RAPID METHODS

RAPID METHODS FOR ANALYSIS OF FOOD AND FOOD RAW MATERIAL

Edited By

Werner Baltes

TECHNOMIC
PUBLISHING CO., INC.
LANCASTER · BASEL

Rapid Methods for Analysis of Food and Food Raw Material

a **TECHNOMIC**® publication

Published in the Western Hemisphere by
Technomic Publishing Company, Inc.
851 New Holland Avenue
Box 3535
Lancaster, Pennsylvania 17604 U.S.A.

Distributed in the Rest of the World by
Technomic Publishing AG

Printed in the United States of America
10 9 8 7 6 5 4 3 2 1

Main entry under title:
  Rapid Methods for Analysis of Food and Food Raw Material

A Technomic Publishing Company book
Bibliography: p.

Library of Congress Card No. 90-71262
ISBN No. 87762-794-0

Preface

The increasing availability of physical methods for food analysis has also stimulated the development of rapid methods for the quick determination of specific parameters in food. Rapid methods for food analysis are required not only for industrial process control but also for obtaining a fast overall view of the state of a food. Last, but not east, the efficiency of the official food control is dependent on the possibility to proceed rapid tests for special ingredients, contaminants or additives. On the other hand, it must be stated that rapid methods are sometimes less specific or less accurate, therefore they are applicable for very specific problems only.

The idea of compiling a book on rapid methods in food chemistry came up during a symposium initiated by the Behr's Verlag. The market has accepted this event very well. This encouraged the Behr's Verlag to publish the book in English language, too. With this translation the editor anticipates to make a contribution to an international program of food analytical methods that, among others, can also support the world-wide trade of food products.

The editor wishes to express his gratefulness to the Behr's Verlag for the cooperation and understanding. If the efforts of authors, publishing company and editor result in the propagation of rapid methods in food analysis this book has accomplished its purpose.

Werner Baltes

Authors

Prof. Dr. W. Baltes (Editor)
Institut für Lebensmittelchemie
der Technischen Universität Berlin
Straße des 17. Juni 135
D-1000 Berlin 12

Dr. P. Barker
BHP Melbourne Research Laboratories
245–273 Wellington Road
Mulgrave, Victoria
Australia 3170

Prof. Dr. H. Büning-Pfaue
Rheinische Friedrich-Wilhelm-Universität Bonn
Endenicher Allee 11–13
D-5300 Bonn 1

Prof. Dr. K. Cammann
Anorganisch-Chemisches Institut
Lehrstuhl für Analytische Chemie
Wilhelm-Klemm-Straße 8
D-4400 Münster

Prof. Dr. K. Eichner
Institut für Lebensmittelchemie
der Westfälischen Wilhelms-Universität
Piusallee 7
D-4400 Münster

Prof. Dr. H. Engelhardt
Fachbereich Physikalische Chemie
Universität des Saarlandes
D-6600 Saarbrücken

Prof. A. Fricker
Ringelberghohl 12
D-7500 Karlsruhe 41

O. Fröhlich
Institut für Lebensmittelchemie
Universität Würzburg
Am Hubland
D-8700 Würzburg

Dr. Herbert O. Günther
Labor Dr. S. Ehrensdorfer
Gögginger Straße 78
D-8900 Augsburg

Dr. G. Henniger
Boehringer Mannheim GmbH
Biochemica Werke Tutzing
Bahnhofstraße 10
D-8132 Tutzing

Dr. H.-J. Hoffmann
Bayr. Staatsministerium für
Landesentwicklung und Umweltfragen
Postfach 81 01 40
D-8000 München 81

Dr. rer.nat. Frank Honold
Wissenschaftlich-Technisches Werkstätten GmbH
Trifthofstraße 57a
D-8120 Weilheim

Prof. Dr. H. Jork
Fachbereich Haushalts- und
Ernährungswissenschaften
Universität des Saarlandes
D-6600 Saarbrücken

Priv.-Doz. Dr. R. Matissek
Lebensmittelchemisches Institut des
Bundesverbandes der Deutschen
Süßwarenindustrie e.V.
Adamsstraße 52–54
D-5000 Köln 80

Petra Offizorz
c/o RCC
In den Leppsteinwiesen 19
D-6101 Rosdorf

Dr. L. Rudzik
c/o Milchwirtschaftliche Lehr- und
Untersuchungsanstalt der
Landwirtschaftkammer Hannover
Harenbergstr. 130
D-3000 Hannover 91

Dr. K. Schmidt
c/o E. Merck
Postfach 41 19
D-6100 Darmstadt 1

Prof.Dr. P. Schreier
c/o Institut für Lebensmittelchemie
Universität Würzburg
Am Hubland
D-8700 Würzburg

K.H. Torkler
c/o Foss-Electric
Waidmannstr. 12 B
D-2000 Hamburg 50

Dr. E. Windhab
c/o Deutsches Institut für
Lebensmitteltechnik e.V.
Postfach 11 65
D-4570 Quakenbrück

Dr. R. Wittkowski
Max von Pettenhofer Institut
des Bundesgesundheitsamtes
Thielallee 88–92
D-1000 Berlin 33

Dr. G. Zaeschmar
c/o Allgäuer Alpenmilch AG
Industriestraße 3–5
D-2940 Leer

Dipl.-Biol. Regina Zschaler
c/o NATEC GmbH
Behringstraße 50
D-2000 Hamburg 54

Table of Contents

3   Equipment for Rapid Methods in Food Quality Control

K. H. Torkler, Hamburg

4 Ion-Selective Electrodes

F. Honold, Weilheim and K. Cammann, Münster

5 Rapid Determination of Metallic Contaminants in Food and Food Raw Materials by ICP-AES and AAS

H.-J. Hoffmann, München

6   Thin-Layer Chromatography – A Screening Method for Food Analysis

H. Jork, Saarbrücken

7   High Performance Liquid Chromatography (HPLC) Reflections on Application in Food Analysis

H. Engelhardt, Saarbrücken

8 Characterization of Changes during Processing and Storage by Chromatographic Determination of Tracer Substances

K.Eichner, Münster

9 Rapid Sample Preparation for Instrumental Analysis

P. Schreier and O. Fröhlich, Würzburg

10 Fast Quality Control by Headspace Analysis

R. Wittkowski, Berlin

11 Application of Infra-Red Spectroscopic Methods

L. Rudzik, Hannover

12 Application of NIR to Analysis of Dairy Products

G. Zaeschmar, Leer

13 Low Resolution NMR

P. J. Barker, Melbourne

14 Rapid Determination Methods for Drugs and Fattening Substances in Animals

H. Büning-Pfaue, Bonn

15   Enzymatic Rapid Methods

G. Henniger, Tutzing

16   Immunochemical Methods

H.O. Günther, Augsburg

17 Theory and Application of Isotachophoresis in Food Analysis

P. Offizorz, Berlin

18 "Rapid Methods" in Sensory Analysis of Food

A. Fricker, Karlsruhe

19   Physical Methods for Rheology, Consistency and Particle Size Measurements

E. Windhab, Osnabrück

20   Microbiological Rapid Methods

R. Zschaler, Hamburg

Rapid Methods
for Food Analysis
– Possibilities and Limits –

R. Matissek

1 Rapid Methods for Food Analysis – Possibilities and Limits

R. Matissek, Köln

Summary

Rapid methods for examination and evaluation of food and associated raw materials belong to the actual topics of the application oriented analytical chemistry. Due to progress in development of modern equipment and instrumental analysis an increasing interest in such methods can be seen, even though no clear or uniform ideas exist on the actual character of a rapid method. This survey attempts to characterize the term and to determine its position in the area of conflict between possibilities and limits. According to the above criteria two categories of rapid methods can be distinguished: the substitution type and the organization type. Finally, rapid methods for determination of relevant parameters in food production and food control are described using method-oriented representations and are explained using some selected examples.

1.1 Introduction

The term "Rapid Methods" has not yet been given a valid definition. Definition and limitation are very complex and difficult because various interpretations are to be taken into account, all of them more or less relative to each other but coming from different philosophies. Additionally, the adjective "rapid" is preferably used today as argument for advertising purposes – even in the field of analysis – and with that it is subject to a certain erosion and lack of conceptual clarity.

Two selected examples of modern food chemical analysis, both with the key word "rapid methods", define the problems here at the beginning and serve to introduce the subject.

Figure 1 contains a short description of all necessary steps for a rapid method to determine Strontium-90 (Sr-90) in milk [1]. In principle, it is an extraction of the daughter product Yttrium-90 (Y-90) from Sr-90 in the incinerated sample after treatment with tributylphosphate, subsequent intermediate precipitation as hydroxide, transformation into oxalate, and measurement of the ß-activity with an anticoincidence counter with very low background

---

● **Method for Determination of SR-90 in Milk
– Rapid Method –**

Principles: Extraction from ash, chemical clean-up, ß-measurement
Source: Chemisches Untersuchungsamt, Hannover

☞   "The methods has – compared to 'common methods' – the
         advantage of requiring significantly less working and time
         effort.

         One employee can carry out at least 3–4 parallel analysis
         of Sr-90 in 1.5 days."

Fig. 1 Rapid Method – Example 1

count rate (<0.008 s^{-1}). At a quick glance, the time requirement (Figure 1) seems to be very long for a rapid method. For better evaluation it is important to know that sensitive and exact Sr-90 analyses are extremely work- and time-consuming. And they require extensive chemical know-how. The common Sr-90 determination takes 2–3 weeks, most of it being necessary for adjustment of a good Sr-90/Y-90 ratio for the measurement [2].

The second example of a rapid method, shown in Figure 2, for the determination of diethylene glycol (DEG) in wine [3]. The method is based on extraction of DEG with chloroform and subsequent determination after capillary gaschromatographic separation by means of mass spectrometry (Selected Ion Monitoring, SIM-technique). This method is, besides being very fast, extremely sensitive as well as specific. The detection limit is 0.2 mg DEG/l wine.

Other methods [4–6] are either less sensitive and are thereby less useful or more time-consuming. The explanation of the author of this work regarding its classification as rapid method, as quoted in excerpts in Figure 2, is remarkable. It mentions not only the highly organized course of action but the special status the team work has here.

These examples are very interesting regarding method theory and problem solving but here we cannot go further into this subject. They stand for all other so-called "rapid methods" and show that limitation and classification are manifold and heterogeneous. Consequently, the following points are very important and should be kept in mind:

- **Method for Determination of Diethylene Glycol in Wine**

Principles: Extraction, capillary gaschromatographiy/mass spectrometry

Source: Kuhlmann, F. (1986) Dtsch. Lebensm.-Rdsch. 82:84

☞ "With the described method a team can determine diethylene glycol in wine in a large series in short time.

Organization of sample reception, course of analysis in
the laboratory and the following indicidual evaluation are
very important."
Time requirement: approximately 0.5 h!

Fig. 2 Rapid Method – Example 2

1. Even though rapid methods are characterized by time, the term does not
   ad definitionem include the condition of a time limit.

2. "Rapid" is not only a question of methodology and instrumentation but
   also a question of organization up to the level of automatization.

3. Rapid methods are increasingly employed in food control activities and
   not only, as originally planned, for quality control in industrial food production processes. This means that a change in the problem perception
   has taken place.

4. The extent and diversity of analytical parameters are increasing daily,
   with constantly more exacting requirements regarding the qualityof analytical statements being made especially in the important field of trace
   analysis.

For characterizing the term "rapid method" as well as for localizing it in the
area of contention between possibilities and limitations three basic questions arise, as stated below:

1. What is a "rapid method?
   – Definition –

2. Why and where are rapid method employed?
   – Possibilities –

3. Which parameters should or could be determined by rapid methods?
   – Limits –

In this survey only such methods will be considered in which chemical and physical-chemical principles of measurement are used. This means that microbiological and purely physical measurements are excluded. Such a survey can, of course, not claim to be complete.

1.2   What is a "Rapid Method"?

Figure 3 shows the attempt to characterize and define "Rapid Methods" in a schematic diagram.

1.2.1 Explanation of the Term "Method"

A method is a procedure based on a regulated system and serves to obtain knowledge or practical results [7]. For an analysis this is an examination carried out according to a certain plan. Usually this takes place in a laboratory or a laboratory-like facility and includes sample taking, sample preparation, the actual measurement, and evaluation of the results. This is the so-called laboratory method.

Some methods, due to their simplicity, can be carried out independent of laboratory facilities. Among these are potentiometric determination of pH-values, radiometric density measurements, refractometric extract determinations, test-strip reflectometry, NIR- and NMR-measurements, test-kits, etc. If non-destructive measurements are carried out on-line and integrated into the production process, they can serve as continuous process controls.

This type of measurement is well-suited for quality control in manufacturing and/or processing plants. This is an example of the ideal, but rare case, of a "rapid method", which should be more exactly described as an "immediate method".

Analysis methods can be of varied significance depending on the problem, concept and execution. They can be differentiated according to Figure 3 into:

1. Detection and screening methods

2. Tests or limiting value methods

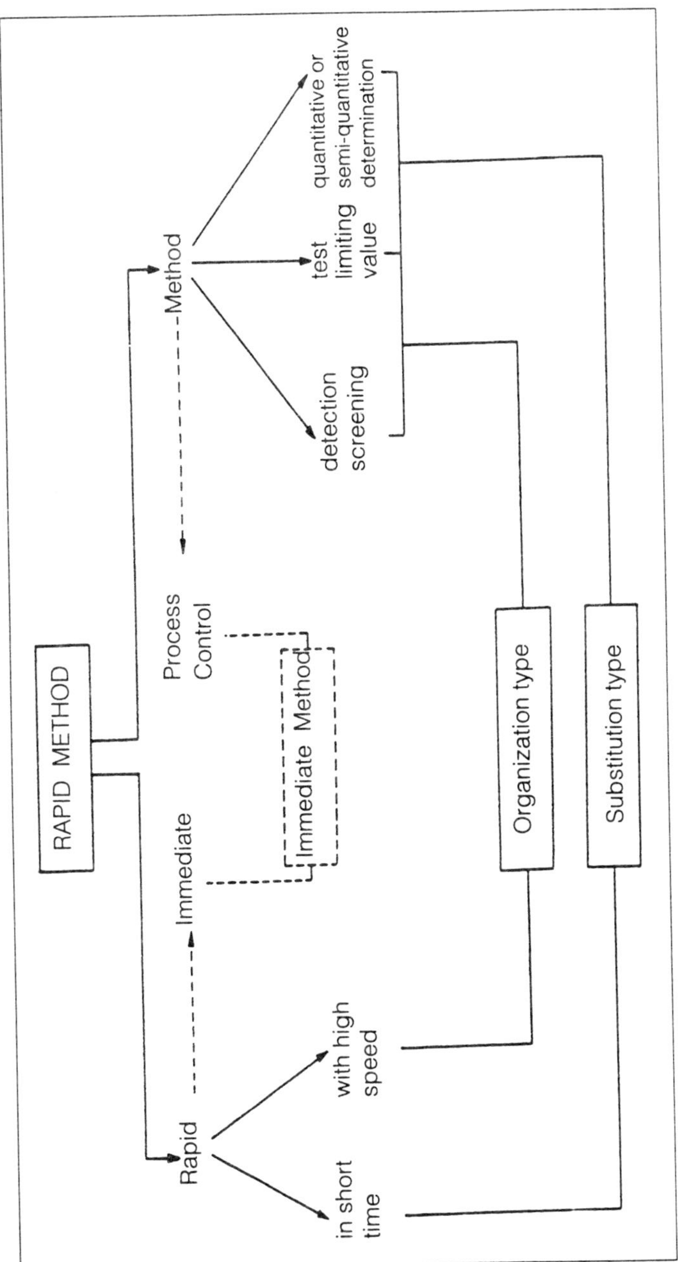

Fig. 3 Schematic Diagram for Characterization and Definition of "Rapid Method"

3. Quantitative or semi-quantitative determination methods

4. Methods of process control.

1.2.2 Explanation of the Term "Rapid"

The adjective "rapid" is based on two important meanings:

1. within short time

   and/or

2. with relatively high speed [8].

For rapid methods a division into two types can be made in relation to the common, time-consuming, often official analysis methods which are convention or reference methods. It must be distinguished between:

1. methods which quickly lead to a result within the shortest possible time by exchanging or avoiding time-consuming work or process steps, usually by introducing a different measuring principle

   or

2. methods with which the result can be obtained faster because the tests can be carried out with higher speed, even if the basic measuring principle is retained.

1.2.3 Category of Rapid Methods

The first rapid method mentioned can be called the "substitution type" and the second one the "organization type". Even though both types are aimed at rapidity, they are basically different. Transitions and combinations are possible and quite common. In practice a mixed type is often used. So-called multicomponent methods take a special position, they show apparent rapidity only because they produce a high number of parallel results.

The characteristics mentioned are not only significant because they classify and define rapid methods but because they show the generally different level of analytical significance of rapid methods depending on the type of rapid method involved.

1.3 Why and Where Are Rapid Methods Employed?

Some year ago, when the equipment for instrumental analysis was not developed to the present extent, classical wet-chemical methods which required much analytical experience and much time were preferred in the laboratories. Rapid methods were employed because they gave results in a shorter time, usually due to simplified handling and by avoiding common measuring principles as well as simplification in regard to sample preparation. Practically only methods of the so-called substitution type were employed.

Figure 4 gives a schematic presentation on two significant fields of application for rapid method analysis.

1.3.1 Food Production

The necessity of employing rapid methods was seen, for the in-plant quality control in the industrial food production, because biological raw materials with natural fluctuation in composition were processed. By rapid determination of valuable constituents in raw materials, intermediates and end products, it became possible to influence the production in order to obtain optimal economy and quality. Quality is the measure of agreement between intention and implementation. Today, the question of the quick determination of possibly undesirable substances is of increasing significance for large food production plants.

1.3.2 Food Inspection

The increasing interest in rapid methods in the area of food quality control is due to the increasing number of samples and simultaneously increasing number of analysis parameters. This is true not only for the official state laboratories but also for the independent commercial laboratories. Legal or toxicological aspects often require rapid information as to the presence or absence of one or more substances either exceeding or falling below a given limiting value. Furthermore, the institutes are forced to save time and material since resources are short. On the other hand, effective consumer protection must be guaranteed. Some examples have been published lately, for example the detection of diethylene glycol in wine.

In the field of food inspection, including, in the broader sense the environmental analysis, the determination of defined individual substances – singularly or simultaneously – has become very prevalent in the last years.

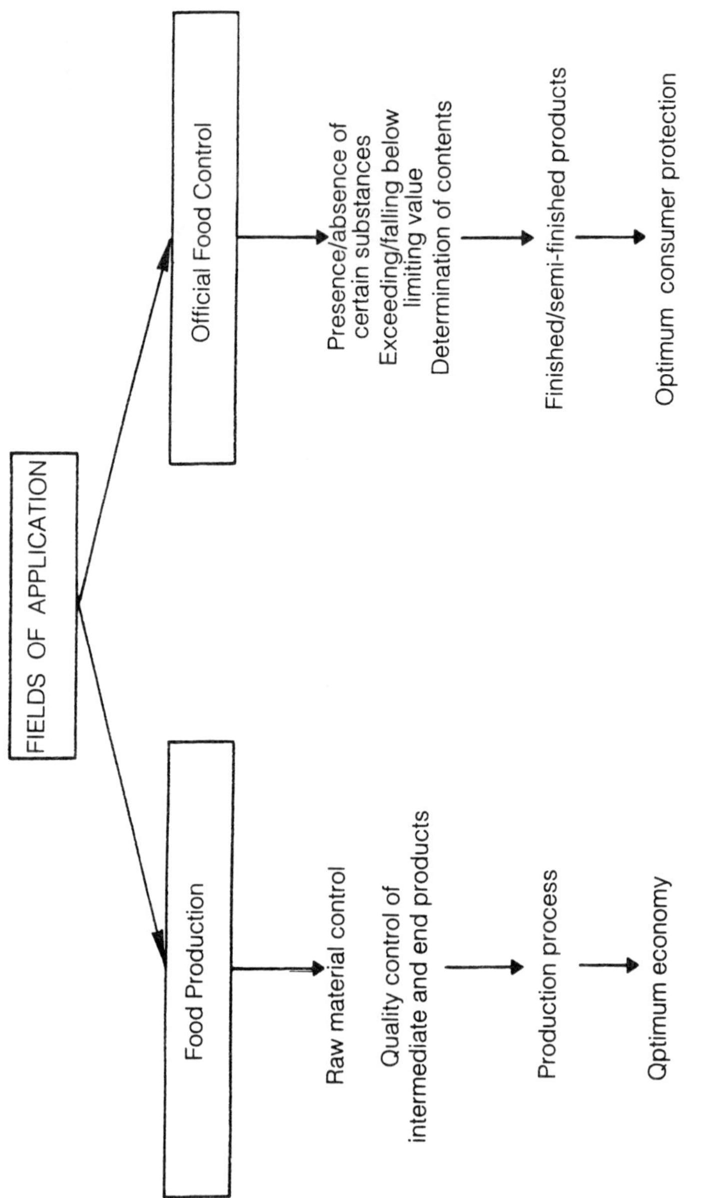

Fig. 4 Fields of Application and Objectives for Employment of Rapid Method Analysis

For these problems, modern instrumental methods which can be adjusted by time-optimizing automatization or better organization of known measuring techniques to allow faster, more exact and sensitive determinations are of primary interest. These rapid methods belong mainly to the organization type.

1.4 Which Parameter Should or Could be Determined by Rapid Methods?

In this survey a comprehensive answer to this question is not possible. This field is continuously developing and a corresponding catalogue would be extremely extensive. Furthermore, there are often several ways for determining parameters of interest. An individual listing does therefore not seem very appropriate. A method-oriented representation is favored in the following. Not the object to be analyzed is considered, but rather the method.

1.4.1 Substitution Type

For quality control in a production plant, the determination of valuable constituents of food raw material and food products is most important. From the analytical point of view, however, the main constituents such as fat, protein, solid matter, carbohydrates, etc. are aggregate parameters which are defined by application of reference methods. A faster analysis of aggregate parameters with an identical measuring principle can be practiced only in exceptional cases with specially developed equipment. Such equipment for aggregate parameter determination is usually developed from simplified standard or more simple alternative methods. Table 1 shows a list of the main quality characteristics to be determined by analysis as well as the measuring principles of the methods employed. The rapid methods listed are mainly of the substitution type. The following text contains explanations with descriptions on the main fields of application. This article cannot go into more detail, however, the extensive literature citations offer the possibility of further special method studies.

1.4.1.1 Determination of Fat Content

Reference methods are gravimetric determinations after extraction with a suitable solvent, carried out with or without a previous acid- or alkaline decomposition, depending on the sample material. These methods are known

Tab. 1 Rapid Methods of the Substitution Type – A Comparison

Parameter	Measuring Principle	
	Reference-/Convention Method	Rapid Method
Fat	Gravimetry after solvent extraction with or without acid/alkaline decomposition: Weilbull-Stoldt, Röse-Gottlieb, Schmid-Bondzynski-Ratzlaff, Soxhlet	Densitometry Refractometry Volumetry X-ray absorption Radiometry NIR/MIR CM-extraction NMR
Protein	Volumetric analysis after decomposition and distillation: Kjeldahl	Direct distillation Color binding methods NIR/MIR NMR
Solid matter/ Extract/ Density	Gravimetry after thermal drying in drying chamber	Gravimetry after IR-or microwave drying Refractometry Radiometry Conductometry Dielectrometry NIR
Carbo-hydrates	Reduction methods: Luff-Schoorl, Potterat-Eschmann; Polarimetry, Enymatic methods	NIR/MIR HPLC (Enzymatic methods)
Further quality attributes	Various special methods	NIR Enzymatic methods Test-strip reflectometry Ion-selective electrodes etc.

under the name of their developers, e.g. Weilbull-Stoldt, Rîse-Gottlieb, Schmid-Bondzynski-Ratzlaff, Soxhlet [9–12].

There are many different rapid methods for fat determination:

(1) **Densitometry**: The density of a solvent-fat-mixture is determined by means of a buoyant body (food, meat products) [13].

(2) **Refractometry**: The refractive index of a ‡-bromonaphthalene fat mixture is determined (cocoa products, confectionaries, meat products) [14].

(3) **Volumetry**: Determination of fat content by means of the volume in minced meat after thermal rendering [15]. A rapid method, developed for the practice by Gerber in 1892 and modified since then, for determination of fat content in milk and milk products belongs to this category, too. Here, the fat is set free by means of a sulphuric acid decomposition and subsequent centrifugation and is then determined in a butyrometer [16].

(4) **X-ray absorption**: This method is used for meat "as grown" (approximately 7 kg sample). The absorption of X-rays through the sample is measured. The mineral content of fat is low, thus absorption is lower the higher the fat content is in the sample [17,18].

(5) **Radiometry**: The ratio of inelastic and elastic scattered radiation (Compton-, Rayleigh scattering) by means of an Am-241 radiator ("soft gamma-rays")is determined (meat products, fat-containing food) [19].

(6) **NIR/MIR**: Reflection measurement of non-absorbed radiation of defined wave length in near (NIR) or medium (MIR) infra-red range (various food) [20–28].

(7) **CM-Extraction**: Isolation of the total fat in food by means of chloroform-methanol extraction and subsequent gravimetric determination [29].

(8) **NMR**: The presence of paramagnetic atomic nuclei, usually protons, is determined with a nuclear impulse resonance spectrometer. Relaxation times give information on solid and liquid fat contents in a sample [30–35].

1.4.1.2 Determination of Protein Content

Reference method is the volumetric determination of released ammonia after sulphuric acid decomposition according to Kjeldahl and after distillation [9–12].

Rapid methods to be mentioned are:

(1) **Direct Distillation**: That portion of protein-ammonia is determined which is released without decomposition in an alkaline distillation of the sample (milk, grain, seeds) [36].

(2) **Color binding methods**: Proteins have the ability to bind colorants. This characteristic is used here. It is a photometric determination of excess color, not bound by proteins. This analysis method is known as the amido black method for milk testing [37].

(3) **NIR/MIR**: [20–28]

(4) **NMR**: After addition of so-called "relaxation reagents" relaxation times are measured by means of nuclear impulse resonance spectrometers and the times are correlated to the protein content [38].

1.4.1.3   Determination of Solid Matter Content/Extract Content/ Density

Reference methods are the gravimetric determinations after thermal drying in a drying chamber]9–12].

Rapid methods are:

(1) **Gravimetric difference measurement**: The heat transfer by means of thermal conduction and convection usually occurring in the drying chamber in convention methods, is substituted by more rapid heating using microwave or infrared radiation (compare [10] Vol. II/2, page 17).

(2) **Refractometry**: Determination of refractive index and calculation of extract or solid matter content (solutions, liquids) [39].

(3) **Radiometry**: Determination of density by absorption measurement of gamma quanta. Source is Am-241 (sugar-, beverage-, milk-, starch industry; process control) [40,41].

(4) **Conductometry**: Determination of water content by measurement of electric conductivity of samples (compare [10] Vol. II/2, page 34).

(5) **Dielectrometry**: Measurement of the dielectric constant for the determination of the water content in samples [42] (compare [10] Vol. II/2, page 34).

(6) **NIR**: [20–26, 27].

1.4.1.4 Determination of Carbohydrate Content

Reference methods are reduction methods (e.g. according to Luff-Schoorl or Potterat-Eschmann), polarimetric methods and enzymatic methods [9–12].

The following rapid methods are of importance:

(1) **NIR/MIR**: [20–28].

(2) **HPLC**: By high performance liquid chromatography the carbohydrates (sugars) are separated and detected individually [43–46].

(3) **Enzymatic methods**: In some cases, rapid methods are reference methods [9]. These are the so-called UV-test-kits. The change of UV-absorption after an enzymatic transformation reaction is measured photometrically [47–48].

1.4.1.5 Further Quality Attributes

For determination of several other quality features some very different methods can be employed. In the following a limited selection of rapid methods are listed:

(1) **NIR**: From NIR-measurements various quality attributes can be derived, e.g. evaluating the malt quality of barley, evaluation of baking quality of wheat as well as evaluation of degree of starch degradation in flour [21]. NIR can also be used for determination of dietary fiber of wheat bran [49].

(2) **Special enzymatic test-kits**: (compare [47,48]).

(3) **Test strip reflectometry**: The color intensity of a test strip is measured reflectometrically. A study on the determination of the following constituents and additives in juices had been carried out: ascorbic acid, potassium, nitrite/nitrate, sulfite [50].

(4) **Ion-selective electrodes**, although not explicitly mentioned, are especially suitable for determination of ions in water or aqueous solutions. They too, as well as many other measuring methods, belong to this category.

When comparing the results of common methods with the results of rapid methods of the substitution type, system inherent differences in accuracy and sensitivity can be verified.

If the measuring errors ("measuring differences") are known in regard to kind and size, this can be taken into consideration through correction or calibration. Rapid methods usually have narrow or special fields of application and are often limited to certain concentration ranges. For quality control in food processing this is not disadvantageous because matrices of the raw materials and the products are known and show a certain consistency. It must be considered that some methods may be limited by concurrent reactions. For example, when applying heat radiation drying (IR) to carbohydrate-containing foods the inter- and intramolecular water tends to hydrolyze and may cause possible local overheating [51].

1.4.2 Organization Type

The food inspection agencies require the determination of many different parameters. The list of substances to be determined is extensive, heterogeneous and includes, in addition to the main food constituents, minor constituents such as vitamins, enzymes, and minerals as well as additives and contaminates in the trace range. As already known, in special cases the application of a reference method is mandatory. However, for many areas a quick analysis is necessary. It is, of course, not possible to make an extensive list for the complete area of food analysis, but some important methods should be mentioned.

Table 2 lists rapid methods of the organization type. The methods are sorted according to the measuring principle. In the center column some rapid method are listed, in the right-hand column an example of their application is given. Within this survey not all listed subjects can be discussed in detail.

1.4.2.1 Chromatography

(1) The term **Very-High-Speed Liquid Chromatography** indicates in the name the high speed used in carrying out this method. Separations, i.e. analyses are finished within a few minutes [52–54].

(2) **Automated HPLC-systems** are available from several companies. Special systems for rapid analysis of amino acids are offered [55,56]. HPLC-systems can be universally employed (45,46,57,58). The ion-chromatography with its variants belongs in this category [59,60].

(3) The **Head-Space Gaschromatography** allows rapid analysis because the "gas portion" of a sample is directly injected into the chromatographic

Tab. 2 Rapid Method of the Organization Type – Survey and Examples

Basis Principle	Rapid Method	Examples of Application
Chromatograhpy	Very high speed LC	Preservatives
	Automated HPLC-systems	Amino acids
	Head space GC	Aroma substances, volatile compounds
	Flash chromatography	Isolation/sample preparation
Spectrometry	Fourier-transformation-technique	Identification/spectra
	Diode-array-technique	Identity/spectra
	ICP-AES	Multi-elemental analysis
	GC/MS-organization	Diethylene glycol
Photometry-combination with enzymatic/ color reactions, etc.	Centrifugal analyzer	Organic acids/carbohydrates
	Flow injection analysis (FIA)	Various
Colorimetry	Test-kits (color-cards/ color comparators)	Water analysis
Amperometry	Titrating automates (e.g. KF)	Water content
Kjeldahl method	Khel-Foss-Automatic	Protein content

separation system and usually the obligatory sample preparation is not necessary [61].

(4) The **Flash Chromatography** – "fast as a flash", as the name indicates – is an optimized low pressure column chromatography very suitable for quick sample preparation for several problems [62,63].

1.4.2.2 Spectrometry

(1) The **Fourier-Transform-Technique** allows significantly faster spectra scans since it is not longer necessary to scan the total measuring range but only to calculate the spectrum from a short measuring signal [22,64]. Equally fast, but based on a different principle, are diode-array-photometers [65].

(2) The **Atomic Emission Spectrometry** with inductively coupled plasma allows simultaneous detection of many elements. This methods produces a large number of results in short time (multi-component method) [66,67].

(3) The **Gas Chromatography/Mass Spectrometry** (GC-MS) has been described previously using the example of DEG-determination in wine in the introduction [39].

1.4.2.3 Photometry

(1) For the photometric rapid methods, the **combination with enzymatic methods** is of great interest. Centrifugal analyzers allow automatization of photometric measurements and are especially suitable for enzymatic analysis [68]. The principle is based on centrifugal mixing and subsequent longitudinal measurement in rotating cuvettes. The examples, listed in Table 2, are from the field of fruit juice analysis [58].

(2) The **combination of flow injection analysis** (FIA) with photometers as detectors belongs to the photometric rapid methods, too. Reactions and transformations take place in thin tubes (capillaries). The FIA is suitable for high sample throughput [58,69,70]. In this way enzymatic or chemical color reactions can be automated. The FIA can also be coupled with ion-selective electrodes or AAS for detection.

These methods originated in the clinical chemistry, but they have quickly entered the field of rapid food analysis.

1.4.2.4 Colorimetry

In colorimetry the test kits for quick characterization or evaluation of water quality are of primary interest. Here, color cards or color comparators are often employed for obtaining (semi-)quantitative statements [71,72].

1.4.2.5 Other Methods

Furthermore, titration automates as, for example, the Karl-Fischer-Automates for water determination are important. Finally, the automated Kjeldahl method for rapid protein determination belongs to the methods of the organization type (KJEL-FOSS-Automatic).

1.5 Conclusion and Outlook

Figure 5 shows in an allegory, symbolically the most important steps of computer development which can also be applied to the development of rapid methods.

This figure shows how a complex mathematical problem is solved with: 1. paper and pencil, 2. slide-rule and 3. electronic calculator.

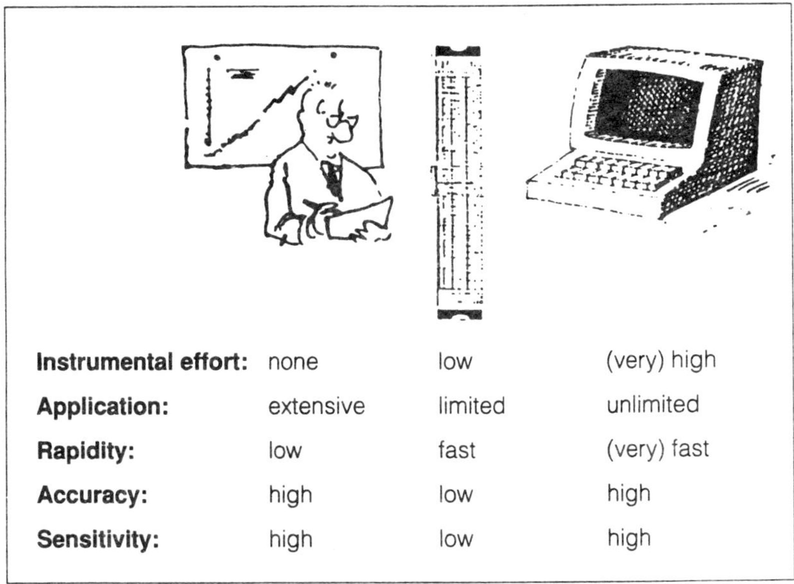

Instrumental effort:	none	low	(very) high
Application:	extensive	limited	unlimited
Rapidity:	low	fast	(very) fast
Accuracy:	high	low	high
Sensitivity:	high	low	high

Fig. 5 Allegory – Development of Calculating Methods/Rapid Methods

Ad 1: The paper and pencil needed for this calculation can be compared on a level with the laboratory test tube and Bunsen burner. The instrumental effort is relatively low. The field of application is not directly limited but rapidity is low while accuracy and sensitivity are high with careful working manner.

Ad 2: Many special units for rapid determinations are similar to the state of development of a slide-rule. Of course, there are exceptions. Since the concept was designed for special applications, the instrumental effort is significantly higher but not too large. The methods are rather quick, but relatively inaccurate and not very sensitive. It should be noticed that these instruments belong to the substitution type.

Ad 3: Due to further development in technology and increasing implementation of electronic calculators it is possible to solve and automate difficult analysis problems very quickly, accurately and sensitively. The instrumental requirements may be very high. In principle, the fields of application are unlimited. These systems belong predominantly to the category of the organization type.

Assuming this comparison for future possibilities in the field of rapid methods, it is evident that due to high-tech progress in many analytical sectors, the development tends more and more towards the organization type. It should be made clear that each analysis method has its limits, in regard to selectivity and flexibility, as well as to detection limit and reproducibility. These factors should be considered and weighed against each other prior to selecting an analysis method. Another basic consideration, in addition to the time requirement, which is here the focal point of interest, is the cost effort.

REFERENCES

1. Chemisches Untersuchungsamt Hannover (1984): Verfahren zur Bestimmung von Strontium-90 in Milch—Schnellmethode; Methode E-Sr 90-Milch-02-01

2. MORFELD, U. (1986): Private Mitteilungen

3. KUHLMANN, F. (1986): Schnellmethode zur Bestimmung von Diethylenglykol in Wein; Dtsch. Lebensm.-Rdsch. **82**:84

4. LEHMANN, G., GANZ, J. (1985): Nachweis von Diethylengykol in Wein; Z. Lebensm. Unters. Forsch. **181**:362

5. FÜHRLING, D., WOLLENBERG, H. (1985): Zur Bestimmung kleiner Mengen von Diethylenglykol in Wein 1.; Dtsch. Lebensm.-Rdsch. **81**:325

6. LITTMANN, S. (1985): Zur Bestimmung kleiner Mengen von Diethylenglykol in Wein 11., Dtsch. Lebensm.-Rdsch. **81**:329

7. DROSDOWSKI, G. (Hrsg.) (1978): Duden—Das große Wörterbuch der deutschen Sprache; Bibliographisches Institut, Mannheim, Bd.4

8. DROSDOWSKI, G. (Hrsg.) (1980): Duden—Das große Worterbuch der deutschen Sprache; Bibliographisches Institut, Mannheim, Bd.5

9. Bundesgesundheitsamt (Hrsg.): Amtliche Sammlung von Untersuchungsverfahren nach §35 LMBG; Beuth Verlag, Berlin, Köln

10. SCHORMÜLLER, J. (Hrsg.) (1965): Handbuch der Lebensmittelchemie; Springer-Verlag, Berlin

11. RAUSCHER, K., ENGST, R., FREIMUTH, U. (1986): Untersuchung von Lebensmitteln; VEB-Fachbuchverlag Leipzig

12. BEYTHIEN, A., DIEMAIR, W. (1970): Laboratoriumsbuch für den Lebensmittelchemiker; Verlag Liedl, München

13. MONTAG, A.(1973): Zur densitometrischen Fettbestimmung in milcheiweißreichen Erzeugnissen mit dem FOSS-LET-Gerät; Dtsch. Lebensm.-Rdsch. **69**:470

14. FINCKE, A. (1965): Handbuch der Kakaoerzeugnisse; Springer-Verlag. Berlin, S. 428

15. Hobart Maschinen, Offenburg (Hrsg.) (1978): Fett-Tester Mod. F-101; Informationsschrift PR-1074-080

16. DIN 10310 (1970): Bestimmung des Fettgehaltes von Milch nach dem Gerber-Verfahren; vgl. [9] L01.00-8 (1981)

17. Seffelaar & Looyen, Odenzaal Holland (Hrsg.) (1978): Anyl-Ray/Bestimmung des Fettgehalts eines Fleischmusters mit X-Strahlen; Selo Bulletin Ablage 12 5-5-71-1

18. MADIGAN, J.J. (1978): Grundlagen für Fleischanalysen und Standardisierung des Rohmaterials; Informationsschrift Anyl-Ray Corporation, Waltham, Massachusetts, USA

19. SCHÄTZLER, H.P., KUHN, W. (1978): Zerstörungsfreie Schnellbestimmung des Fettgehaltes von Fleisch und anderen fetthaltigen Nahrungsmitteln; Lebensmittelchem. Gerichtl. Chem. 32:126

20. KORTÜM, G. (1969): Reflexionsspektroskopie; Springer-Verlag. Berlin

21. OSBORNE, N.G. (1981): Principles and practice of near infra-red (NIR) reflectance analysis; J. Fd. Technol. **16**:13

22. RUDZIK, L. (1986): MIR-NIR in Lebensmitteln; GIT Supplement **2/86** Lebensmittelchemie, S. 30

23. Technicon Instruments, Tarrytown USA (Hrsg.): Infra Alyzer 400—Schnellbestimmung von Inhaltsstoffen durch Reflektionsmessung im NIR; Informationsschrift

24. Technicon Instruments, Bad Vilbel (Hrsg.) (1979): Infra Alyzer zur Konzentrationsmessung von Inhaltsstoffen; Seifen Öle Fette Wachse **105**:439

25. MEYHACK, U. (1986): NIR—Die Ulbrichtsche Kugel sammelt das Licht; Chem. Rdsch. 39: Nr.**9**, S.16

26. Berwind Instruments, Wheldrake England (Hrsg.) (1981): Schnelles Infrarot- Analysengerät für Milch und Molkereiprodukte, Kurzbericht; Z. Lebensm. Technol. Verfahrenstechn. **32**:152

27. PRÜFER, H. (1986): NIR-Spektroskopie—Analytik in nahen Infrarot zur Qualitäts- und Prozeßkontrolle; Kontrolle Bd. 5 Juli/August, S. 11

28. BUCKENHÜSKES, H., HARRER, M. GIERSCHNER, K. (1986): Quantitative Bestimmung von Inhaltsstoffen in Schokoladenmassen mit Hiffe von NIR-Messungen; Lebensmittelchem. Gerichl. Chem. **40**:40

29. HALLERMAYER, R. (1976): Eine Schnellmethode zur Bestimmung des Fettgehaltes in Lebensmitteln; Dtsch. Lebensm.-Rdsch. **72**:356

30. Bruker Analytische Meßtechnik, Rheinstetten (Hrsg.) (1979): Minispec. p 20; Informationsschrift

31. WEISSER, H. (1979): NMR spectroscopy in the food industry; Bruker Application Note

32. KOHN, R. (1966): Anwendung der Kernresonanz-Spektroskopie in der Lebensmittelanalytik; Fette Seifen Anstrichm. **68**:795

33. McCARTEN, J. (1979): Fat analysis in margarine base products; Bruker Application Note

34. VAN PUTTE, K. (1979): Pulse NMR as a routine method in the fat and margarine industry I–111; Bruker Application Notes

35. McCARTEN, J. (1979): Determination of the fat content and solid-to-liquid ratio in cocoa products with the minispec p 20 NMR Process Analyzer; Bruker Application Note

36. SCHMÜTZ, W., Do, Q.N. (1979): Eiweiß-Schnellbestimmung mit dem Direktdestillations-Verfahren; Dtsch. Lebensm.-Rdsch. **75**:398

37. KIERMEIER, F., LECHNER, E. (1973): Milch und Milcherzeugnisse; Verlag Paul Parey, Berlin, S. 256

38. Newport Instruments, England (Hrsg.) (1980): Protein Analyzer P—100, Kurzbericht; Z. Lebensmittel. Technol. Verfahrenstechn. **31**:369

39. MILLIES, K., BÜRKIN, M. (1984): Anwendung der Refraktometrie zur Produktkontrolle in Fruchtsaftbetrieben—Meßwertkorrekturen; Flüssiges Obst **51**:629

40. SPRISSLER, G. (1978): Genehmigungsfreie radiometrische Dichtemessung mit Am-241; Z. Lebensm. Technol. Verfahrenstechn. **29**:315

41. Berthold Kernstrahlungsmeßgeräte für Industrie, Wissenschaft und Medizin, Wildbad (Hrsg.) (1980—84): Radiometrische Dichtemeßanlagen; Informationsschriften

42. AHLGRIMM, H. J. (1g77): Zur dielektrischen Bestimmung des Feuchtegehaltes von Lebensmitteln; Z. Lebensm. Technol. Verfahrenstechn. **28**:305

43. BAUER, H., QUAST, H., SHALAEY, A., ROCEK, P. (1986): HPLC von Kohlenhydraten; LaborPraxis Spezial '86 Chromatographie Spektroskopie, S. 64

44. Millipore Corporation, Eschborn (Hrsg.) (1986): Kohlenhydratanalyse in sechs Minuten; Waters Aktuell Ausgabe Nr. 6, S. 4

45. MACRAE, R. (1980): Applications of high pressure liquid chromatography to food analysis; J. Fd. Technol. **15**:93

46. Fogy, I., Fioresi, C., Schmid, E. R., Markl, P. (1980): HPLC in der Lebensmittelanalytik; Kontakte (Merck) Nr. 2, S. 24

47. Boehringer, Mannheim (Hrsg.) (1980): Methoden der enzymatischen Lebensmittelanalytik; Informationsschrift

48. Boehringer, Mannheim (Hrsg.): Die enzymatische Lebensmittelanalytik; Heft 1-6/4

49. Horvath, I., Norris, K. H., Horvath-Mosonyi, M., Rigo, J., Hegedüs-Völgnesi, E. (1984): Untersuchung zur Bestimmung der Ballaststoffe von Weizenkleie nach dem NIR-Verfahren; Acta Alimenta **13**:355; Ref. Z. Lebensm. Unters. Forsch. **182**:76

50. Schwedt, G. (1986): Teststäbchen-Reflektometrie fur Ascorbinsäure, Kalium, Nitrat/Nitrit und Sulfit in Lebensmitteln; Dtsch. Lebensm.-Rdsch. **82**:111

51. Untze, W. (1982): Zur Einrichtung eines Betriebslabors und zur Durchführung von lebensmittelchemischen Untersuchungen; in Florin, U., Lehmann, D. (Hrsg.): BDL-Spectrum 1, Behr's Verlag, Hamburg, S. 151ff

52. DiCesare, J. L., Dong, M. W., Ettre, L. S. (1981): Very-high-speed liquid chromatography 1. The system and selected applications; Chromatographia **14**:257

53. DiCesare, J. L., Dong, M. W., Atwood, J. G. (1981): Very-high-speed liquid chromatography 11. Some instrumental factors influencing performance; J. Chromatogr. **217**:369

54. Dong, M. W., DiCesare, J. L. (1982): Very-high-speed liquid chromatography 111. Quantitative Analysis of Parabens in Cosmetic Products; J. Chromatogr. Sci. **20**:49

55. Millipore Corporation, Eschborn (Hrsg.) (1986): Hochempfindliche Analyse physiologischer Aminosäuren; Waters Aktuell, Ausgabe Nr. 6, S. 4

56. Hewlett-Packard, Bad Homburg (Hrsg.) (1986): Automatisierte Methode für die Bestimmung von Aminosäuren; PEAK Bd. 8; Nr. 213, S. 5

57. Schwedt, G. (1984): Spezielle HPLC-Techniken in der Lebensmittelanalytik; Labor-Praxis **8**:1028

58. Scheuermann, C., Fischer-Ayloff-Cook, K. P., Hofsommer, H. J. (1986): Ausgewählte moderne Laborgeräte fur Fruchtsaftbetriebe; Confructa Studien **30**:162

59. Franklin, G. O. (1985): Development and applications of ion chromatography; International Laboratory Vol. **15**, No. 6, P. 56

60. Harrison, K., Beckham, W. C., Yates, T., Carr, C. D. (1986): Rapid, cost-effective ion chromatography; International Laboratory Vol. **16**, No. 3, P. 90

61. Kolb, B. (1982): Anwendung der Headspace Analysis; LaborPraxis **6**:4

62. Still, W. C., Kahn, M., Mitra, A. (1978): Rapid chromatographic technique for preparative separations with moderate resolution; J. Org. Chem. **43**:2923

63. Matissek, R., Häussler, M. (1985): Flash-Chromatographie—Eine schnelle Niederdruck-Säulenchromatographie zur Trennung präparativer Probemengen; Dtsch. Lebensm.-Rdsch. **81**:50

64. Oelichmann, J., Rau, A., (1986): Mikro-Methoden in der FTIR-Spektroskopie; LaborPraxis **10**:958

65. OWEN, T., EMMERT, J. (1986): Diodenarray-Technik für Routineanalysen; LaborPraxis **10**:1512

66. SCHRADER, W., GROBENSKI, Z, SCHULZE, H. (1981): Einführung in die AES mit induktiv gekoppeltem Plasma (ICP); Angewandte Atom-Spektroskopie (Perkin-Elmer), Heft 28

67. KAHN, H. L. (1983): MS oder ICP—Jede Technik hat ihre eigenen Vorteile; Kontrolle Bd. **2** Juni, S. 11

68. Hoffmann-La Roche, Basel (Hrsg.) (1984): Cobas FARA—Eine neue Dimension der Zentrifugalanalyse; Informationschrift

69. MÖLLER, J. (1982): FIA—Eine neue Analysenmethode; LaborPraxis **6**:278

70. HANSEN, E. H. (1986): Recent advances in flow injection analysis; International Laboratory Vol.**15**, No. 8, P. 14

71. Merck, Darmstadt (Hrsg.) (1985): Schnelltest Handbuch; Informationsschrift

72. KOCH, E. (1985): Schnelltests zur Umweltanalytik; in: Fresenius, W., Günzler, H., Huber, W., Lüderwald, 1., Tölg, G.. Wisser, H. (Hrsg.): Analytiker-Taschenbuch, Springer-Verlag, Berlin, Bd. **5**, S. 183

Rapid Tests
for the Analysis
of Water and Food

K. G. Schmidt

2 Rapid Tests for the Analysis of Water and Food

K.G.Schmidt, Darmstadt

2.1 Introduction

The continuously growing request for chemical-analytical information from industry and trade, as well as in various fields of the everyday life, cannot alone be fulfilled by "official analysis" represented by DIN or reference methods. For the last 20 years, rapid methods have been developed based mainly on titrimetry and colorimetry, allowing chemical-analytical information to be obtained directly on-site.

Further developments of these methods have led to high-quality rapid tests, alternatives to reference methods, because they provide comparable results within tolerable limits with low effort. Alternative methods compared to reference methods require a full quantitative detection which today is done almost exclusively with modern small photometers.

So-called semi-quanitative methods with test strips and color scales or simple titration tests, under the condition that they fulfill certain quality standards, can be used for exploratory testing or as screening methods for selection of samples that must be examined more closely.

Regardless of considerations for rapid test applications in analytical laboratories, these tests can also be applied for on-site examinations when mainly semi-quantitative methods are asked for. These tests enable even chemical amateurs such as farmers, gardeners, housewives, plumbers, police men, and customs officers, cooks, bakers, engineers, etc. in their special fields to analyze, for example water, soil, or fat in a simplified way.

The industry supplying those tests has a high responsibility since the use of the tests must be safe and ecologically harmless. On the other hand, the user of these tests must be given precise instructions for correct interpretation of the analytical values determined. Today, not all desirable requirements are fulfilled. But there still is a fast development in the field of rapid testing, so that the fulfillment of open requirements can be expected in the near future.

2.2 Test Strips

Simple test- and reagent papers such as lead acetate for detection of sulfides, potassium iodide starch paper for detection of oxidizers, or curcumin paper for detection boric acid or borates have been known for a long time. Additionally, indicator papers and non-bleeding indicator strips for determination of the hydrogen ion concentration in water belong to this group of tests.

A more recent development is the semi-quantitative indicating test strips, e.g. Merckoquant®. Compared to the purely qualitatively indicating reagent papers, these strips additionally indicate the concentration range of the detected substance. The test strips consist of a reactive test zone firmly bonded to a plastic backing. The test zones are impregnated with reagents, buffers and other substances.

To test a sample, the test zone is immersed into the water sample for 1–2 seconds and then compared to the color scale on the package (visual colorimetry) (Fig. 1).

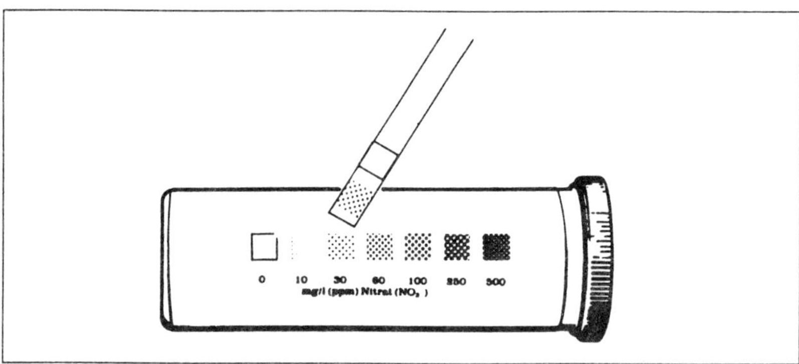

Fig. 1   Color Comparison of the Test Strip Reaction Zone with the Color Scale

The test strips allow a semi-quantitative analysis and are used for rapid exploratory testing of substance concentrations as low as 1 mg/l (ppm). In case of nitrate determination the color scale is graded as follows: 0 – 10 – 25 – 50 – 100 – 250 – 500 mg/l (ppm). On the test stick there is also a so-called nitrite warning zone. If this zone indicates the presence of nitrite, amido-sulphonic-acid must be added to the water sample to destroy the nitrite prior to the nitrate detection.

A finer graduation of the color scale would be desirable. From the print technical point of view this is possible, but the visual resolution is not improved and a non-existent accuracy would be the result. However, there are two methods, that can be used to improve the reading significantly. In the first method, known as the "limiting simultaneous measurements", the reaction zone of the test stick after immersion is not compared to the color scale printed on the package but to other test strips whose reaction zones have been colored in standard solutions with known contents.

It is advisable first to measure the content range by comparing the test stick to the printed scale and then make further measurements above and below this value. In the range between 10–100 ppm a graduation of approx. 10 ppm is obtain-able. This principle can generally be used with test strips.

A second possibility for improvement of the reading accuracy is the method of reflectrometry with the a Nitracheck reflectometer which is of increasing importance for practical determination of nitrate in water, plants, and soil.

2.3 Titration Tests

Rapid titrimetric methods are suitable, for example, for determination of acidity, alkalinity, carbonate and total hardness, oxygen content, calcium- and chloride ion concentrations. These tests are mainly intended for analysis of boiler feed water and coolant water in power stations as well as spring and surface water. The rapid tests available on the market as e.g. Aquamerck® are packed into sturdy plastic boxes or hinged boxes (Fig. 2).

The test kits contain ready-made reagent solutions, a graduated plastic titrating vessel as well as a precision dropping pipette with a scale which gives the concentration or the measuring value directly. For example, for the total hardness determination of water with Aquamerck® 8039, one graduation mark on the titrating pipette corresponds to 0.2 °dH (German degree of hardness) (Fig. 3).

2.4 Color Card Tests

The color card test is visual colorimetry applied in its simplest form. Colors on a card, each valid for a concentration range, are compared by the incident light method to the color of a test solution after addition of reagents (Fig. 4).

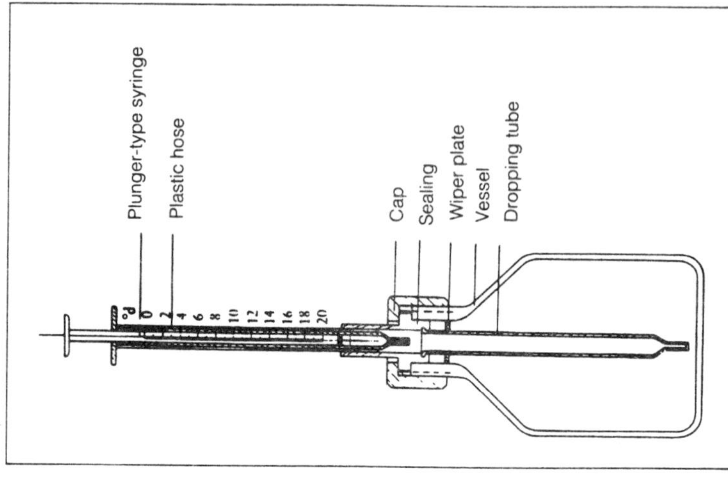

Fig. 3   Titrating Pipette

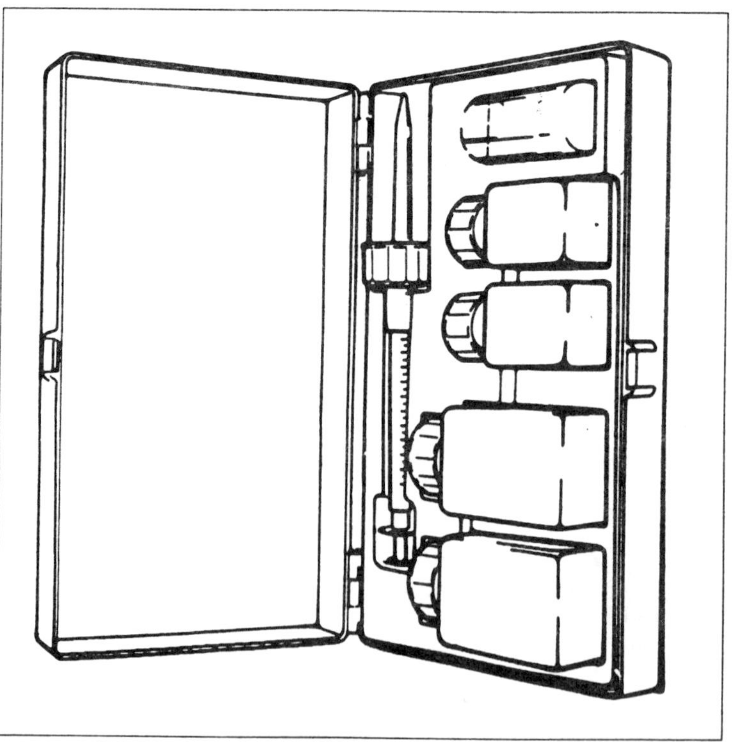

Fig. 2   Reagent Kit with Titrating Pipette

Fig. 4   Test Kit with Color Card

2.5 Tests with Sliding Color Comparator

A significant increase in detection sensitivity is obtained by increasing the layer thickness (test glasses with a fill level up to 80 mm) in a sliding color

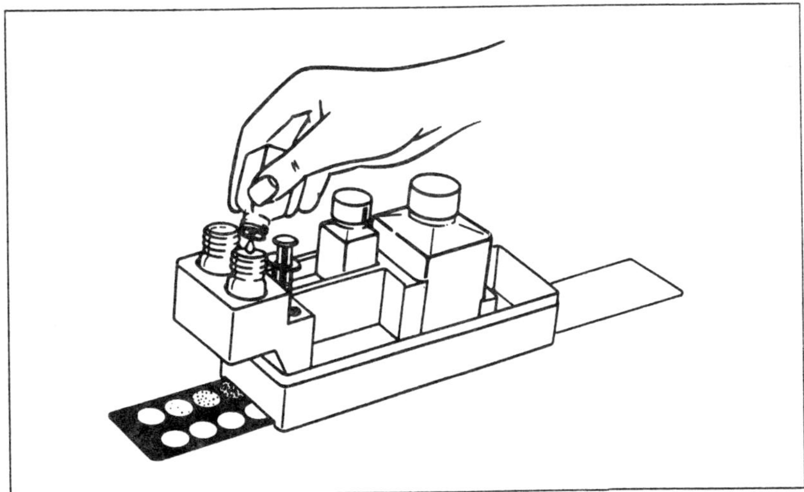

Fig. 5   Test Kit with Sliding Color scale Comparator

comparator. To perform a test, two tubes are filled with water samples. Add reagent to the tube located above the white spot or spots of the same color. Match the color by sliding the color scale under the tubes so that when viewed from above both tubes appear to be of the same color. A series of colored spots allow differentiated graduation of the concentration ranges (Fig. 5). The Aquaquant® sliding color comparator is especially suitable for trace analysis. Concentrations of different parameters up to 1 ppb (0.001 mg/l) can be detected.

2.6 Test Vessels for Colorimetric Measurements

The test vessels consist usually of three rectangular chambers. The center chamber is the reaction vessel for the color reaction, the two outer vessels are for color comparison standards. The color comparison standards consist of plastic cubes which are color-balanced to the different concentrations (Fig. 6).

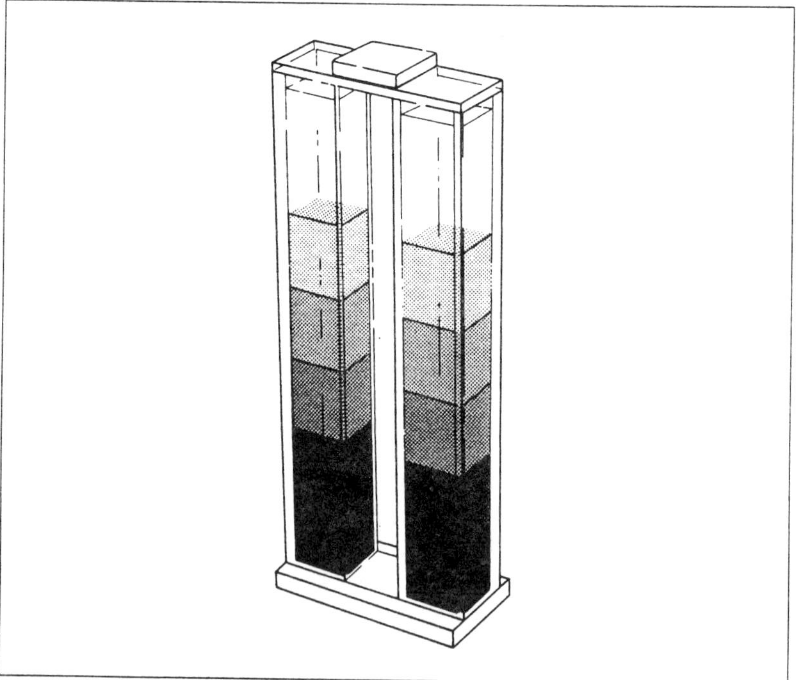

Fig. 6   Test Vessel (Color Comparator)

2.7 Rotating Disk Comparator

There is often a need for rapid colorimetric tests in a medium sensitivity range for determination of colored or turbid samples. For this purpose rotating comparator discs operating with transmitted light are well suited (Fig. 7).

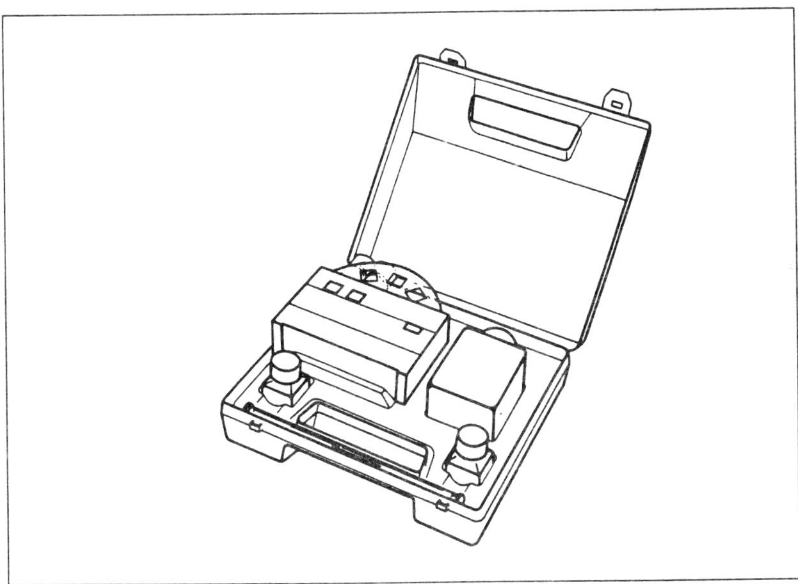

Fig. 7 Reagent Kit with Rotating Comparator Disc

This analytical method contains e.g. in the case of Microquant® a 10-stage color comparator disc with colored plastic windows. Two cuvettes are inserted into the comparator. The second cuvette (reference cuvette) is used for compensation of the intrinsic color of a water sample.

The special advantages of this system are:

1. Compact and sturdy design, ideal for rough operating conditions (e.g. field studies).

2. Measurements of turbid samples possible, e.g. photometers cannot distinguish between turbidity and coloring.

3. Measurements of samples with intrinsic color possible by using a comparison cuvette.

2.8 Spectroquant® Analysis System for Photometric Rapid Analysis

In most cases, rapid visual colorimetric methods provide sufficient sensitivity, are easy to use and low priced. Certain disadvantages, however, should not be disregarded. Lack of objectivity due to changing light conditions and vision defects, and limitation to distinct color values which allow only rough allocation of each concentration are the most notable. For some time now, photometric practices in the laboratories have been simplified in a way that the photometer can also be applied to rapid analysis. This is only possible in combination with reagent kits, specially optimized and working according to advanced methods. Such test kits contain all reagents necessary for the photometric analyses of one parameter each (Fig. 8).

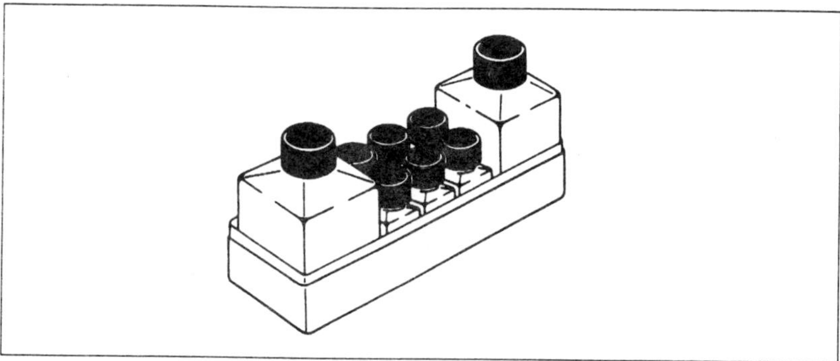

Fig. 8   SPECTROQUANT® Reagent Kit for Photometric Analysis

Liquid and powder concentrates with long shelf-lives cut the time requirements to a minimum compared to classical methods. Photometric determinations can be done, independent of well-equipped laboratories and, for field studies on-site, in combination with suitable photometers.

2.9 Application of Test Kits in Food Analysis

Soon after introduction of Merckoquant® test strips their suitability for rapid exploratory testing in food analysis was examined. In 1973, R. Musche and W. Lucas published "Orientierende Vorprüfung auf Schwermetallgehalt in flüssigen Lebensmitteln" (Exploratory pretesting of heavy metal content in liquid foods). It was found that heavy metals such as iron, cobalt, copper and zinc in liquid foods could be detected as far as those food did not show too much inherent color. In 1985, G.Schwedt and W. Altkofer published "Zur Leistungsfähigkeit von Teststäbchen in der Lebensmitteluntersuchung" (Test stick capacity for food examination). They reported that semi-quantitative analyses with test strips could be done to determine pH-value, potassium, nitrate, sulfite, tartaric acid, and ascorbic acid in various foods. Therefore, test strips can be applied for "purposeful pre-selection of analysis samples as well as for the determination of concentration ranges (e.g. for the production of serially diluted solutions) in combination with quantitative analyses". According to Schwedt and Altkofer, test sticks cannot and should not replace quantitative instrumental analysis, but "they are an important help within a method of industrial analysis, a so-called process step, because they allow localization and with that decrease of cost as well as a closer supervision of special food constituents".

In addition to the previously mentioned parameters, the following substances which can be determined using test strips are also of importance: aluminium, arsenic, calcium, total hardness, nitrite, peroxide, and tin.

The test strips have been emphasized here because they are used extensively in the food industry. The use of the more costly color comparator test as well as full quantitative tests, e.g. Spectroquant® are becoming more and more widespread.

Also well known are the fat tests Fritest and Oxifrit-test which allow users of small deep-frying equipment themselves to examine the quality of their fats. These tests are also used as a screening method for quality control.

More sensitive determinations of nitrite and phosphate in food can be done, after corresponding sample preparation, with Aquaquant®, Microquant®, or Spectroquant®.

Finally it should be pointed out that the detection of disinfection agents is possible. For this purpose, formaldehyde- and chlorine tests are also offered. The possibilities of analyzing spring, drinking, and mineral water with the described systems should be mentioned.

Rapid Determination Equipment for Food Quality Control

K. H. Torkler

3 Rapid Determination Equipment for Food Quality Control

K. H. Torkler, Pinneberg

3.1 Introduction

Food of animal or vegetable origin sometimes show significant fluctuations regarding valuable constituents. Quality control is aimed at a routine inspection of production processes to achieve a standardized, defined product quality and to exclude defective products.

The food manufacturer wants to recognize the constituent contents as well as the variations in his processing materials. Data as actual and reliable as possible are asked for to adjust the final product exactly to the desired quality.

The official food authorities monitor how this quality corresponds to the relevant legal norms.

For economical reasons, the manufacturer uses rapid determination equipment for quality control, the official food authorities use this equipment to assure effective consumer protection. In both cases semi-automated or automated units for rapid determination of valuable constituents are used, sometimes based on different test methods.

It is necessary, however, to delimit the term "rapid determination". For example, instruments offering a simplified operation of standard methods could be considered to be rapid methods because it is possible to execute more tests within the same time. The same is valid for equipment in which several parameters during a longer test time can be determined. Thereby a rational personnel employment is realized but the reporting of constituents is not speeded up.

What is rapid determination equipment?
This question can only be answered in comparison with a reference value. In this case the classical standard methods of the official food authorities are the criteria. They are also reference points for evaluation of measurement accuracy. If constituents can be detected a many times quicker than with classical methods, then we have rapid determination equipment or rapid determination methods, respectively.

The better the alternative methods correspond to the conventional methods, the more effective is the application of those procedures. For the official food authorities, the measuring accuracy is used for evaluating to what extent results obtained during routine checks by such equipment must be counterchecked by post-determinations with standard methods. The food producer can take the measuring accuracy as an indicator of the economic success which results from employment of the rapid determination equipment.

This article is limited to the most important constituents but the interested user can choose from a variety of different equipment and select the best for his application. The different rapid determination equipment can be employed for one or more parameters. The corresponding equipment techniques are distinguished according to the rapid determination of

1. the moisture content

2. the protein content

3. the fat content and

4. other constituents.

Explanation of these techniques and their application, respectively, are limited here to the basics. For a better overall view, test period (from sample taking to result), measuring accuracy, and measuring range (as far as known) are listed. But not all rapid determination methods can be mentioned. On the one hand, due to low distribution of the methods not described here. On the other hand, it shows the difficulties in observing this fast developing market and shows how few extensive reports are available on this subject. Besides the equipment described in the following, there are further rapid determination equipment such as remote temperature measuring units for monitoring raw material storage, germinability determinator or methods developed for special applications.

3.2   Instrumentation for Rapid Determination of Moisture Content

3.2.1 Infra-red Drying

Infra-red rays penetrate the product to be dried and dry it. With this technique it is necessary to assure a sufficient distance between infra-red radiator and sample material to avoid surface carbonization. The equipment is often available with integrated weighing system.

Sample weight: 2 – 10 g
Test period: 15 – 30 min
Test range: 0 – 90%
Equipment: table model as infra-red oven or as infra-red incident
                    light lamp
Application: food, in general

3.2.2 Microwave-Drying

In a closed chamber the test substance is exposed to a microwave radiation of previously determined energy level. The microwaves causes certain molecules inside the test substance to rotate. The product heats up and dries out. For adjusting the energy level as well as the drying period exact conditions determined for that specific product by comparative measurements must be retained to avoid carbonization. If carbonization occurs this means that, in addition to water, other constituents have become volatile.

Sample weight: 5 – 70 g
Test period: 3 – 10 min
Equipment: table or stand model with or without integrated balance
Application: food, in general

3.2.3 Measuring the Electrical Conductivity

Corresponding to the ability of substances to transmit electrical current, current is lead through a product and the product resistance is measured. The specific conductivity for a substance changes under influence of the bound water content. The temperature influences the conductivity by 1 – 3 °C, therefore, a temperature compensation during measurement is necessary. Special calibrations have to be prepared for different products. Constantly changing mixed portions influence the measured result negatively, because with each portion change the specific conductivity changes also.

Sample weight: approx. 40 g
Test period: approx. 2 min
Measuring accuracy: ± 0.5% absolute (grain)
Equipment: table model or continuously measuring and controlling equipment in the production
Application: ground grain or granulates as well as powdery products

3.2.4 Measuring the Dielectric Constant

Measurement of the dielectric constant is based on the precise measurement of a capacity. The dielectric constant of a substance is the capacity ratio between the condensator, filled with the substance, and the empty condensator. The more water bound in a product, the higher the dielectric constant. The dielectric constant is dependent on the temperature, and therefore, a temperature compensation is necessary. This measuring method is also negatively influenced by changing the mixture proportions. Calibrations have to correspond to the product.

Sample weight:	200 – 300 g
Test period:	30 sec
Measuring accuracy:	$s_d = 0.25$ to $s_d = 0.40^*$
Test range:	1 – 45%
Equipment:	portable model as well as table model
Application:	grain or granulates without previous milling

$$^* \; s_d = \sqrt{\frac{\Sigma \, (x_1 - x_2)^2}{n \; - \; 1}}$$

s_d = Comparability of results of a measuring method with measurement values obtained by standard method of the official food authorities

x_1 = Individual measuring value of the standard method

x_2 = Individual measuring value of the comparison method

n = Number of comparisons

3.2.5 Refractometric Determination

This method is based on measurement of the refraction index in liquids. This is an indirect measurement. The extract content of aqueous solutions is determined and recalculated with regard to the water content of the solution. The substance is dissolved, if needed, and the refraction at 20 °C is read with an immersion refractometer. Tables are used for recalculating the water content.

Application: Confectionery, marmalade, milk products and beverages

3.3 Instrumentation for Rapid Determination of Protein Content

3.3.1 Automated Kjeldahl Method (Kjel-Foss-Automatic)

After depositing the sample in the Kjel-Foss-Automatic equipment the whole classical Kjeldahl method is fully automated. The equipment is arranged as a carousel and the individual process steps are divided into six interlocking automatic steps. The positions have the following purpose:

Position 1 and *2* are for sample disintegration.

At *position 3*, the hydrolysate cools off and is diluted with water.

Position 4 starts with sample alkalization and distills then the free ammonia into a simultaneously running titration device in which the color change is determined photometrically.

The Kjeldahl flask is emptied automatically at *position 5* where the catalyst used precipitates as a sulphide.

Finally, at *position 6* the flask is ready for receiving another sample.

Each step takes about 3 minutes. The desired result is available after 12 minutes, that is after *position 5* has been reached. Due to the carousel arrangement it is possible to examine a new sample every 3 minutes.

Sample weight: 1 g (Protein content > 50% = 0.5 g)
Test period: 15 min (20 samples/hour)
Measuring accuracy: as standard method
Test range: 0 – 100%
Equipment: stand model
Application: food of all kinds

3.3.2 Direct Distillation

Ammonia is liberated from protein when heated in alkaline solution. This ammonia comes mainly from hydrolysis of the protein components glutamine and asparagine, which is completed after a few minutes. Other amino acids are destroyed more slowly and therefore, only partly responsible for ammonia formation. The free ammonia is distilled for 5 minutes into a common receiver and quantitatively determined by titration. The ammonia quantity freed by direct-distillation stands in close correlation to the total nitrogen content. For determination of reliable results the examination of an extensive number of comparison samples with the common Kjeldahl method is necessary. It is important to have always the same conditions, otherwise different protein components may be hydrolyzed and freed as ammonia.

Sample weight: approx. 5 g
Test period: 15 min
Test range: 0 – 25% protein
Equipment: stand model
Application: grain and grain products, meat and meat products

3.3.3 Color Binding Methods

These methods are based on the fact that proteins can bind certain coloring matters, for example amido black 10-B buffer solution. First, the color intensity of the original buffer solution is determined photometrically. Then the test substance is treated with this buffer solution. For liquids this is done directly, while for grain products a longer treatment (approx. 10 min) in special color reaction vessels is necessary. Loss of color intensity due to color binding on the protein can be observed. The intensity loss is determined photometrically and set in relation to the total protein content of the sample.

Sample weight: 1 g
Test period: 1 min (milk)
                        12 min (grain)
Measuring accuracy: s_d = 0.1 (milk)
                     s_d = 0.3 (grain)
Test range: 0 – 20% protein
Equipment: table or stand model
Application: milk and grain

3.3.4 Measuring the Thermal Conductivity

For this method, organic nitrogen compounds are converted into gaseous nitrogen and the amount of nitrogen is determined by means of its thermal conductivity.

The samples are weighed in quartz crucibles and incinerated in an induction oven with pure oxygen at approx. 1000 °C. The incineration converts all organic compounds into gases, the bound nitrogen is almost exclusively converted into N S42 T. Humidity and salts are condensated. The nitrogen oxides are reduced with copper at 650 °C to elementary nitrogen and disturbing gases are selectively absorbed. The tota elementary nitrogen is registered in a thermal conductivity detector and displayed as protein with the desired conversion factor.

Sample weight:	0.125 – 1.0 g
Test period:	approx. 10 min
Test range:	0 – 55% protein
Equipment:	automatic operating table models
Application:	grain products, other foods and feedstuffs

3.4 Instrumentation for Rapid Determination of the Fat Content

3.4.1 Density Determination of a Solvent-Fat-Mixture (Foss-Let)

Fat, which has a relatively low density, is dissolved in a solvent (perchlorethylene) with high density.

45 g sample material are weighed in special preparation chambers and 120 ml solvent are added. For binding the humidity in the sample 1 – 2 tablespoons gypsum are necessary. The chamber is placed inside an homogenizer in which the fat is extracted within 2 minutes. The chamber content is then filtered into a measuring unit until completely filled.

When the measuring temperature has been reached the buoyancy of a calibrated floating body inside the measuring chamber is determined automatically. The buoyancy point of the floating body is related to the density of the solvent; which is related to the fat content displayed digitally.

Sample weight:	45 g
Test period:	5 – 10 min
Test range:	0 –100%
Equipment:	table model consisting of dispenser, homogenizer, and measuring unit
Application:	food, in general

3.4.2 Density Determination of Raw Meat

This method is based on the principle that lean meat has a different volume compared to fatty meat. A 6 kilogram meat sample is placed in a special cylindric container made of sturdy stainless steel. A piston plate equal to container's diameter holds the meat under constant hydraulic pressure. The measured impression depth is related to the fat content and displayed electronically as fat content.

Sample weight: 6 kg
Test period: 3 min
Measuring accuracy $s_d = 1.0\%$
Test range: 2 – 90%
Equipment: stand model
Application: raw meat

3.4.3 X-Ray Absorption

Minerals absorb X-rays. An approx. 7 kilogram coarsely ground meat sample is placed into a container that is then clamped into a measuring chamber. In it the sample is trans-illuminated by X-rays, the absorption of the minerals in the meat is measured, respectively. The mineral proportion in meat varies slightly, approx. 1.5% absolute. Using the mutual relationship between fat, protein, water and the mineral content in raw meat, the fat content can be calculated. The result is displayed in % fat.

Sample weight: 6.5 kg
Test period: 4 min
Test range: 2 – 90%
Equipment: stand model
Application: raw meat

3.4.4 Gerber – Determination of Milk Fat

A butyrometer is filled consecutively with 10 ml sulphuric acid (90%), 10.75 ml milk, and 1 ml amyl alcohol. The butyrometer is centrifuged for 5 minutes at 1200 rpm and then placed into a water bath at 65 °C for 5 minutes. For raw milk the fat content can be read as gram fat, while for homogenized milk the procedure must be repeated, starting a centrifugation, prior to reading the result of the test .

Sample volume: 10.75 ml
Test period: 15 min (raw milk)
                    25 min (homogenized milk)
Test range: 2 – approx. 60%
Equipment: calibrated butyrometer and centrifuge
Application: milk and milk products

3.4.5 Turbidity Measuring (Milko-Tester)

This rapid method is based on photometric measurement of the fat globule concentration. The casein turbidity in the sample is removed by "versene solution".

Test period: 45 sec
Test range: 1 – 55% fat (with increasing fat content the accuracy
                    decreases)
Equipment: table model, semi- or fully automatic, also as monitor-
                    ing unit integrated in the production process

3.5 Instrumentation for Simultaneous Rapid Determination of Several Constituents

General: Various equipment is available, which utilizes the above mentioned methods, for different simultaneous or consecutive determinations with one or more integrated modules. Because these techniques have already been discussed, and they are only a combination of these methods, they are not mentioned here again.

3.5.1 Infra-red Spectroscopy

Infra-red light is an electromagnetic radiation following the visible spectral range toward the longer wavelengths. In general, infra-red has a wave length of 760 nm – 0.5 mm (1 mm). It is divided in:

    Near infra-red   (NIR)   760   nm–   2 .5 µm
    Medium infra-red (MIR)   2 .5 µm– 25   µm
    Far infra-red   (FIR)   25   µm–500   µm

As already described in section 3.2 (Rapid Determination of Moisture Content), infra-red radiation causes heat development because infra-red light excites molecular bonds causing vibration. Simply said, part of the infra-red radiation energy is consumed by vibration excitation and therefore, absorbed. If only specific wave lengths are allowed to pass through a filter, only defined molecular bonds, characteristic for a certain constituent, are excited. The resulting absorptions can be measured using the techniques, described in the following.

3.5.1.1   NIR-Reflection Measurement

This technique is based on light absorption in the near infra-red range. The proportion of absorbed near infra-red light at a defined wave length compared to an equipment related reference value is determined. For each measurement the radiation quantity reflected from the surface of a substance is determined. To obtain sufficient accuracy for determination of an constituent with this technique, it is absolutely necessary to measure at several different wave lengths. Reflection measurement is purely a surface measurement and therefore, a consistent, exacting preparation of the samples is important.

Prior to obtaining comparable results with such equipment, extensive comparative measurements with the standard methods of the official food authorities are necessary. The resulting calibration is very product specific. Samples not considered in the calibration, cannot be measured exactly and lead, as do changing mixture ratios, to higher deviations than could be expected according to the comparison values of the calibration when compared to the standard methods.

Test period: 3 – 10 min
Constituents: water, fat, proteins, carbohydrates – non-infra-red light absorbing components such as crude fiber, ash or grain hardness cannot be measured, but can be recalculated using logarithmic data.
Equipment: table models; for powdery products production process integrated by-pass measuring devices
Application: solid and pasty products as milk, grain, and meat products as well as raw materials for production control

3.5.1.2   MIR-Transmission Measurement

The measuring principle is based on the light absorption in the medium infra-red range. The absorption maximum (measuring filter) of each constituent is measured at a wavelength significant for that special substance and compared to the corresponding absorption minimum (reference filter) by transmitted light method. The following wave lengths show specific absorption maxima and absorption minima, respectively:

Fat measuring filter (A)	5.73 µm (ester groups)
Fat reference filter (A)	5.50 µm
Protein measuring filter	6.46 µm (peptide bonds)
Protein reference filter	6.80 µm
Carbohydrate measuring filter	9.60 µm (primary alcohol groups)
Carbohydrate reference filter	7.80 µm
Fat measuring filter (B)	3.40 (C-H bonds)
Fat reference filter (B)	3.60 µm

The liquid or liquefied sample is transferred to in an infra-red inactive cuvette and trans-illuminated by infra-red light. For measuring solid or pasty products those have to be liquefied. This is done with a special solution giving within 4 minutes of mechanical homogenization a stable, finely distributed emulsion. To measure the mentioned products, it is necessary to calibrate the total measuring range of the equipment by comparative measurements with standard methods. Newer developments allow fully automatic calibration under microprocessor control. For different product groups, such as milk, meat, and fish products independent calibrations are necessary due to the different protein components.

Test period:	10 sec (liquid products)
	15 min (liquefied products)
Measuring accuracy:	$s_d = 0.02$ (liquids up to 10%)and
	$s_d = 0.10$ (liquids above 10%)
(Soxhlet)	$s_d = 0.60$ (liquefied products – not milk)
(Kjeldahl)	$s_d = 0.40$ (liquefied products – not milk)
(Drying chamber)	$s_d = 0.80$ (liquefied products – not milk)
Constituents:	fat, proteins, carbohydrates, and water
Equipment:	table and monitor models integrated in the production process
Application:	delivered milk (payment samples), milk performance tests (individual animal samples for breeding evaluations), milk, meat, and fish products

3.6　Importance of Equipment for Rapid Methods in Food Quality Control

Following the survey of different rapid determination methods the importance of the rapid determination equipment for food quality control should be given further attention.

Such methods are very popular in the official food analysis because, on the one hand, they are an effective contribution towards consumer protection due to their ability to make a prompt statement in regard to food quality. This is especially important in a quick-turning and widely spread distribution systems. On the other hand, the increased costs, the increased control needs with high turnover rates, cause implementation of methods which require less personnel and expensive operating budgets.

Compared to that, the interests of the food manufacturer are different. The raw material is the highest cost factor. Therefore, the manufacturer needs at the time of raw material delivery a quick evaluation of valuable constituents that, at the same time, forms the basis for the acceptance/rejection decision. In fact, in most cases a neutral examination prior to delivery is done, however, this must not necessarily correspond with the actual quality delivered.

To avoid defective products it is important to determine the constituent variations for quality assurance. Only if the variations are known it can be determined what proportion of a total batch should be used in order not to exceed a fixed error when making a at determination of average content. Variations, which occur in the raw materials, necessitates taking of raw material samples. Together with the measuring accuracy of the analysis system employed, sampling is very important for evaluation and combination of raw materials.

Employment of modern computer systems has pushed the recipe creation and recipe optimization to the fore. The cost evaluation of the individual raw materials is, in addition to the delivery price, only as good as analytical data concerning the raw material are present. The more frequent and the more available data are, the more realistic the evaluation is. Standard methods and rapid determination methods are suitable for determination of constituents. But, standard methods require extensively more qualified laboratory personnel or are more expensive if done externally.

Because recipe optimization for production assurance operates with flexible averages, the possibility in an actual situation of using additional raw material reserves is given by employing rapid determination equipment. At the same time this method is used to control the desired quality in the individual production steps, in order to be able to draw conclusions from productions errors at any time.

Despite intense efforts of equipment manufacturers, instruments for immediate determinations are the exception due to the associated inaccuracy. The food manufacturer buys in large amounts for production and distributes them in smaller quantities into the production process. If he wants to optimize the raw materials economically, a small material buffer is needed at the time of actual evaluation. It is clear that shorter test periods result in shorter intermediate storage periods. But it is too early to conclude that the quickest result is the best result, because often the measuring accuracy influences the application ability even more. More exact methods usually give rise to longer intermediate storage periods but nonetheless favor the economic success more than the costs arising from increased storage time. Therefore, for employment of such rapid determination equipment, one is always looking for a compromise between production reality, test period, measuring accuracy, and economic success, including evaluation of results. At the same time, it should be stated that for utilization of all economic possibilities a certain adaptation of the production process to these rapid determination instruments seems to be unavoidable.

Ion-Selective Electrodes

F. Honold
K. Cammann

4 Ion-Selective Electrodes

F. Honold, Weilheim and K. Cammann, Münster

4.1 Introduction

The users of analytical examination methods need techniques which are simple, reliable and universally applicable. They should be automatized, have a high sample throughput and must be low-priced.

For years now, the ion-selective potentiometry, i.e. the concentration determination by means of potential measurements, fulfills many of these requirements. In addition to its low costs and low operating expenses, it requires only minimal sample preparation and can be applied to small sample quantities. It plays an important role in process control, in supervision of industrial processes or in the operating room where vital parameters must be monitored continuously. In Figure 1 different ions and substances that can be determined are listed.

With enzyme electrodes acetylcholine, 5'-adenosine-monophosphate, albumin, amygdalin, L-arginine, asparagine, catechol, creatinine, cyanide, L-cysteine, diphenylcarbamylfluoride, glucose, glutamine, L-leucine, D-methionine, nitrite, penicillin, L-phenylalanine, sulphate, thiosulphate, tyrosine, and urea, among others, can be measured.

Recently, new developments allow the above mentioned requirements to be fulfilled even better. Additionally, the spectrum of ions and neutral compounds detectable by means of ion-selective electrodes (ISE) has been expanded significantly due to application of new electrode design and materials. This is especially important in biology and medicine. On the other hand, the data handling is much easier due to application of microprocessor controlled ionmeters and completely automatic analysis systems. New techniques such as flow injection analysis allow more rapid and disturbance-free operation.

4.2 Measuring with Ion-Selective Electrodes

A measuring cell as shown in Figure 2 consists of an ion-selective electrode, a reference electrode, and the measuring device [1]. The electrodes are immersed into a solution which contains the ion to be determined.

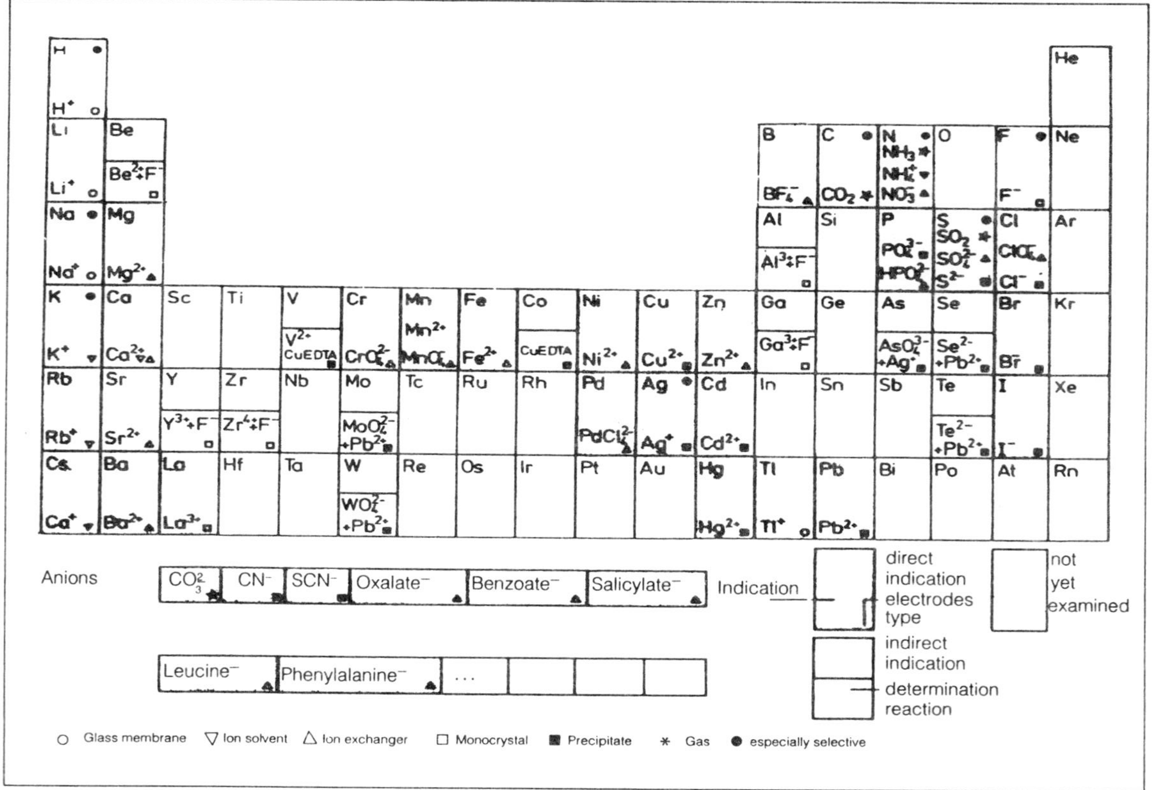

Fig. 1   Survey of Some Ions and Neutral Substances Detectable by means of Direct Potentiometry (listing not complete)

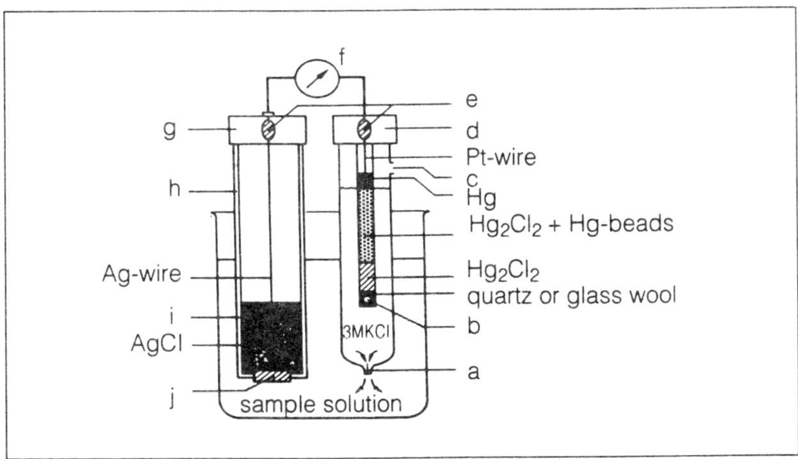

Fig. 2   Ion-Selective Measuring Cell Consisting of a Membrane Electrode and a Calomel Reference Electrode

a   asbestos fiber, metal wire, frit or ground sleeve as diaphragm
b   opening
c   opening for filling KCl-solution
d   calomel electrode
e   soldered joint
f   measuring device
g   ion-selective measuring electrode
h   electrode body
i   internal solution
j   ion-selective membrane

The ion-selective electrode has an ion-sensitive layer, normally a membrane. It can consist of glass, a low-soluble solid substance, a non-reacting organic phase fixed to a porous backing, or plastic. In the ideal case the membrane exchanges only **one** defined ion species [2] with the measuring solution at its phase boundary. The equilibrium of the exchange is given by:

$$\text{Ion}_{\text{solution}} \rightleftharpoons \text{Ion}_{\text{membrane}} \tag{1}$$

and must lie on the right side. Otherwise the electrode will stop functioning or only poor detection will result. Furthermore, the exchange must be fast. If not, drift and insufficient selectivity will be the result. The charges transported during this process constitute a potential difference at the boundary layer whose value depends on the ion activity in the solution. It follows the Nernst-equation:

$$\Delta\phi = \Delta\phi_0 \pm \frac{RT}{z_i F} \ln a_i \tag{2}$$

with

$\Delta\phi$ potential difference at the phase boundary in reference to the standard hydrogen electrode

$\Delta\phi_0$ potential difference at standard conditions

R   gas constant

T   absolute temperature

z_i ionic charge of the species i

F   Faraday constant

a_i activity of species i

$\pm$ polarity sign depending on charge

To measure this potential difference the electric circuit must be closed by means of the second electrode, the reference electrode. This electrode is, as the name indicates, the reference point for the ISE. Its potential, therefore, does not depend on the composition of the measuring solution, including zero. Any constant value is allowed. Usually, electrodes of the second order are employed, for example, the widely used calomel electrode, the thalamide electrode for ISE with a strongly negative potential ($\Delta\phi_0 = -575$ mV at 25 °C and a 3.5 M KCl-electrolyte), or for higher temperatures (up to approx. 130 °C) the Ag/AgCl electrode. As electrodes of the second order, they must be immersed into a solution with constant chloride ion activity. Because the measuring solution very often does not assure this, the connection to the solution is made by using a salt bridge consisting of a strong electrolyte with a constant chloride concentration. Contact with the measuring solution is established by means of ceramic diaphragm, a bundle of platinum wires, or a ground sleeve. At this junction variable diffusion potentials can occur which disturb the measurement. These are caused by the different migration velocities of the ions present. These effects are reduced by a salt bridge electrolyte with anions and cations having almost equal migration velocities, e.g. KCl or NH_4Cl, especially at higher concentrations (1–3 mol/l). Other disturbances occur if the measuring solution and the salt bridge electrolyte are incompatible. If, for example, a KCl-salt bridge is employed in a measuring solution containing silver ions, precipitation on the diaphragm can occur. If one used a double salt bridge reference electrode these difficulties can be avoided. The inner salt bridge electrolyte supplies the necessary chloride ions. The outer salt bridge electrolyte is chosen so that no interference occurs in the measuring solution. The total voltage of the measurement cell, which is then measured, is

$$E = E_0 + \frac{RT}{z_iF} \ln a_i \qquad \text{whereby} \qquad\qquad (3)$$

E   cell voltage of measuring cell

E_0 cell voltage at measuring cell at standard conditions including $\Delta\phi$ of the reference electrode and the diffusion potential

The term $\ln(10) \cdot RT/zF$ is also called the Nernst-slope S. For monovalent ions at 25 °C this has the value 59.16 mV/activity decade. A calibration function, as actually found for pNa glass electrodes, is shown in Figure 3.

The concentration is plotted on a logarithmic scale. The linear part of the curve in the figure can be described by equation (3). The electrode slope actually determined deviates slightly from the theoretical Nernst-slope. An electrode is considered still good if the slope of the calibration line has about 95% of the Nernst-slope. During the life of the electrode the slope decreases. The durability depends on the electrode type. Glass electrodes may last several years, liquid membrane electrodes last only a few weeks. If the concentration present is less than 10^{-3} mol/l, the deviations between concentration and activity may be neglected. For very low concentrations the calibration curve will be horizontal. According to the IUPAC (International Union of Pure and Applied Chemistry) the detection limit for ISE is de-

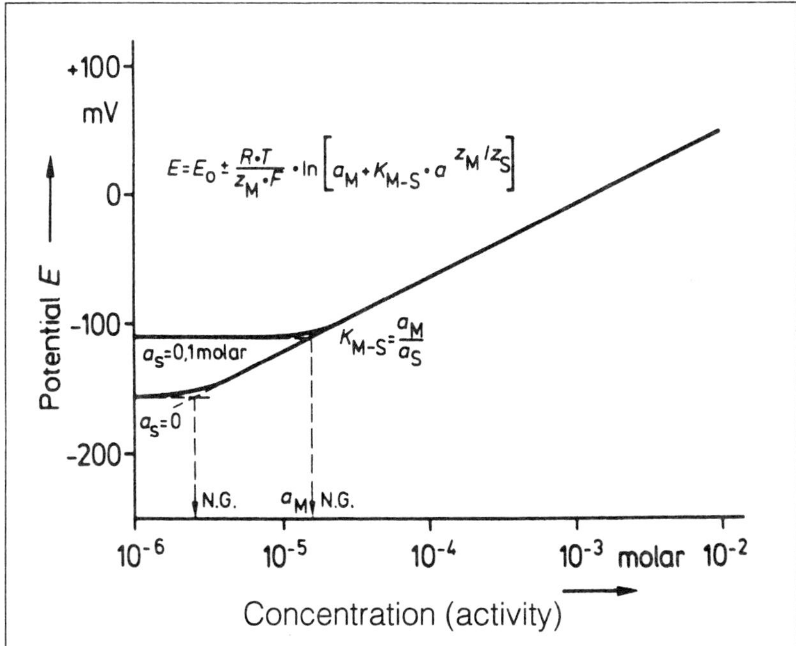

Fig. 3 Calibration Curve of a pNa-Electrode.

Without interfering ions ($a_S = 0$) the detection limit is at approx. $2.5 \cdot 10^{-6}$ mol/l. In the presence of interfering ions ($a_S = 0.1$ mol/l KCl) the detection limit increases to $1.5 \cdot 10^{-5}$ mol/l. The selectivity coefficient K_{Na-K} has the value $1.5 \cdot 10^{-4}$

fined as the point where the deviation of the extrapolation of the linear part of the calibration curve from the actual curve is equal to $18/z_i$ mV (refer to Figure 3).

Usually, ISE can be employed for concentrations down to 10^{-5}–10^{-7} mol/l. Quantitative determinations can be done in the linear as well as the non-linear part of the calibration curve. In the second case, however, more calibration points are to be taken and larger errors may occur. Modern equipment supports the evaluation in this case by suitable interpolation.

Unfortunately, ion-selective electrodes do not respond to only one species of ions, but more or less, with other ions with similar characteristics [3]. In order to describe their behaviour, equation (1) is extended to the Nicolsky-equation:

$$E = E_0 + \frac{RT}{z_i F} \ln (a_i + K_{ij}\, a_i^{zi/zj} + K_{ik}\, a_k^{zi/zk} \dots) \tag{4}$$

Here K_{ij}, K_{ik} ... are the so-called selectivity coefficients of the electrode in reference to the interfering ions j, k... . The value $K = 1$ indicates that the interfering ion and the sample ion are of equivalent magnitude. Good selectivity coefficients are less than 0.001. The interfering ion must be present in a 1000-fold concentration compared to the ion to be determined to produce the same result. pH-glass electrodes have the best selectivity coefficient against alkali ions. As an example, the value of $K_{H\text{-}Na}$ of less than 10^{-12} should be mentioned.

In principle, the measuring device is a pH/mV-meter. The measuring technique requires the input connection with a very high Ohm value ($> 10^{12}$ Ohm). It may also be called an electrometer input. This is necessary because the signal source, e.g. a glass electrode, may have a membrane resistance itself of several hundred megaohm. In simple voltmeters this can lead to significant errors. The wiring to the ISE usually is a shielded cable. The shielding can, at the same time, serve as reference electrode connection. This is the case in combination measuring cells. These are equipped with an electrode body which contains the ISE as well as the reference electrode.

Simple measuring devices operate analog and indicate the measuring cell voltage in mV or the pX-value (negative decimal logarithm of the activity of the ion X). More elaborate units operate with a microprocessor which execute integrated programs. They allow calibration of the unit in dialogue, thus avoiding operating errors. Some analysis methods, e.g. standard additions, can be done in this way easily and reliably. In addition to connecting to a laboratory recorder, it is possible to output the data to a printer or to transfer the data to a computer for further processing.

4.3 Types of Electrodes

The membrane composition is decisive for the ion or the compound which is indicated by the electrode. One must differentiate between solid-state membranes and liquid membranes. Among the detectable non-ionic compounds are dissolved gases in equilibrium with an ion, and substrates which form ions during the transformation by enzymes.

Among the solid-state membranes, the glass membrane is known best and has been used for the longest time (Figure 4a). Electrodes made of this material respond almost specifically to H^+-ions. Even today, the glass electrode is the most widely used ion-selective electrode for determination of pH-values, although there are polymer membranes available for H^+-ions. For high (pH >10) and low (pH <1) pH-values deviations from the calibration line can occur depending on the glass composition. The error in the alkali range is dependent on the concentration of alkali-ions present (mainly Li^+, Na^+, K^+). This observation led to the development of glasses that can be used specifically for determination of these ions (refer to Figure 3). Therefore, if a suitable sample matrix is present, this is an alternative to atomic absorption spectroscopy. Glass electrodes are often designed as combination electrodes.

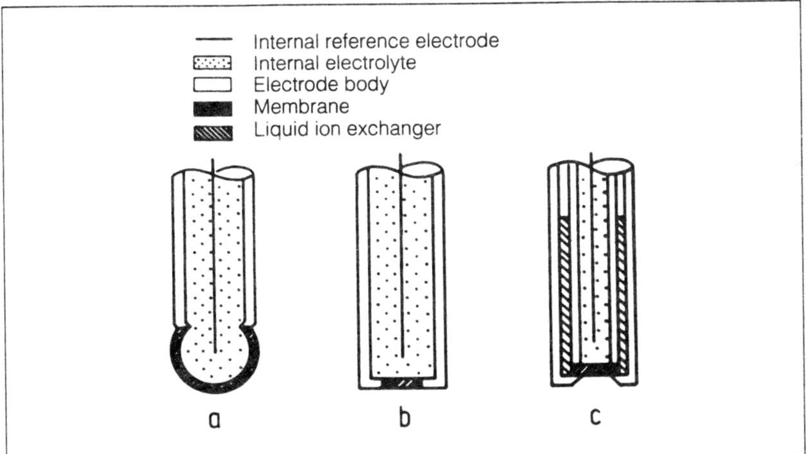

Fig. 4   Ion-selective Electrodes
a) glass electrode,
b) solid-state electrode,
c) liquid membrane electrode

In the 1960's the development of solid-state electrodes (Figure 4b) started with membranes consisting of a low-soluble precipitate of the element in question. With single crystals of AgCl, AgBr, AgJ, and Ag_2S e.g. silver cations as well as the corresponding anions Cl^-, Br^-, J^-, and S^{2-} can be detected. The best known type is the fluoride electrode on the basis of lanthanum fluoride (LaF_3) which is doped with europium to assure high ion mobility for F^-. Other ions such as Cd^{2+}, Cu^{2+}, Pb^{2+}, and CN-ions can also be determined with solid-state electrodes.

Liquid membrane electrodes (Figure 4c) consist of a with water non-miscible organic solvent in which liquid ion exchangers or ionophores are dissolved. Solution carriers are porous plastic discs which absorb the liquid. The ion exchangers can be either cation exchangers such as substituted

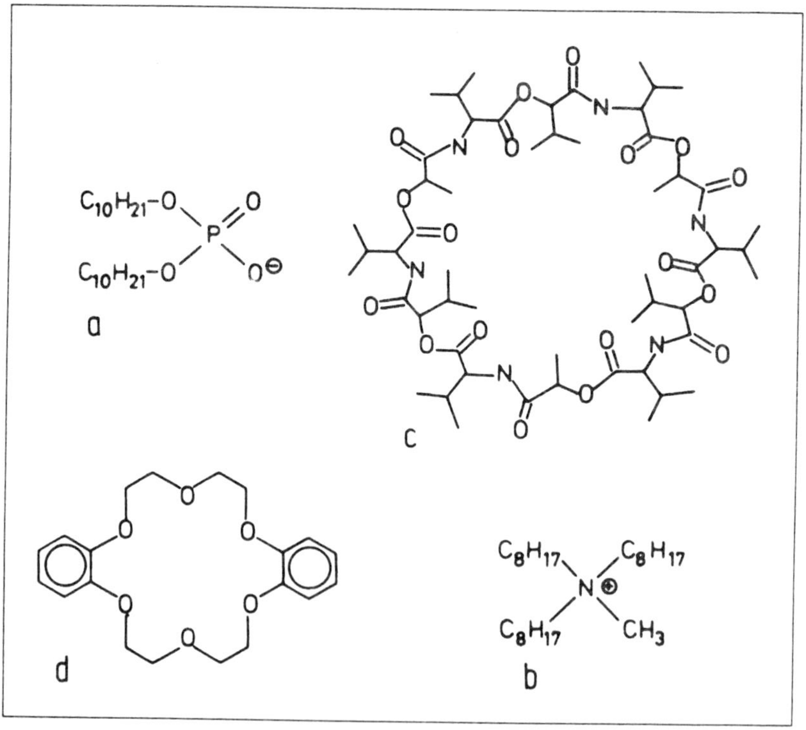

Fig. 5   Ion-active Compounds
        a) Cation exchanger b) Anion exchanger
        c) Ionophore d) Crown ether

phosphates responding to Ca^{2+} or Zn^{2+} (Figure 5a), or anion exchangers such as e.g. the tricapryl-methyl-ammonium-ion (Figure 5b). Ionophores are complex-forming active substances with high selectivity for individual cations. Examples are compounds as the naturally occurring valinomycin (Figure 5c) which is used for excellent potassium electrodes, or the synthetic molecules as the crown ethers (Figure 5d) or the so-called "Simon compounds" for Ca^{2+}, Li^{2+}, and Ba^{2+}.

Interesting modifications of the liquid membranes are the PVC-membranes. This type of membrane consists of an homogeneous mixture of organic solvent, electro-active ion exchanger or ionophore and high-molecular polyvinylchloride. Due to the stabilisation with PVC, this form of liquid membrane has a significantly longer life time.

The liquid membrane electrodes can be produced in a special form as microelectrodes [4]. An ionactive liquid phase is filled into the tip (diameter approx. 1 µm) of a conic glass tube. Such combination electrodes are called double barrel microelectrodes. Due to the extremely small contact areas the electric resistance of such electrodes is very high. This requires special measuring techniques to screen them from outside interferences.

Figure 4 also shows that all ion selective electrodes possess an inner reference electrode. This is often accomplished by using a silver wire covered with AgCl. The inner solution must of course contain chloride and the ion to be determined must also be present, because a stabile potential difference should form at the boundary layer membrane/inner solution. The inner solution is constant in its composition and so are the related potential differences.

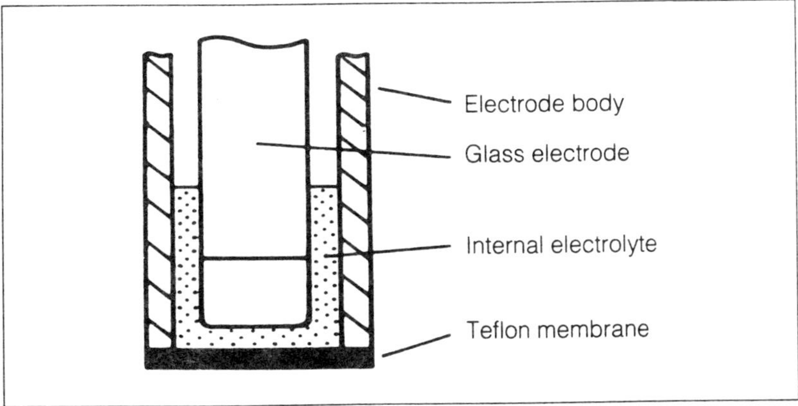

Fig. 6   Ammonia Electrode

Other electrode types are the gas-sensitive electrodes. These are common ion-selective electrodes responding to non-ionic compounds due to a pre-existent equilibrium between a gas and the solution in which ions are produced. The ammonia electrode (Figure 6) is explained as an example. The ammonia in the analysis solution diffuses through a teflon membrane, through which ionic compounds can not pass, into the inner electrolyte. There it causes a change of the pH-value according to

$$NH_3 + H_2O \rightleftharpoons NH_4^+ + OH^- \tag{5}$$

This pH-value change is indicated by a pH-meter. In the outer measuring solution the same reaction occurs. Fluctuations in the pH-value can shift the ratio NH_3/NH_4^+ there. For this reason conc. NaOH is added, resulting in a pH-value above 12. At this point the equilibrium is almost completely on the side of the ammonia. For an analogous carbon dioxide membrane acid must be applied. In this way other gases, such as NO_2, H_2S, SO_2, HF, etc. can be determined.

Enzyme electrodes are similarly designed [5]. An enzyme reacting with specific substrates releases only defined ions in a pre-reaction. Those ions can be detected by an ion selective electrode [6]. Depending on the enzyme,

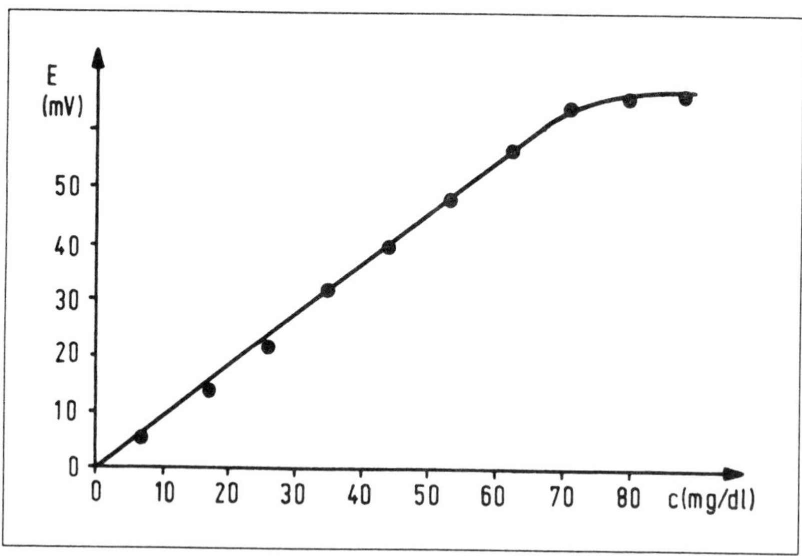

Fig. 7 Calibration Curve of a Glucose Electrode

the resulting electrodes are very selective. Combinations of enzymes may also be employed. Enzyme electrodes can be produced on amperometric basis. Among the electrodes operating according to potentiometric principles, the electrode for determination of glucose is explained here. According to the reaction

$$\text{Glucose} + O_2 + H_2O \xrightarrow{\text{GOD}} \text{Gluconic acid} + H_2O_2 \tag{6}$$

Glucose is oxidized to gluconic acid in the presence of oxygen and under the influence of the enzyme glucose oxidase (GOD). The enzyme is immobilized in a membrane permeable by the involved substances. Local changes in the pH-value can be detected by means of a pH-electrode. It can be seen that the logarithmic Nernst-relationship between concentration and cell voltage of the measuring cell is no longer valid. Instead the cell voltage, within a certain concentration range, is proportional to the concentration. This is caused by the buffer effect of the basic solution which has neutralized part of the gluconic acid.

4.4 Analysis Techniques

In principle, there are four different possibilities for working with ion-selective electrodes:

4.4.1 Direct Potentiometry

The direct potentiometry requires constructing a calibration curve (see Figure 3) with solutions of known content. Normally, measuring of two calibration solutions is sufficient to define the line. The concentration of the analysis solution to be determined can be read from this curve as long as its voltage response E is known. Most of the instruments available on the market operate according to this principle and the pX-value can be read directly (negative decimal logarithm of the concentration of the ion X). Other instruments are designed for indicating the result in mol/l. It must be noted that the direct potentiometric measuring always is an activity determination. If the concentration

$$c_i = \frac{a_i}{f_i} \qquad \text{with} \tag{7}$$

c_i concentration of ion species i
a_i activity of ion species i
f_i activity coefficient of ion species i

is to be determined, the activity coefficient f_i must be known. It is, however, possible to determine the concentration indirectly by measuring calibration and measuring solutions under addition of an excess of non-detectable ions. The ionic strength and with it the activity coefficient remains constant in the solution. All solutions are measured with the same activity coefficient which is thereby eliminated and the results are de facto concentrations.

For exact determinations it is important that measuring and calibration solution are similarly composed. This does not only decrease the influence of various activity coefficients, but influences on the reference electrode occur in the same way for both determinations. For this purpose, one of several TISAB (Total Ionic Strength Adjustment Buffer) may be added to the solutions prior to the analysis in equal amounts. Additionally, they reliably stabilize the pH-value since only narrow pH-ranges are allowed for certain electrodes.

4.4.2 Standard Addition

If the standard addition techniques are employed no calibration curve is necessary. The sample solution is measured and the resulting value is

$$E_1 = E' + S \cdot \log (C_x) \tag{8}$$

Then, a defined amount of analyte ions is added which causes a defined increase C_s of the analyte ion concentration at known volume. This results in the value

$$E_2 = E' + S \cdot \log (C_x + C_s) \tag{9}$$

The difference E2 – E1 after transformation of the evaluation formula leads to

$$C_x = \frac{C_s}{10^{(E_2 - E_1)/s} - 1} \tag{10}$$

The dilution effect caused by the added standard volume is not yet taken into consideration yet. Care must be taken paid that both measurements are made at the same diffusion potential value of the reference electrode. No change in the activity coefficient, whether due to over dilution or excess addition of standard solution with high ionic strength, may occur. In addition to the advantages of not having to make a calibration curve, the second measurement following the standard addition is carried out in the same solution with the same matrix components as in the first measurement. For this reason, this technique allows, under consideration of certain preconditions the determination of complex bound ions. In such a case, the complexing agents must be present in excess, so that the ratio of complexing agents to the free analyte ions is constant even after standard addition.

For this method as described here, the electrode slope must be known. It is, however, possible, to carry out a second standard addition thus eliminating the parameter S of the determination equation.

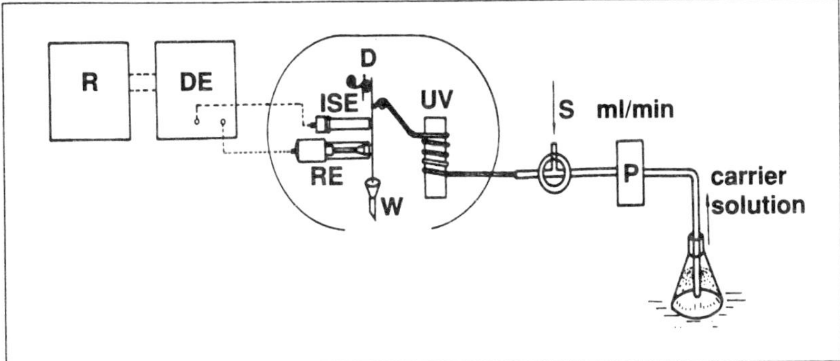

Fig. 8 FIA-Analysis Unit
P Peristaltic pump; *S* Injection block; *D* Detector unit with ion-selective electrode (ISE), reference eletrode (RE), waste (W) and optional UV-lamp; *DE* mV-meter; *R* laboratory strip chart recorder

4.4.3 Potentiometric Titration

Finally, the potentiometric titration is another possibility. Here, the ion-selective electrode is only an aid for determining the end point of titrations. The advantage of this combination is that on the one hand titration is an analysis technique with high accuracy (error < 0.5%) and one the other hand the end point determination by means of ion-selective electrodes is much more precise than with commonly employed indicators. For an exact determination of the end point the Gran-method can be used. It linearizes the relation between electrode signal and concentration, but contrary to the Nernst diagram according to Figure 3, the concentration curve is linear and the ordinate with $10^{E/S}$ exponential. This allows extrapolations to a theoretical concentration of zero (at the end point). The instruments employed today titrate and evaluate automatically.

4.4.4 Flow-Injection

The flow-injection analysis (FIA) is a very young technique which is employed advantageously if the selectivity coefficients are too large for an error-free measurement [7]. Figure 8 shows a suitable design with simplest means.

The carrier electrolytes continuously transported from the pump are injected into the injector block through micro-liter syringe. If analyte ions are present when the sample passes the electrodes a signal peak is obtained. The advantages of a better selectivity are paid for by a loss of electrode slope. This is explained by the fact that a steady-state equilibrium is not reached for time reasons. With an UV-lamp, organic compounds which under heavy radiation release an analyte ion are detectable [8]. Figure 9 shows a calibration curve for p-chlor-nitrobenzene and a chloride selective monocrystal electrode.

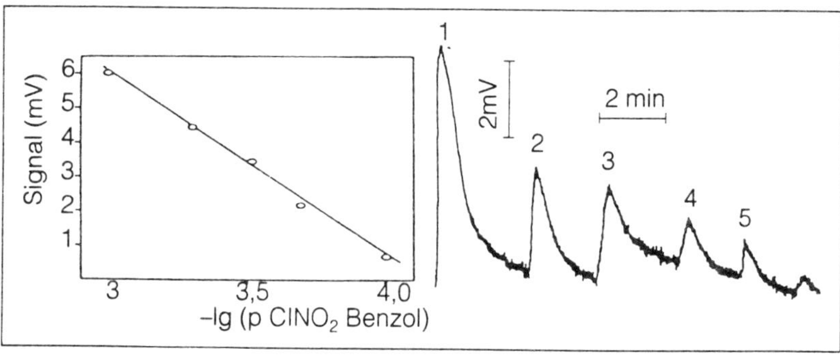

Fig. 9 FIA with UV-Destruction of p-Cl-nitrobenzene
50 µl injections of solutions with concentrations (1) 10^{-2}M; (2) 10^{-3}M; (3) $5 \cdot 10^{-4}$M; (4) $3 \cdot 10^{-4}$M; (5) $2 \cdot 10^{-4}$M

4.5 Applications

Many publications regarding the application of ionselective electrodes are available. Only a few examples should be mentioned here.

ISE are used for water control. Carbonate, which to a significant degree causes water hardness and is characteristic for natural waters, can be determined with a special liquid membrane electrode. Alternatively, a gassensitive electrode can be employed to measure CO_2 development after acidification. In water analysis, the chloride electrode can replace the titration with $AgNO_3$ and chromate indicator which have a very poorly recognizable end point. For both methods higher halides are interfering factors. In common waters, however, they are present in traces only. Waste water must also be examined. Some industries must determine the cyanide content in this matrix. This is done with direct potentiometry using a cyanide solid-state electrode.

There are ions which are not easy to measure with ISE. Sometimes this problem can be solved by indirect determination. Take aluminium, for example. The aluminium sample is added to a fluoride solution in which is the fluoride was determined by a F^- electrode. Al^{3+} will form a complex with F^- and the concentration of free fluoride in the solution will decrease. From the decrease as indicated by the electrode, the aluminium can be calculated. A phosphate determination works similarly. The phosphate is precipitated with La^{3+} and the excess lanthanum is back-titrated with fluoride.

Ion-selective electrodes can also be used in food analysis. The nitrate values of spinach and other vegetables can, after fine grinding and leaching with water, be measured comfortably with a NO_3 selective electrode. pH-values are often determined with electrodes in various shapes. With pointed electrodes fruit and meat can be measured. Flat glass electrodes are suitable for surface measurements.

Many applications are possible in medicine. Na^+ and K^+ are important body electrolytes which can be determined by direct potentiometry of the serum. Determination of ammonia in urine is carried out with a gas-sensitive electrode. In the field of dental medicine determination of fluoride are done by dissolving tooth material in perchloric acid and measurement with a LaF_3 solid-state electrode. The Ca-electrode (liquid membrane with dodecylphosphate) is used for measuring the content of free Ca^{2+}. This value is often more significant than the total Ca-amount which includes calcium bound to low molecular complexing agents and proteins. The total Ca-content can be determined by atomic absorption spectrometry. In this context it should be mentioned that heparine, usually added in whole blood analysis can complex analyte ions and lead to false low values. Today, analytical instruments are available which are able to execute automatically blood analyses for Na^+, K^+, Ca^{2+}, and Cl^- within a few minutes with an accuracy of 2%. Combined with an automatic sampler which takes several prepared analysis samples from a sample plate this unit is very suitable for routine determinations in hospitals.

The previously mentioned microelectrodes are employed for examinations of nerve excitation, based on rapid activity changes of certain ions (Na^+, K^+)

in biological cells. With these devices either as glass or liquid electrodes it is possible to take measurements inside individual cells.

The medical sector is the main area of use for enzyme electrodes. They are also called biosensors because they can detect biologically active molecules. With enzyme electrodes with a design based on ISE e.g. urea can be determined. The enzyme urease which produces ammonia from urea is employed together with a NH_3 electrode. The combination of urease with arginase allows additional determination of the amino acid arginine. Another amino acid, tyrosine, is quantified with tyrosine decarboxilase which produces carbon dioxide. This is then detected by a CO_2 sensor. Penicillin can be detected and determined when penicillinase transforms the substrate to penicillic acid which is then detected by a pH-electrode. The latest development are electrodes which use the highly specific antigen-antibody-reactions as selection mechanisms.

The ion-selective potentiometry covers a broad spectrum of analytical applications. More than 60 % of the main and secondary group elements of the periodic system as well as a large number of non ionic compounds can, in one form or another, be measured with ISE. The development of IS-electrodes is moving towards more selectivity, less sample preparation and increased application possibilities. Especially in determination of biologically active substances, which can be indicated by a sensor without prior separation, developments can be expected.

REFERENCES

1. CAMMANN, K.: Elektrochemische Untersuchungsverfahren, in "Untersuchungsmethoden in der Chemie", Hrg. H. Naumer, W. Heller, Thieme Verlag, Stuttgart, New York (1986)

2. CAMMANN, K.: Das Arbeiten mit ionenselektiven Elektroden, Springer Verlag, Berlin, Heidelberg, New York (1977)

3. CAMMANN, K.: Fehlerquellen bei Messungen mit Ionenselektiven Elektroden, in Analytiker Taschenbuch Bd. 1, 245—267, Springer Verlag, Berlin, Heidelberg, New York (1980)

4. HAVAS, J.: Ion- and Molecule-Selective Electrodes in Biological Systems, Springer Verlag, Berlin, Heidelberg, New York (1985)

5. CAMMANN, K.: Fresenius Z. Anal. Chem. (1977), 287, 1—9

6. SCHINDLER, J. G., SCHINDLER, M. M.: Bioelektrochemische Membranelektroden, de Gruyter, Berlin, New York (1983)

7. ILCHEVA, L, CAMMANN, K.: Fresenius Z. Anal. Chem. (1985) 322, 323—326

8. ILCHEVA, L, CAMMANN, K.: Fresenius Z. Anal. Chem. (1986) 325 11–14

Rapid Determination of Metallic Contaminants in Food and Food Raw Material by ICP-AES and AAS

H.-J. Hoffmann

5 Rapid Determination of Metallic Contaminants in Food and Food Raw Material by ICP-AES and AAS

H.-J. Hoffmann, München

5.1 Introduction

Metals play an ambivalent role in biology. Many of them are regarded as essential trace elements for humans, animals, and plants. But it is indisputable that due to anthropogenic activities the heavy metal concentrations in some areas of the biosphere have increased and in part reached alarming levels. Most metals exhibit their toxic effects in the trace range. Their determination can only be accomplished using the trace analysis techniques. Generally, the analysis of extremely low metal contents is contrary to the need for high sample throughput.

Due to the recent and rapid development in instrumental analysis, metal analysis techniques have made significant progress. Today, metallic contaminants, even in low concentrations, can be determined quickly using methods of atomic spectrometry. In practice, the atomic absorption spectrometry (AAS) and within the last years, the atomic emission spectrometry with inductively coupled plasma (ICP-AES) have been employed successfully in the laboratories. Both methods are based on the principle of solution analysis. The sample preparations as described in the following are valid for both.

5.2 Sample Preparation

Size reduction and homogenization are important pre-requirements for obtaining a representative sample of the test material. Prior to this pretreatment the sample is often dried. Usually, the water is removed by heating (e.g. in the drying chamber) or by freeze drying. Furthermore, the sample must be chosen carefully with regard to the problems in question. For example, for ecological problems it is advantageous to examine the socalled "detoxification organs" of the fish, such as kidneys and liver. They allow, due to their high accumulation ability, conclusions to be drawn regarding the heavy metal load of the respective waters. On the other hand, samples of the significantly less contaminated fish fillets are suitable to answer questions regarding human consumption.

The decomposition serves to transform the sample from the solid state into a dissolved state. This is one of the most important requirements of a solution analysis. The following demands must be met by the decomposition method:

1) All substances must be decomposed completely (this requirement can be limited for convention methods)

2) The elements to be determined must quantitatively remain in the digestion vessel (this is especially difficult to achieve with volatile elements, e.g. Hg).

3) Decomposing reagents such as acids, for dissolving incineration residues, should only be used in low quantities. With that the blank values as well as the dilution of lower heavy metals can be kept in certain limits.

4) Interaction between the decomposition material and the material of the digestion vessel should be reduced to a minimum. This requirement leads to the use of smaller vessel surfaces, lower decomposition temperatures and inert vessel materials (e.g. teflon or silica glass).

5) Of significant economic importance is the employment of digestion vessels that can be reused several times without becoming contaminated.

6) For simplification and acceleration of trace analysis as many elements as possible should be determined from one decomposition solution.

Practically, these demands can only partly be fulfilled with a reasonable effort regarding instrumentation and personnel. The following decomposition methods for fish and plants are used successfully, for example: the digestion vessel is a mini-autoclave with a teflon cup (V = 50 ml). The sample weight varies between 0.1 and 1.0 g, depending on the sample type and the gas development expected. The sample is decomposed with 1–3 ml conc. nitric acid under pressure at approx. 140 °C. The decomposition period depends on the matrix type, usually, 2–4 hours are sufficient. For complete destruction of some plant matrices the addition of a trace of hydrofluoric acid (e.g. 0.1 ml) is sometimes necessary.

This decomposition method has the advantage that, in addition to almost all heavy metals, some volatile elements as mercury and sulfur can be determined from one solution.

Experience show that one must be very cautious when decomposing samples with a high fat content.

Basically, the direct analysis of solid matters and oils is possible with AAS and ICP-AES. This technique has not yet been widely accepted for solid matter analysis because the problems occurring with the necessary micro-sample weights and calibration of such very complex solid matrices have not been sufficiently resolved.

5.3 Method Selection

Selection of a suitable determination method is mainly dependent on the element to be determined and the matrix. In the following the most important atomic spectrometric methods are described.

- Atomic absorption spectrometry (AAS)

- Atomic emission spectrometry with inductively coupled plasma (ICP-AES)

5.3.1 Atomic Absorption Spectrometry

In the last years the atomic absorption spectrometry has been developed to become the standard method in almost all laboratories which are concerned with analysis of metals. This method offers many large advantages, as there are:

- high selectivity

- favorable price/performance ratios
  (cost of investment + operating cost compared to attainable sample throughput)

- perfected technique

- low detection limits

- universal application for almost all materials and sample types

- comprehensive literature and methods
  (e.g. DIN- and VDI-methods for the legal field)

- the possibility of determining an extensive element range from one sample.

For practical applications the atomic absorption spectrometry is divided into the following variants:

- Flame AAS (e.g. air/acetylene flame)

- Furnace AAS (e.g. graphite furnace technique)

- Cold-vapour AAS technique

- Hydride AAS technique

5.3.1.1   Flame AAS

Spraying a solution into a flame for atomizing the metal in a sample is a well-known technique in the atomic absorption spectrometry. Aerosol formation, necessary for this technique, is mostly done with a pneumatic nebulizer. Contrary to the graphite tube- or hydride technique a time independent signal is produced whose height is proportional to the content of the element to be determined in the solution for measurement.

5.3.1.1.1   Air/Acetylene Flame

An air/acetylene flame is the form of excitation used mostly in AAS. With a temperature of approx. 2300 °C it is for many elements suitable as an excitation source with a noticeable self-absorption only below 230 nm. The following metals can be determined with sufficient sensitivity using the air/acetylene flame:

silver	cadmium	copper	cobalt	manganese
nickel	zinc	lead	iron	

The practical operating ranges for these elements start usually at concentrations around 50 to 100 µg/l.

5.3.1.1.2   Nitrous Oxide/Acetylene Flame

Medium temperatures of approx. 2750 °C can be obtained with the nitrous oxide/acetylene flame. Therefore, it is suitable for analysis of metals that can not be determined without interference when using the air/acetylene flame:

aluminium	chromium	barium	beryllium
vanadium	calcium	tungsten	molybdenum
titanium	zirconium		

In practice, determination limits of about 100 µg/l can be achieved.

5.3.1.2 Furnace AAS

The graphite furnace technique is a highly efficient alternative to the flame AAS because atomization of the metals is done electrothermally. Due to longer retention periods of the atoms in the narrow furnace, the usual determination limits for metals are about 100-times lower than for the flame technique. But it must be noted that interferences increase almost to the same extent. Newer background correction techniques such as Zeeman- and Modifier-additives can eliminate some of the causes for measurement errors. Furthermore, the addition technique offers valuable possibilities for measuring metal traces even in difficult matrices by means of the graphite furnace technique. In practice, the graphite furnace AAS (e.g. PE 5000) equipped with a Zeeman-underground correction has proven itself. This instrument (PE 5000) is equipped with a lamp revolver with six lamps. Theoretically a sample can be examined for six metals using this instrumentation. This can only be realized if the samples are very similar in matrix and composition.

5.3.1.3 Cold-Vapour AAS Technique

For determination of mercury other techniques, e.g the cold-vapour technique, must be applied because mercury is insensitive to flame AAS. For the cold-vapour technique, the unique characteristic of this heavy metal, Hg exists in atomic form at room temperature, is used. In spite of several chemical reactions, with their potential susceptibility to interferences, the cold-vapour AAS technique has not yet been replaced by other atomic spectrometric determination methods. For practical application in the laboratory low-priced measuring devices that can be operated independently from the "normal" AAS-unit have been proven best.

5.3.1.4 Hydride AAS Technique

It is known that the elements arsenic, selenium, antimony, etc. cannot be detected with the necessary sensitivity by the previously described techniques. This problem can be solved easily by employing the hydride technique. This variant of the AAS uses the tendency of some elements of the IV., V., and VI. main group of the periodic system to form volatile hydrides. The advantage of volatilization of hydrides lies in the separation and concentration of the elements to be determined. For producing the hydrides usually sodium hydridoborate in hydrochloric acid solution is used.

Normally, the hydride technique requires special decomposition methods to transform e.g. arsenic or selenium into a suitable form, followed by a com-

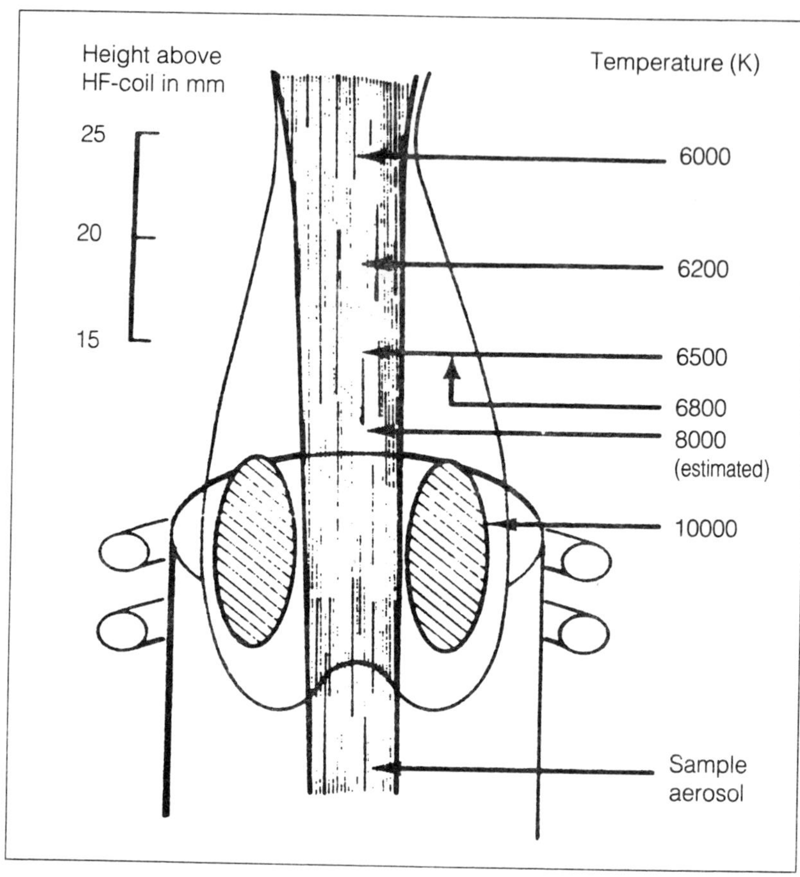

Fig. 1   Temperature Profile of ICP-Plasma

plete reaction to AsH_3/AsH_5 or SeH_4. As examples the $MgO/MgNO_3$ – or H_2SO_4/H_2O_2 – decomposition can be mentioned.

5.3.2 Inductively Coupled Plasma – Atomic Emission Spectrometry (ICP-AES)

While the term "rapid" for the AAS could be used with limitation only, it is valid for the ICP-AES-method without any limitations.

The inductively coupled plasma (ICP) as excitation source is highly significant for plasma emission spectrometry. This is due to the following factors:

- the favourable characteristics of the ICP-source allow a suitable compromise of conditions for the simultaneous excitation of almost all elements

- the high temperatures in the plasma (Figure 1) allow hardly any interferences

- rapid determination by optical emission spectrometry with photomultipliers as highly sensitive detectors.

These characteristics make application in almost all fields of the elemental analysis possible.

As already mentioned in the beginning, the ICP-AES is also a solution analysis method. Sample preparation techniques of the AAS can usually be directly taken over. Depending on the type of spectrometer the ICP can be employed in sequential and simultaneous spectrometers.

5.3.2.1 ICP Sequential Spectrometer

Usually a sequential spectrometer is supplied with a rapid scanning monochromator (Figure 2). Larger parts of the spectrum must be scanned in intervals.

For practical applications this measuring arrangement shows advantages regarding flexibility in selection of elements to be determined and surface correction. Disadvantageous are the often extended measuring periods.

5.3.2.2 ICP Simultaneous Spectrometer

The ability of ICP-AES to produce simultaneous analyses of 20–40 elements essentially in the same time as a single element analysis is clearly one of the most fundamental advantages of the method. Most spectrometer of this type are equipped with a polychromator system in Paschen-Runge mounting in the geometry of the Rowland circle (Figure 3).

With that the ICP abilities as an excitation source for the atomic spectrometry are shown to advantage. This is especially true regarding the relatively

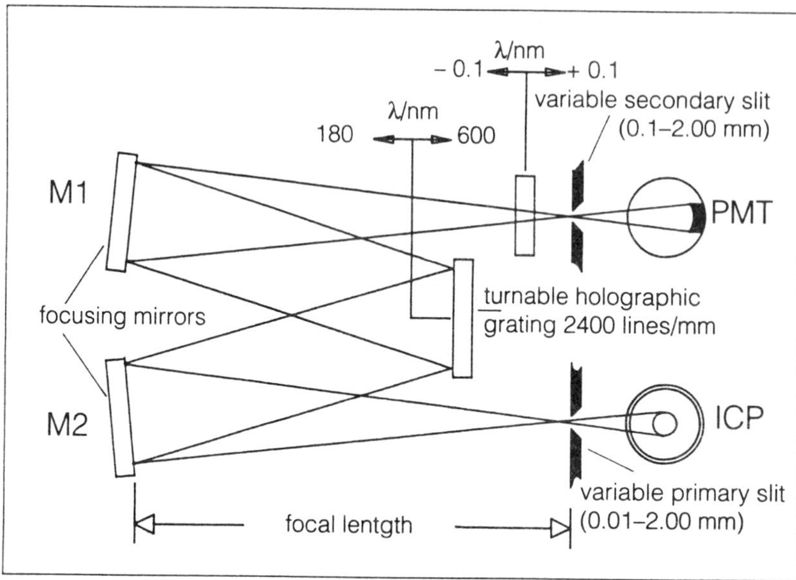

Fig. 2   Monochromatic Arrangement according to Czerny/Turner for Sequential Emission Analysis

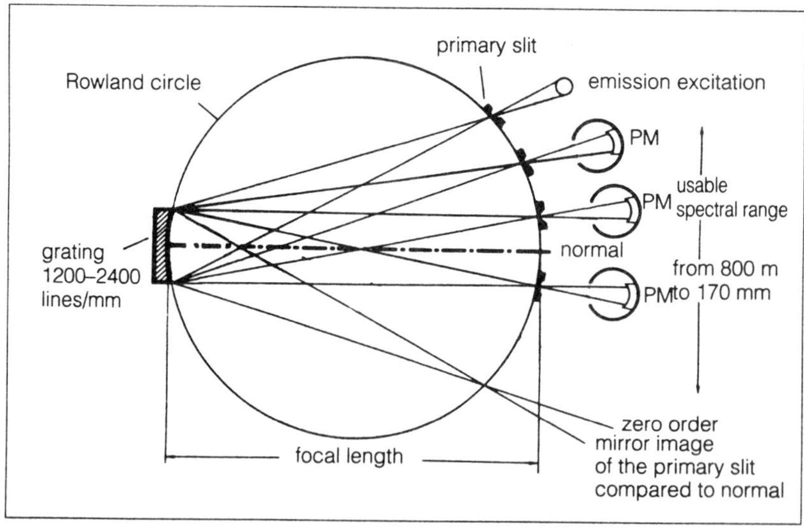

Fig. 3   Spectrometer Arrangement according to Paschen-Runge

simple setting-up of "compromise conditions" for excitation of almost all metals. With this measuring arrangement (Figure 4) extremely low analysis periods of a few seconds for 30–40 elements can be obtained. Disadvantageous is, however, the limitation to a fixed element selection, but generally this can be extended.

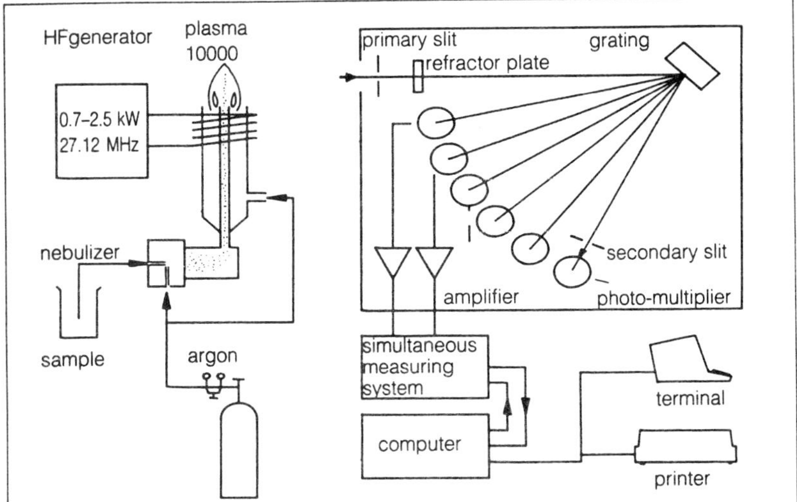

Fig. 4   Schematic Diagram of a Simultaneous ICP Spectrometer

5.3.2.3   Characteristics of ICP-AES

The combination of well-established spectrometer technique with the new excitation source ICP and sensitive detection systems has led to a highly efficient determination method for most trace elements.

5.3.2.3.1 Linear Dynamic Range

For many elements it can be operated within a wide linear dynamic range over three to four powers of ten.

5.3.2.3.2 Detection Limits

Even under practical conditions, ICE-AES determination limits are usually between those of flame and graphite furnace techniques. Table 1 lists the lower operating range limits for some elements. They are from DIN 38 406 part 22. The data were determined experimentally and confirmed statistically.

Tab. 1:   Limits of Determination of ICP-AES (in solution for measurement)

Element	Wavelength (nm)	Limit of determination (mg/l)	Element	Wavelength (nm)	Limit of determination (mg/l)
Ag	328,068	0,02	Mo	202,030	0,03
	338,289	0,02		204,598	0,05
Al	308,215	0,1	Na	589,592	0,1
	396,152	0,1	Ni	231,604	0,02
As	193,696	0,1	P	178,287	0,5
	197,197	0,1		213,618	0,1
B	208,959	0,005		214,914	0,1
	249,678	0,006	Pb	220,353	0,1
	249,773	0,01	S	182,036	0,5
Ba	233,527	0,004	Sb	206,833	0,1
	455,403	0,002		217,581	0,1
	493,409	0,003	Se	196,026	0,1
Be	313,042	0,002		203,985	0,1
	234,861	0,005	Si	251,611	0,02
Bi	223,061	0,04		212,412	0,02
	306,772	0,08		288,158	0,03
Ca	315,887	0,1	Sn	235,848	0,1
	317,933	0,01		189,980	0,1
	393,366	0,0002	Sr	407,771	0,0005
Cd	214,438	0,01		421,552	0,01
	226,502	0,01		460,733	0,1
	228,802	0,01	Ti	334,941	0,005
Co	228,616	0,01		336,121	0,01
Cr	205,552	0,01		337,280	0,01
	267,716	0,01		368,520	0,01
	283,563	0,01	V	290,882	0,01
	284,325	0,01		292,402	0,01
Cu	324,754	0,01		310,230	0,01
	327,396	0,01		311,071	0,01
Fe	259,940	0,02	W	207,911	0,03
K	766,490	2		209,860	0,06
Li	460,286	0,9		239,709	0,06
Mg	279,079	0,03		222,589	0,06
	279,553	0,0005		202,998	0,08
Mn	257,610	0,002	Zn	206,191	0,01
	293,306	0,02		213,856	0,005
			Zr	343,823	0,01

5.3.2.3.3 High Speed Measurements

Very high measuring speeds are a feature of ICP-AES . According to our experience it is possible to analyze in simultaneous measuring operations up to 100 decomposed solid matter samples for 20–30 elements per working day (=2000 to 3000 measured values).

5.3.3 ICP-MS

Finally, as a very modern method for trace analysis of metals, the ICP-MS, should be introduced here. For this analysis system the ideal excitation concept of the ICP has been combined with the very sensitive mass spectrometric detection.

Table 2 shows the fantastic determination limits of this new method for different elements. They are about one or two decades below the usual for ICP-AES.

Tab. 2   Limits of Determination of ICP-MS and ICP-AES (µg/l)

ICP-MS			ICP-AES			
0,1–0		1–10	0,1–1	1–10		10–100
Ag	Mg	B	Na	Ag	Mo	Al
Al	Mn	Ge	Be	B	Ni	As
Ba	Mo	Fe	Ca	Cd	Sr	K
Be	Na	Li	Mn	Ce	Ti	Na
Ca	Sn	Zn	Mg	Co	V	Pb
Cd	Sr		Sc	Cr	Zn	Se
Ce	Ti			Cu		Si
Co	Ti			Fe		Sn
Cr	V		La		Te	
Cs	W					
Hg						

5.4   Conclusion

The rapid determination of metallic contaminants in food and food raw materials is carried out successfully by combining several atomic spectrometric methods as e.g.:

ICP-AES →graphite furnace AAS → cold-vapour AAS → hydride AAS

In this way, time consuming sample preparations (size reduction, homogenization, decomposition) can be combined optimally with a rapid instrumental analysis. If ICP-AES is employed, then the flame AAS technique usually is not.

For the following elements German standard methods (AAS or ICP) are available:

Cd   DIN 38 406 part 19 (AAS)   Hg   DIN 38 406 part 12 (AAS)
Cr   DIN 38 406 part 10 (AAS)   Pb   DIN 38 406 part 6 (AAS)
Zn   DIN 38 406 part 8 (AAS)   As   DIN 38 405 part 18 (AAS)

and DIN 38 406 part 22 (ICP) for the elements Ag, Al, As, B, Ba, Be, Bi, Ca, Cd, Co, Cr, Cu, Fe, K, Li, Mg, Mn, Mo, Na, Ni, P, Pb, S, Sb, Se, Si, Sn, Sr, Ti, V, W, Zn, Zr.

REFERENCES

WELZ, B.: Atomabsorptionsspektrometrie, Verlag Chemie Weinheim, Deerlield Beach, Basel 1983

HOFFMANN, H.-J.; ROHL, R.: Plasma-Emissions-Spektrometrie, Analytiker-Taschenbuch, Springer Verlag Berlin, Heidelberg, New York, Tokyo, Band 5 (1985)

Thin-Layer Chromatography
– A Screening Method for Food Analysis

H. Jork

6 Thin-Layer Chromatography – A Screening Method for Food Analysis

H. Jork, Saarbrücken

In a time with daily new press releases on toxic substances in air, water, or food, control methods are increasingly asked for; screening methods for analyzing, reliably and quickly, as many samples as possible. The detection sensitivity should be good and the total methods should not require too many processing steps.

Screening test means selection test. This includes

- that several samples need to be examined,

- that each matrix has a similar composition for analogous pre-chromatographic sample preparation, and

- that parameters have been defined as selection criteria.

An analysis method that fulfills these preconditions optimally and that can be adapted specifically to these requirements is the thin-layer chromatography.

6.1   Preconditions for TLC as a Screening Method

Thin-layer chromatography (TLC) and its further development the "High Performance Thin-Layer Chromatography (HPTLC)" are separation techniques, based on the principles of liquid chromatography, which can be found in every food chemistry laboratory. The TLC is:

- *easy to learn and operate,*

- *rapid,* especially as HPTLC method,

- *always ready for use,* because pre-coated plates can be used without further pre-treatment,

- *easy to control,* because the total chromatogram can be visually controlled on one view.

- There is *no forced elution* for the individual components. This is also valid for substances remaining at the starting point.

- TLC and HPTLC-plates are *disposable*.

- The method can be employed *cost efficiently in routine operations* due to the low solvent consumption and minimal disposal problems. With a few milliliters of mobile phase, up to 70 samples and authentic comparison substances can be chromatographed next to each other in the horizontal chamber [1].

- For solvent selection there are *no restrictions regarding detection.* For sample separation those chromatographic conditions can be chosen which give the best separation results. At the time of detection no solvent is present in the chromatogram.

- As mobile phase acid, alkali or pure aqueous systems can be employed as well as neutral lipophile liquids. This offers not only large variability for the stationary phases but the *total chromatographic system is flexible enough to adjust optimally to each problem* [2].

The detection possibilities can be evaluated positively, too. In principle, physical, micro-chemical, and biological-physiological detection methods can be employed [3].

Among the *physical methods* are the photometric methods of absorption and fluorescence measurements, the phosphorescence quenching as well as the detection of radioactively labelled substances by means of auto-radiographical techniques, scintillation measurements or other radiometric methods.

Micro-chemical reactions belong either to the field of universal reagents or to those detection methods in which defined functional groups are converted (group characterizing detection). Substance specific reactions are practically unknown [4].

The *biological-physiological detection methods* make use neither of the physical nor of the chemical properties of a compound, but they detect the biological efficiency of the separated components [5]. Of course, these methods show the usual fluctuations, as in biology and some manual skills are necessary for this examination. Despite of certain disadvantages, employment of those biological-physiological determination methods, which sometime are very work-intensive, is justified for screening tests [6,7]:

– The methods are highly specific,

– Ineffective components do not interfere with the detection, therefore, cleaning-up techniques are not necessary.

– The detection range of the substance to be determined may be so low that even classical detection and determination methods yield no better results.

Such methods have been employed for detection and determination of antibiotics [8–32] and analogous effective alkaloids [33–36], insecticides [38–42] and fungicides [43–48], mycotoxins [49], vitamins [50–54], bitter substances [55–59], and saponins [48,56,60–65,68].

The above mentioned three possibilities for substance characterization can be used individually or coupled for detection purposes, and in fact, independently from the actual separation process. No on-line coupling is necessary. Therefore, there exists no temporal or physical connection to the separation process. The TLC/HPTLC-plate can be considered as a diskette on which data are stored that can be called upon depending on the different problems. The thin-layer chromatography offers not only a remarkable efficiency and selectivity for substance separation but also an excellent detection selectivity (Fig. 1). Those are the most important reasons why the thin-layer chromatography is very suitable for screening tests.

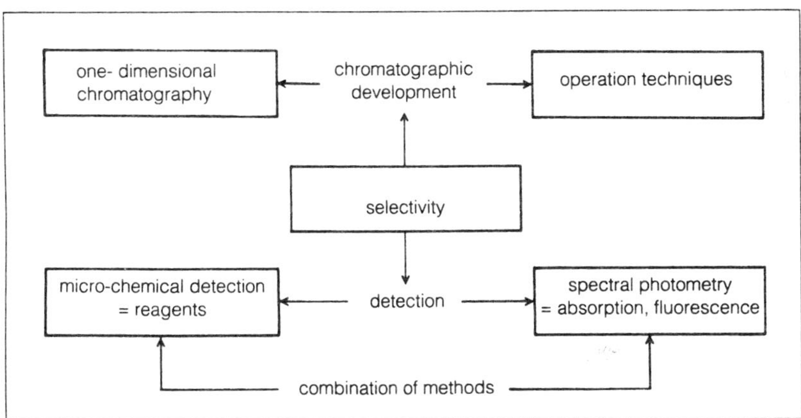

Fig. 1 Selectivity Ranges for Employment of Planar Chromatography as Screening Method for Food Analysis. Bioautographical detection methods are a useful completion (see text).

6.2   Quantitative Determinations

Besides qualitative statements, quantitative determinations can also be carried out. In the beginning, the so-called indirect determination methods [7, 69–73] were used, in which the thin-layer chromatography served practically only as a "clean-up method". The interesting zones were scrapped off the chromatogram and then eluated or directly extracted from the chromatogram with an "Eluchrom" unit [74–84]. Then the solution was optimized.

Direct photometric measurements of the chromatogram are more sensitive and quicker (in-situ measurements [7,72,85,86]). With the instrumentation, available today for laboratories, the total random error for nanogram amounts is usually between VK = ±1 and ±3%.

6.2.1 External Standard Methods

Evaluation by means of an external standard is always employed when compounds in a sample are determined by separately recorded calibration or regression curves (lines). For the recording of one calibration function at least five calibration solutions should be distributed as equidistant as possible over the operating range. For each TLC-plate a new calibration curve must be recorded because external factors may interfere with the allocations of the corresponding value pairs. Therefore, transfer onto other chromatograms is hardly possible. Interfering factors are, e.g.,:

- Vaporizing effects of the mobile phase in relation to time and temperature.

- Deviations in the starting zone (spot size effects) when applying solution.

- Changes of layer thickness of the stationary phase.

The deviations can be up to ±20% [69,71,87–94]. For a new recorded calibration curve the random errors lie between ±1 to ±3% each time.

Often the measuring range can be limited in order to avoid applying the five calibrations per chromatogram (or 10 by double determination). For screening methods this is especially interesting. In this case the so-called **two-standard-method** (two-point calibration) is used allowing more analyses per chromatogram. It is often employed for testing if the concentration of a component in a sample is lower than 90% or higher than 110% of the nominal value [95–99]. For routine analysis and quality control this is especially interesting. Interpolation between both limiting values is done in order to determine the concentration in a sample. A pre-requirement for this method is that the calibration lines intersect the point of origin of the coordinates. Extrapolation above the measuring range limit is not allowed.

The same pre-requirements are valid for operating according to the **one-standard-method** [100–115]. The origin of coordinates is assumed to be the second measuring value through which the calibration line is running. The one-standard method is used for evaluations in the field of quality control. If the content of a component in a preparation is known, but must be checked (stability and storage testings), a standard solution with the same content is made and chromatographed parallel to the sample solution. The concentrations of the standard and the sample solution should be similar. Both chromatogram paths are registered and the peak area values are compared to each other. Calculation of the content is done using "the rule of three" because according to the pre-testing the component contents deviate only minimally (<5%) from the nominal value and the calibration curves or lines will have almost the same values. Figure 2 shows the application schemes for the corresponding evaluations. Two sample solutions each can be applied on HPTLC-plates followed by the standard solution, because the reproducibility of the plate material is higher than for the usual TLC-plates.

Considering the two- and one-standard methods, the question of the placing of sample and standard solutions arises. When working with external standards the **double value application** is most useful. It reduces the application error for the factor $1/\sqrt{2}$. Nonetheless, additional systematic and random errors can occur, such as

- small differences in the distance of the solvent front from the starting point resulting in changes in the hRf-value (100 times retention factor),

- edge effects during development in chromatographic chambers without sufficient chamber saturation,

- different depth distribution of the substance in the layer,

- various chromatogram zone formations, etc.

It is useful – in order to detect those, too, – to apply two matching application zones separated by one-half TLC-plate size (Fig. 3). For evaluation, mean values of the data pairs are taken (**Data-Pair-Technique** [97]) thus reducing the application error as well as the chromatographic error.

In this way, the reproducibility of the measurement can be increased significantly. This is valid for peak area- as well as peak height evaluations.

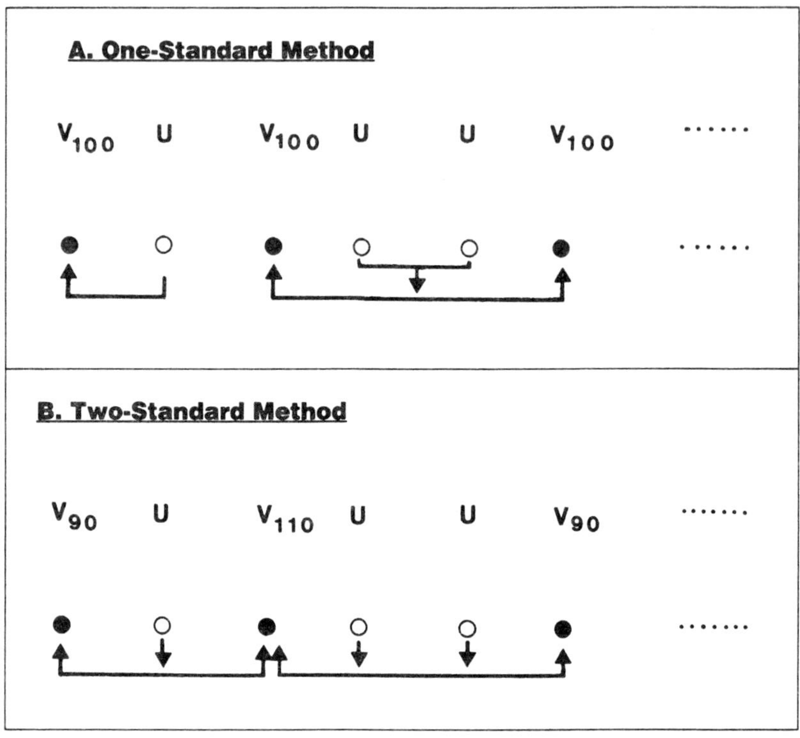

Fig. 2 Application Scheme for One- and Two-Standard-Method. The index numbers correspond to percentages; V = Standard solution, U = Sample solution (For further details refer to text)

The data-pair technique, however, looses importance with increased improvement of chromatography and increased method automatization. Up to that point, however, it should be preferred to all other application and evaluation schemes. Several publications confirm this statement, such as, e.g., in the field of food control and cosmetics [116–118], environmental analysis [119–121], pharmacy and medicine [98, 122–129] as well as industry and engineering [79,130–131].

6.2.2 Internal Standard Method

Evaluation using internal standard substances is well-known for gas- and high pressure liquid chromatography. Selected reference substances are employed – this is also valid for thin-layer chromatography – to

- exclude application errors,

- eliminate the effects of chromatographic gradients (solvent- or saturation gradients) on the measured result as much as possible

- determine recovery rates more exactly because volume transference errors can be eliminated,

- Increase the number of starting zones for analysis solutions available as compared to working with external standards.

This generally positive goal, which also saves time and money because fewer comparison solutions must be prepared, is limited by the requirements placed on internal standard.

Standard substances can only be those compounds which have almost the same chromatographic behaviour as the component to be determined. If possible, *they should be placed directly next to the sample substances on the chromatogram* to assure that the chromatographic influences effect both components equally. Furthermore, measuring time is saved in this way.

Influencing factors are:

- changes in solvent composition along the length of run as well as

- changes in chamber saturation and with that changes in the vaporizing effect in chromatographic chamber systems.

If the substances are spaced widely apart, these influencing factors may cause different zone formations. The spots might not be in the middle of the measuring slit [7] making a re-positioning necessary.

Additionally, the substance distribution in the chromatographic spot may possibly be different. The internal standard substance should not be in the extreme chromatographic ranges, that is, the hRf-value should lie between 30 and 80. For scanner operating according to the double beam principle, possible tailings of solvent fronts (α- or β-fronts) may be compensated, resulting in an increased application range towards the top.

Beside these requirements, which are related to chromatography, the compounds chosen as internal standard substances must have an analytical behaviour similar to the sample substance. For absorbing compounds the absorption maximum should be within the same range of wavelength as for the sample substance. The measuring wavelength should be selected

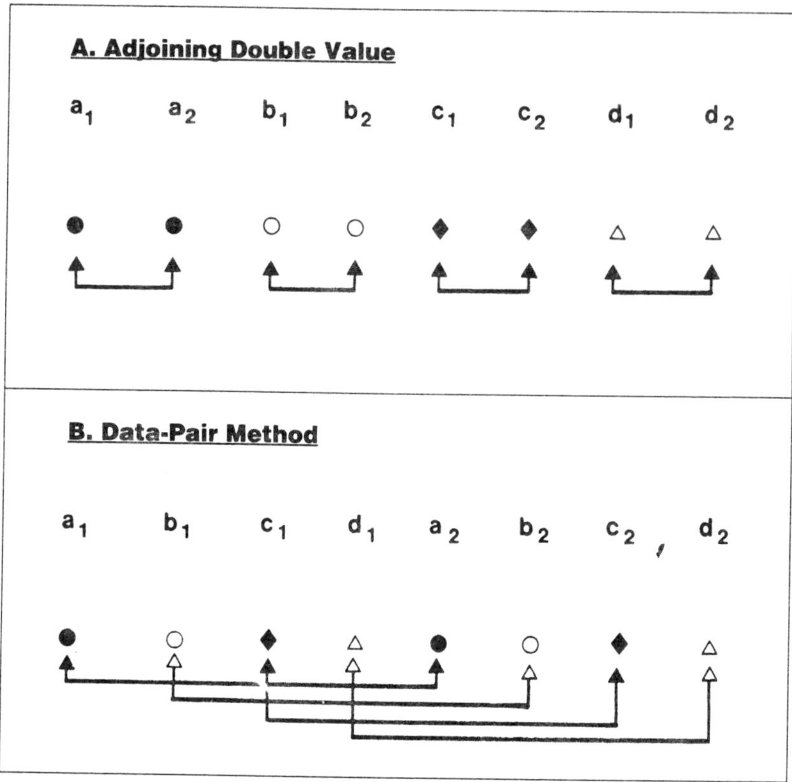

Fig. 3 Application Scheme for Adjoining Double Value and Data-Pair Application (For details refer to text)

optimally to be between both absorption maxima, thus avoiding a wavelength change during registration.

For fluorescent substances the internal standard should not only be a stable compound, but also be able to absorb in the same excitation range as the sample substance. No additional filter change is then required.

The same is valid for emission, if monochromators are used in front of the photomultiplier. If cut-off filters are employed, the wavelength of the fluorescence is no longer so important, as long as the absorption slope of the filters is ahead of the emission range of both substances. By widely spaced fluorescence maxima the differences in the sensitivity of the photomultiplier can become noticeable.

The molare absorption coefficient of the substance used as internal standard should be similar to that of the sample substance. This property is necessary in order to obtain an equivalent measuring signal for similar application amounts (spot sizes). This is similarly valid for the quantum yields of fluorescent compounds, so that the measuring signals for sample and standard substances are of relatively equal size.

If unequal measuring signals still exist, the measuring wavelength is chosen corresponding to the absorption (fluorescence) maximum of the *smaller* peak. For the larger peak the wavelength "slides down" the spectra slope. The absorption coefficient decreases and the measuring signal for this substance is reduced.

If a suitable internal standard is found for the above mentioned requirements (this is the difficulty of TLC, because only few substances can be separated in a single run of, e.g., 10 cm) and the linear dependency of the measuring signal in reference of the applied amount of substance is known, the evaluation is done as follows:

An exactly defined amount of internal standard substance is added to the sample solution. The component to be determined is applied on two or three calibration tracks after addition of the same amount of standard substance. After the chromatographic development, calculation of the sample substance is done according to the following equation:

$$A = B \times \frac{f_{ST}}{f_S} \times \frac{F_S}{F_{ST}}$$

In which:

A = Amount of substance of component to be determined
B = Weight of internal standard substance in the corresponding sample solution
f_{ST} = Peak area of internal standard of the calibration track
f_S = Peak area of the substance to be determined from the calibration track
F_S = Peak area of substance to be determined in analysis track
F_{ST} = Peak area of internal standard in analysis track

The measuring error which can occur during application is included in the quotient f_{ST}/f_S. It is equal in both substances (internal standard and sample substance) and can be cancelled out.

Klaus [133,134] as well as Frei et al. [88, 135] introduced almost simultaneously the quantitative evaluation of absorbing and fluorescent substances through the use of internal standards to thin-layer chromatography. Sample streaks as well as minute spot application techniques were used. The advantage of evaluation by internal standard is the independence from the application volume [92,103,136–139]. The volume does therefore not appear in above equation.

The internal standard must not necessarily be a "foreign substance" which is added to the sample solution. A component of the sample can be chosen which does not decompose during chromatography and must be present in a constant amount. This avoids the time-consuming search for an "open area" in the chromatogram [121] in order to insert the standard. Often this method is used for pharmaceutical formulas, liquids from the field of clinical chemistry as well as cosmetic solutions and lotions as long as the samples are *homogeneous*.

Internal standards can be employed for one-dimensional as well as for two-dimensional chromatography [88]. Attempts to re-evaluate earlier calibration curves by means of the calibration factor f_{ST}/f_S, as proposed by Klaus [134], is possible only for special cases. Usually the calibration factors vary about 5%. It is therefore more favorable to make a new calibration curve or to add addition standards each time as opposed to the further use of earlier values. Ebel et al. [92,140] as well as Glasl and Ihrig [94] have come to the same result. The question of whether the internal standard substance should be in front or below the component to be determined in the chromatogram cannot explicitly be answered. Corresponding tests gave results which do not differ significantly [141].

Compared to working with external standards for which a dilution series for recording a calibration curve must be produced, the evaluation with internal standards is less time-consuming. Only one calibration solution is needed. Additionally, the internal standard is added to the sample solution [92]. In the chromatogram up to three calibration track are provided [137,142]. The other starting zones are available for the standard-sample solution mixture. The effort, compared to the number of analyses performed, is significantly lower than for evaluation by calibration curves. Hagiwara et al. [143] as well as the Lockwood group [144] used this advantage for screening methods. In the field of clinical chemistry, Fenimore [145] recommended this technique while Reh and Jork [146] eliminated transfer difficulties with solution by the addition of a standard substance prior to the chromatography.

The question whether working with external standards makes more sense than the evaluation with internal standards has been answered by Rost [147] and Ebel [138] as follows: *Whenever the application errors are larger than the measuring errors regarding peak height or peak area, the determination of the sample substance by internal standard substances is better.* This statement is derived from the following consideration:

The standard deviation σ for evaluation by height (H) or area (A), respectively, for the external standard method is, according to the law of error propagation, given by the following relationship, if the volume applied is defined as V:

$$\sigma_{ext}^2 = \frac{2\sigma_H^2}{H^2} + \frac{2\sigma_V^2}{V^2} \tag{1}$$

respectively

$$\sigma_{ext}^2 = \frac{2\sigma_A^2}{A^2} + \frac{2\sigma_V^2}{V^2} \tag{2}$$

For the internal standard the following is valid:

$$\sigma_{int}^2 = \frac{4\sigma_H^2}{H^2} \tag{3}$$

respectively

$$\sigma_{int}^2 = \frac{4\sigma_A^2}{A^2} \tag{4}$$

Comparison of the relations (1) and (3) or (2) and (4), respectively, shows: If the measuring error for peak height or peak area determination is smaller than the application error, then the standard deviation for working with an internal standard is smaller.

$$\frac{\sigma_H}{H} < \frac{\sigma_V}{V} \rightarrow \sigma_{int} < \sigma_{ext} \tag{5}$$

But, if the measuring error for peak height or peak area is greater, then the external standard method should be preferred.

$$\frac{\sigma_H}{H} > \frac{\sigma_V}{V} \rightarrow \sigma_{int} > \sigma_{ext} \tag{6}$$

This is usually the case when, e.g., the height and/or the half height band width of registered peaks is measured manually, and when the square of the spot area is included in the calculation. In automatic measuring systems the application and chromatographic errors are often the largest. In such cases working with internal standards becomes more interesting.

6.3 Automatization

Screening methods are characterized by the fact that many similar samples must be analyzed with regard to a common problem. Therefore, employment of automatic analysis systems is recommended. Such systems save time, and by excluding human error, they provide more precise results.

For TLC/HPTLC the entire analysis is not done automatically, but only those steps which are especially susceptible to errors. Consequently, automatization is directed towards application of sample solution, development of chromatogram, homogenous spraying or dipping of solvent-free stationary phase with reagent, and photometric in-situ evaluation.

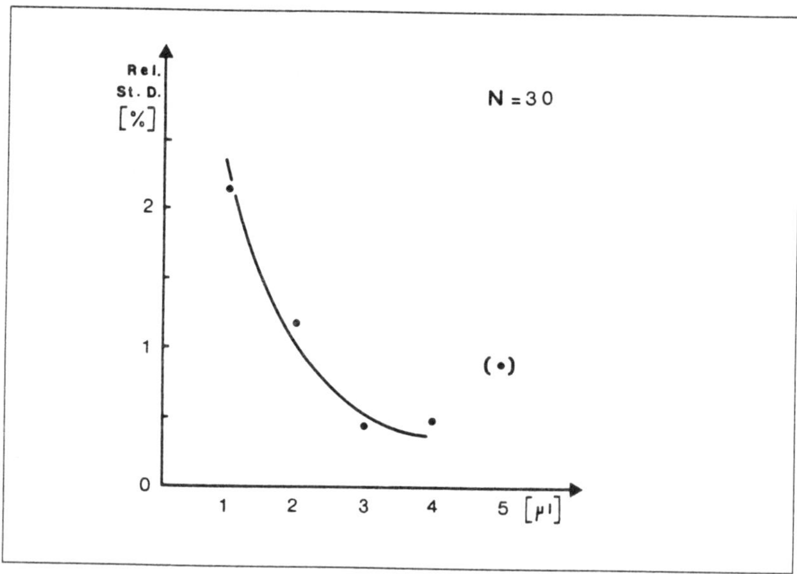

Fig. 4 Determination of Reproducibility for Set Volume and Momentous Application of Solution with a 5 µl Hamilton Syringe (each N = 30)

6.3.1 Application Units

Besides many manual application units for content-uniformity-tests and screening tests, (semi-)automatic systems have been developed (Bayer, Thomae, etc.). For spot-like application of sample solutions the semi-automatic PS-01 Application Unit *) and the Automatic TLC Sampler II**) are commercially available. Band-shaped starting zones can be applied suitably for densitometric chromatography evaluation with the micro-processor controlled Linomat IV**).

For the PS-01 system as well as for the Micro(Nano)Applicator**) a defined application volume is selected and then discharged in one operation. This technique is much more favorable than the application of proportions from a microliter syringe. For example, for a 5 µl Hamilton syringe the single application of 3 or 4 µl solution shows a random error of VK = ±0.4 (N=30). If smaller volumes are drawn up into the syringe and applied, the application error increases (Fig. 4). But it is nonetheless smaller than for applications of 1 µl each five times with a 5 µl syringe (Fig. 5).

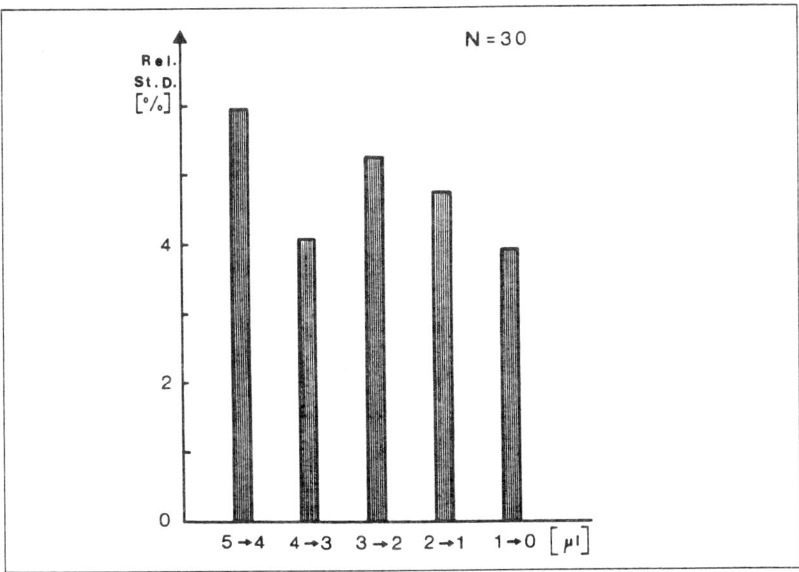

Fig. 5 Determination of Reproducibility for Aliquot Application of 1 µl Sample Solution each with a 5 µl Hamilton Syringe (each N = 30)

*) Manufacturer: Fa. Desaga, Maastraße 26–28, D-6900 Heidelberg, BRD
**) Manufacturer: Fa. Camag, Sonnenmattstra·e 11, CH-4132 Muttenz, Switzerland

With the automatic TLC sampler the starting zone sequence can be programmed freely according to practical needs (e.g., two-standard-, one-standard method, or as data-pair applications). The application volume can be selected between 100 nl and 20 µl (precision > ±1%) [148]. Stepping motors are used for lateral positioning of the application head and for the application syringe drive. The application capillary is made of fused silica and has a steel tube casing.

The Linomat IV is for band-shaped application of larger sample volumes (1 µl to 99 µl). Consequently, 100 µl and 500 µl, respectively, microliter syringes are employed. The table drive (80 steps per mm) and the syringe feed (400 steps per µl) are controlled by stepping motors. The solutions, with nitrogen as carrier gas, are sprayed on in bands. In this way good separation results are obtained for those selected thin-layer chromatographic conditions. Furthermore, certain measuring errors, those which occur during the photometric in situ evaluation, can be reduced.

6.3.2 Developing Systems

The AMD-System (Automated Multiple Development) [149–152] is available for fully automatic development including multiple and stepwise techniques as well as unrestriced selection of a solvent gradient. This system is a further development of the PMD-System (Programmed Multiple Development) which had been postulated by Perry et al. [153–158] in 1973, but was not widely accepted in Germany. The characteristics of the AMD-method are [159]:

– multiple development in one direction,

– each individual run extends over a somewhat longer separation length than the preceding one,

– for chromatographic development over a wide range the elution strength of each run is less than the preceding one,

– between individual runs the solvent is removed completely from the development chamber and the layer is dried by decreasing pressure,

– for each individual run fresh solvent is used.

In this way substances from a large polarity range can be separated in one chromatogram. This is an advantage which also favors the screening techniques since the lengths of run of the individual components are independent of the sample matrix.

6.3.3 Dipping Devices

Often micro-chemical reactions are used for making the separated substances visible. In the past, the reagents had been sprayed on (e.g. spraying automat according to Kreuzig [169]).

More favorable are the dipping units (e.g. Tauchapparatur*) in which the dipping speed of the TLC/HPTLC-plate is constant and the retention time in the reagent bath can be preselected. In this way

- a reproducible dipping process,

- a homogenous layer coating with reagents,

- a spray cabinet can be eliminated thus ensuring a low-pollution working method

free of human error is achieved. Those characteristics benefit the TLC/HPTLC as screening method.

6.3.4 Computer Controlled Chromatogram Evaluation

In 1967, data processing was first used for evaluation of thin-layer chromatograms [161]. Soon operation with minicomputers followed [162,163] which were extended by Ebel et al. (for literature refer to [165]) and finally modified to the computer controlled TLC Scanner II of Camag company.

Compared to manual operation of the scanner, the computer control has the advantage that the chromatogram zone to be determined or to be measured is adjusted by means of stepping motors optimally in the light path before the measuring process is started. The illumination device, the monochromator (wavelength adjustment, slit width) as well as the measuring electronic (sensitivity of the photomultiplier, signal optimization, etc.) are controlled, too. The raw data are stored in order to determine later the optimal basis line by visual screen integration or to re-calculate the result with various other calibration functions.

The necessary communication with the operator is done by means of a screen. After completing the computer controlled chromatogram evaluation, an external protocol is produced by plotter/printer systems. All relevant analysis data are presented [337].

*) Manufacturer: Fa. Baron, Freiherr von Hundbiss Straße 2, D-7752 Insel Reichenau, BRD

The newest development in the field of photometric direct evaluation is the Desaga Denitometer CD 60 which is controlled, e.g., by a IBM-XT control unit. The scanner itself has practically no operating elements anymore. All functions are controlled from keyboard and screen. The menu is in German or English. Per second 400 individual measurements can be carried out (20 mm/sec at 50 μm step length) allowing the measurement of an HPTLC-plate with 30 paths in 5 minutes. The memory capacity is 512 kByte and can be extended at anytime [166]. For evaluation either linear or meander scans with Kubelka-Munk linearizations can be employed. Compensation of interferences due to the plate itself is achieved by two-wavelength measurements.

In the meantime, other scanners are also available commercially).* This allows successful employment of photometric direct evaluation of thin-layer chromatograms especially for screening methods.

6.4 Screening Tests

Before a method can be used as screening method it must be developed exactly in the analysis area and be tested often in practice. The random sample taken must be representative because only an aliquot part is analyzed. The pre-chromatographic sample preparation as well as the clean up should be easy, quick and selectively, if possible, to allow complete, or at least reproducible, isolation of the substance of interest from the "packaging material" meat, plant, food, etc. For the different applications the range of error possibilities and error sizes must be known before the results of a screening method can be used with confidence. Chapter 9 of this book deals with different clean-up methods and, therefore, will not be described here again. Instead several substance groups from various fields of environment and food analysis are listed to supplement earlier surveys, e.g., of Baltes [167], Günther [168], Hezel [169], and Thier [170] and to show the numerous and complex application of TLC/HPTLC. The selection had been made with the main emphasis on the contamination of

- water, soil, air

- meat and meat products

- fruits, vegetables, etc.

including only two fields each (Fig. 6). Finally, the treatment of alkaloids in food is discussed to show that natural genuine constituents are also suitable for screening methods as well as contaminates from the environment.

---

*) Manufacturer: Fa. Biochem, Benzstraße 28 a, D-8039 Puchheim, BRD
Fa. Shimadzu, Albert-Hahn Str. 6–10, D-4100 Duisburg 29, BRD

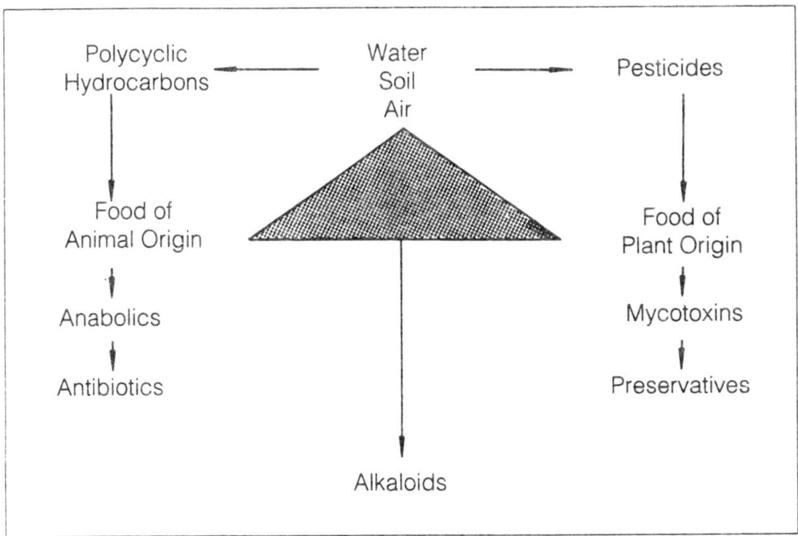

Fig. 6 Selected Fields of Application for TLC/HPTLC as Screening Method in Residue Analysis as well as for Control of Food of Animal and Vegetable Origin

6.4.1 Polycyclic Aromatic Hydrocarbons

With regard to environmental protection, the identification and quantitative determination of polycyclic aromatic hydrocarbons (PAH) is very important. Those substances can be found in soil, plants, tar, soot, exhaust gas, etc. Thereby they may enter the surface water directly and, by the seeping of rain or snow contaminated with PAH from air or dust, the ground water also. This ground water is the basis for our drinking water supply. It should, therefore, contain a maximum of 50 ng PAH/l. Surface waters with low contaminations contain 50–250 ng PAH/l. In heavily contaminated waters the concentration can reach 1 µg/l and up to 0.1 mg/l [7] in waste waters.

In many cases the previously applied summary determination of those cancerogenic substances by means of the IR spectroscopy as a "group determination" is no longer sufficient.

Consequently, the mixture must be separated first and then detected selectively. Figure 7 shows that fluorimetric determination can be applied successfully. By skilful selection of excitation and emission wavelengths, the most important peaks can also be determined individually, even if the chromatographic separation was not optimal. Table 1 lists several publications on the analysis of this substance class in air, water, and food. In form of lipophile substances they are enriched mainly in fat-phases. Comparisons regarding analysis are given among others by Sherma [188], Poole et al. [189] and Rüssel [190].

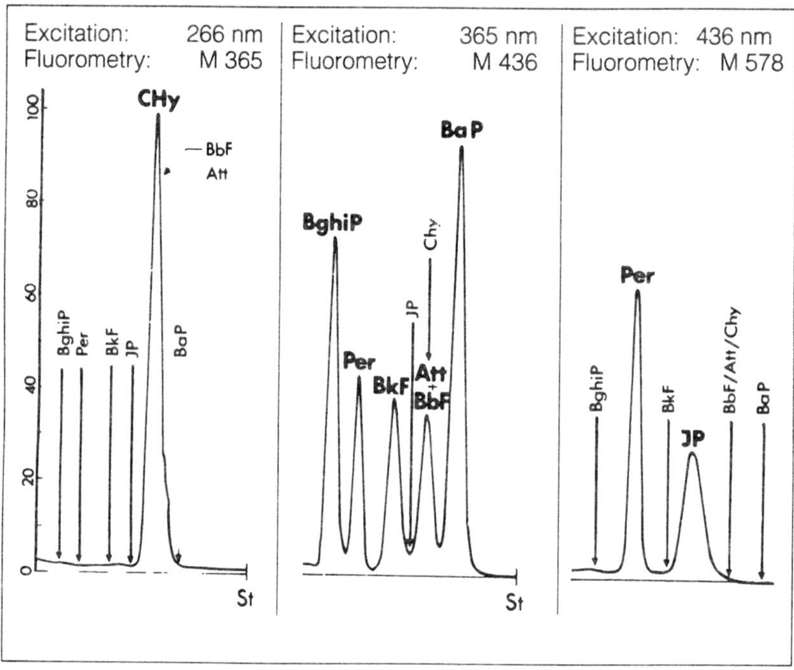

Fig. 7 Example for Selectivity of Fluorimetric In-situ Evaluation for Polycyclic Aromatic Hydrocarbons [177]

Att = Anthanthrene;
BbF = Benzo(b)fluoranthene;
BkF = Benzo (k)fluoranthene;
BaP = Benzo(a)pyrene;
B(ghi)P = Benz(ghi)pyrene,
Chy = Chrysene;
IP = Indenopyrene;
PER = Perylene

Tab. 1 List of some Publications on Polycyclic Aromatic Hydocarbons in Air and Exhaust Gases, Soil Samples, Water, and Food

Substances	Matrix	Clean up	Remarks and Literature
A. Air and Exhaust Gases			
Benzo(a)pyrene Coronen	city aerosols	Sephadex LH-20	detection limit: 16 ng Benzo(a)-pyrene, recovery rate: 95.5% [171]
18 polycyclic aromatic hydrocarbons	city air, laboratory dust	liquid/liquid extraction (DMSO/Cyclohexane)	8 PAHs could be determined in laboratory air The external standard method is more exact than working with internal standard [172]
Benzo(a)pyrene	airborne particles	Cyclohexane-extract no clean up	recovery rate: 102%, random error: VK = ± 2.2% [173]
14 polycyclic aromatic hydrocarbons	exhaust gases, soot	liquid/liquid extraction; vacuum sublimation;silica gel column	between 0.5 and 200 ng, reproducibility of VK = ±1 to 5% [174]
Benzo(a)- and Benzo(e)pyrene	dust particle extracts, water	Cyclohexane-extract no clean up	no TLC separation, but clearly selectable by fluorimetric evaluation [175]
Nitropyrene	exhaust gases, soot	extraction, microchemical reduction	determination limit for aminopyrene: 1 ng [176]
8 polycyclic aromatic hydrocarbons	exhaust gases, air dust filter, water	liquid/liquid extraction (DMSO/Cyclohexan); adsorption chromatography on silica gel; eluate fractioning over Al_2O_3 or Sephadex LH-20 column	reproducibility VK = ±1.4 to ± 6 % recovery rate: 98%, detection limit usually < 10^{-9} [177]

Tab. 1

Substances	Matrix	Clean up	Remarks and Literature
B. Water			
6 polycyclic aromatic hydrocarbons	drinking and ground water	only for colored water samples clean up on Al_2O_3 column	good separation and spot form; recovery better than DIN 38 409, part 13; reproducibility: VK = ±3.3 to ± 5.9 % [178]
Fluoranthene, 3,4- and 11,12-Benzfluoranthene, Benzo(a)pyrene, Indenopyrene, Benzperylene	drinking and industrial water	no clean up	selection test for distinguishing contaminated and uncontaminated water samples [179]
Fluoranthene, Benzo(b)fluoranthene, Benzo(a)pyrene, Indenopyrene, etc.	drinking water	HPTLC-plates RP-18 with concentrating zone	colaborative trial (n=11) confirms suitability for screening methods, limit concentrations 50 ng/l [180, 180a]
6 polycyclic aromatic hydrocarbons of the drinking water regulation	ground-, spring-, reservoir as well as waste water	$CCl_4/CHCl_3$ (1:1) or cyclohexane extract; no clean up	Improvement of indenopyrene detection and decrease of detection limit for more than two of powers of ten. Results are comparable with HPLC [181]
10 polycyclic aromatic hydrocarbons	spring-, well-, lake, or rain water	deactivated silica gel and alkali $Al_2O_{3\,n}$ as clean up column	even without complete separation fluorimetric determination by skilful selection of excitation and measurement wavelength. Example for selective detection [182]
Polycyclic aromatic hydrocarbons	snow in north-eastern Bavaria	—	RP-18 phase chromatography [183]

Tab. 1

Substances	Matrix	Clean up	Remarks and Literature
15 polycyclic aromatic hydrocarbons	bitumen-water extract	liquid/liquid extraction	recovery rate, detection limit and reproducibility were not stated [113]
C. Technical Solvents			
Benzo(a)pyrene, 3-Methylcholanthrene, Chrysene, Dibenz(a,h)-anthracene	technical solvents (n-hexane)	liquid/liquid extraction (Cyclohexane/nitromethane) silica gel column; Al_2O_3-TLC, Cellulose-TLC	the relative error is ±2 to ±5%, recovery rate 36-50% [185]
D. Food			
Benzo(a)pyrene	sardines in oil, streaky bacon, ham, smoked salmon	extraction as watersoluble caffeine complex from petroletheric solution of food and column-chromatographic cleaning	recovery rate: 73-86%; more sensitive than GC less time- and workintensive [186]
Benzo(a)pyrene	milk, butter, smoked streaky bacon	petrolether extraction under presence of triton X-100. Complex formation with caffeine; clean up on silica gel column	recovery rate: 92-98%;low-cost and time-saving method [187]

Tab 2   List of some TLC/HPTLC-Publications on Pesticide Analysis Subdivided into Insecticides, Fungicides, and Herbicides

Substances	Matrix	Clean up	Remarks and Literature
A. Insecticides			
Parathion, Guthion, Methyl-trithion	tap water, blueberries	Florisil	recovery rate: 80-92%, reproducibility: VK = ±1.9-3.6% [132,195]
Cygon, Guthion, Parathion	sea water	no clean up	recovery rate: cygon 87-113% reproducibility at 100-400 ng, guthion: VK = ±4-8% [196]
Sevin	tap water	no clean up	hydrolysis and fluorimetric measurement; recovery rate: 90-94%, reproducibility: VK = ±1-2%, alternative method for present determinations [197]
Guthion, Trithion, Parathion, Proban	water	no clean up	screening method: flavone as spraying reagents, alternative to GC [198]
Mesurol, Landrin, Baygon, Mobam, Sevin	water samples	extraction, hydrolysis and dansylation	recovery rate: 90-106%; reproducibility: VK = ±4.5% (5-500 ng/chromatographic zone),good specification by derivate formation [199]
Landrin, Sevin, Mesurol, Baygon, Mobam, etc.	sea water, clayey ground	hydrolysis, reaction of alkyl-amine with NBD-chloride	recovery rate: 89 ±4%; reproducibility: VK = ±3.3 - 5.6%, detection limit: <1ng/zone, can be used alternatively to GC [200]
Mesurol, Landrin, Sevin, Matacil, etc.	water, soil samples	hydrolysis, reaction of alkyl-amines with dansylchloride	two-dimensional TLC for differentiation of critical pairs [201]
Aminocarb, Mexa-carbat	soil samples	hydrolysis with subsequent dan-sylation, NBD-derivate formation	detection limit: dansyl-derivate: 1 ng/zone; NBD-derivates: 0.5ng/zone; reproducibility: VK = ±3-5% real alternative to GC [202]

Tab 2

Substances	Matrix	Clean up	Remarks and Literature
Dieldrin, DDT, Lindan	model mixture	no clean up	biotest with fruit fly [39]
Sevin, Baygon, Mesurol, Landrin, etc.	methylene chloride solution	no clean up	definition of analytical parameters [203]
Azinphosmethyl, Bayrusil, Cythioat, Dursban and 8 other pesticides	solutions in methylene chloride or ethanol, resp.	no clean up	detection limit for Bayrusil and Maretin: 1-8 ng/chromatogram zone [204]
47 Organothiophosphoric acid pesticides	ethyl acetate solutions	no clean up	fluorimetric detection after hydrolysis, sensitivity: 0.2 - 5 µg [204a]
61 pesticides, among them Aminocarb, Coumaphos, Guthion, Mesurol	chloroform solutions	no clean up	no data on recovery rate, reproducibility and detection limit [193]
Carbaryl	apples, salad, grain	for grain: AOAC-method on TLC-plates	working range: 0.10 ppm, recovery rate >89% [66]

B. Herbicides

Substances	Matrix	Clean up	Remarks and Literature
Chlorpropham, Swep, Barban, Linuron, Diuron, Maloran etc.	sea water	saponification and dansylation	recovery rate: 84-102%; reproducibility: Linuron VK = ±2.3 - 6.8% detection limit: <1ng/zone [206]
Silvex, MCPA, 2,4,5-T; 2,4-D	lake-, river-, industrial water	liquid/liquid extraction for heavily contaminated samples	recovery rate: 97-102% (without clean up); 93-98% with clean up); detection limit: 1ppb, equivalent to GC [207]

Tab 2

Substances	Matrix	Clean up	Remarks and Literature
Monuron, Neburon, Metobromuron, Chlorpropham, Propanil	muddy clay, water	neutral Al_2O_3 column	recovery rate: 75-85%, reproducibility: VK = ±98% [208]
Monolinuron, Linuron, Chlorbromuron, Chloroxuron	carrots	liquid/liquid extraction	recovery rate: 90-98%, reproducibility: VK = ±1.6% (data pair), detection limit: 20-66 ng [209]
Diuron, Linuron, Monolinuron, etc; Propham, Chlorpropham, Swep, etc.	carrots, potatoes wine	liquid/liquid extraction; hydrolysis at the start	recovery rate: 83-106%, detection limit: 1 to 8 ng herbicide [210]
Desmetryn, Prometon, Atrazine, Prometryn, and 10 other triazines	fennel, sage	liquid/liquid extraction, further clean up steps	reproducibility: 1.5-2.2% (TLC); data pair method; detection limit: 3-5 ng(HPTLC); 10-13 ng(TLC) [211]
Malathion, Dimethoat	salad leaves	no clean up for dimethoat determination, florisil-column-clean up for Malathion analysis	recovery rate for 10 ng Dimethoat: 85%; for 100 ng 105% [212]
2-Methyl-4-chlorphenoxy acetic acid(MCPA)	apples	liquid/liquid extraction at various pH-values	detection limit: 0.1 mg/kg, recovery rate: 103-114% at 0.1 to 0.5 mg MCPA/kg [213]
Dimethoat, Bromophos	--	--	[214]
Chlormequat (CCC)	grain, mulch, grapes, raisins, pears, milk, buttermilk, yogurt, etc.	Al_2O_3-column	recovery rate: 70-103%, detection limit: 20 ng/chromatog. zone [215]

Tab 2

Substances	Matrix	Clean up	Remarks and Literature
Chlormequat (CCC)	wheat, oat, various flour samples	Al_2O_3-column	recovery rate at 0.5 ppm or 1 ppm, resp.: 80-90%; detection limit: 0.1 ppm [216]
Chlormequat (CCC)	wheat, oat, wheat- and oat mulch	ion exchanger DOWEX 50W x 8; activated carbon; Al_2O_3, neutral	recovery rate:72-86%; reproducibility: VK = ±5%; detection limit: 50-100 ng/chromatogram zone corresponding to 0.01 ppm [217]
Isouron	sugar cane	liquid/liquid extraction	examination of metabolites, not very suitable for series tests [218]
Atrazine, Propazine, Prometryn, Prometon	methanolic solutions	no clean up	detection limit: 250-400 ng per chromatogram zone [219]
Methabenzthiazuron, Naptalam as well as other insecticides and fungicides	methylene chloride solution	no clean up	detection limit: 2 ng (Metabenzthiazuron); 40 ng (Naptalam) [220]
Phenmedipharm	emulsion concentrate	no clean up	reproducibility: VK = +4%; per day 32 individual determinations possible [162]
Formetonat, Chlorphenamidine	pesticide	no clean up	reproducibility: VK = ±3.4% to 7.7%; detection limit for contaminations: 0.02 to 0.2% [163]
Metoxuron, Diuron	powdery preparation	no clean up	reproducibility: VK = ±1%; detection limit:10-50 ng per chromatogram zone, comparison with HPLC and titration [221]
Atraton, Atrazine, Amethryn, Prometon, Simazine, etc.	pure substances	no clean up	reproducibility: VK = ±3% [222]

Tab 2

Substances	Matrix	Clean up	Remarks and Literature
C. Fungicides			
Captan, Folpet, Captafol	water, apples, salad	Florisil or Chromosorb 102	recovery rate Captan: 81-89%; Folpet: 90–93%; Captafol:84-85%; determination limit: 100 ng [223]
Benomyl	apples, blackberries, grapes, cabbage, sugarbeet	liquid/liquid extraction and hydrolysis	recovery rate: 87%; sensitivity: 0.02ppm (apples), 0.08ppm(cabbage) [224]
Benomyl	cherries	no clean up	the method is much quicker than all others, published until 1974. reproducibility: VK = ±8-13.6% (200 ng); detection limit: approx. 20 ng [225]
Thiourea	citrus fruits	Al_2O_3-column	recovery rate: 76%, reproducibility: VK = ±3% at 100 ppm; detection limit at 1 ppm [226]
Captan, Captafol	apple, potato leaves	Hyflo-super Cel,liquid/liquid extraction	reproducibility: VK = ±7-8% (at 0.2 ppm); detection limit: 0.02 ppm [227]
Benomyl, Coumatetralyl, Fuberidazole, etc.	methylene chloride solutions	no clean up	detection limit: 1-20 ng [228]
Benomyl	ethyl acetate solutions	no clean up	detection limit: 0.6 µg/chromatogram zone [229]
Thiabendazole	--	--	see preservatives, Table 6

6.4.2 Pesticides

The TLC is the "house" method for the residue chemist [190] and most suitable as a screening method. In many cases it can be applied alternatively to GC and HPLC [191]. Of course, not all 7,500 substances, listed in the Pesticide Index [192] can be dealt with here, even though e.g. rodenticides [190,193,194] used for rat control can also lead to significant harm to domestic and wild animals, or vermicides may occur in milk and chicken eggs [205].

In human nutrition, residues coming from insecticides. herbicides, and fungicides play an important role. Table 2 surveys TLC/HPTLC-publications dealing with the subject of residues in human foodstuffs coming from substances from the environment.

6.4.3 Anabolics

In the 1950's, it became known that estrogens (e.g. Diethylstilbestrol) influence the growth of certain animal species [230]. Anabolics caused the protein synthesis in the organisms to be increased significantly [168]. The FAO recommended allowing use of anabolic substances for animal breeding if the following basic conditions were fulfilled:

– knowledge of metabolism

– introduction of safe and routinely applicable control methods to exclude misuse

– knowledge of action in specified large scale use

– meat products on the market must be guaranteed to be free of residues.

These basic conditions must be easy to verify. Consequently, the determination of anabolics must be possible in the micro- and nanogram range. Even today, the sample preparation corresponds in principle with the analysis technique, recommended by Verbeke [231] (Fig. 8).

This technique requires relatively much work, therefore. more and more pre-examinations of faeces and urine samples from calves to be slaughtered are carried out. The sample preparations are simplified [232] and the detection sensitivity is sufficient. Table 3 lists selected publications on this subject.

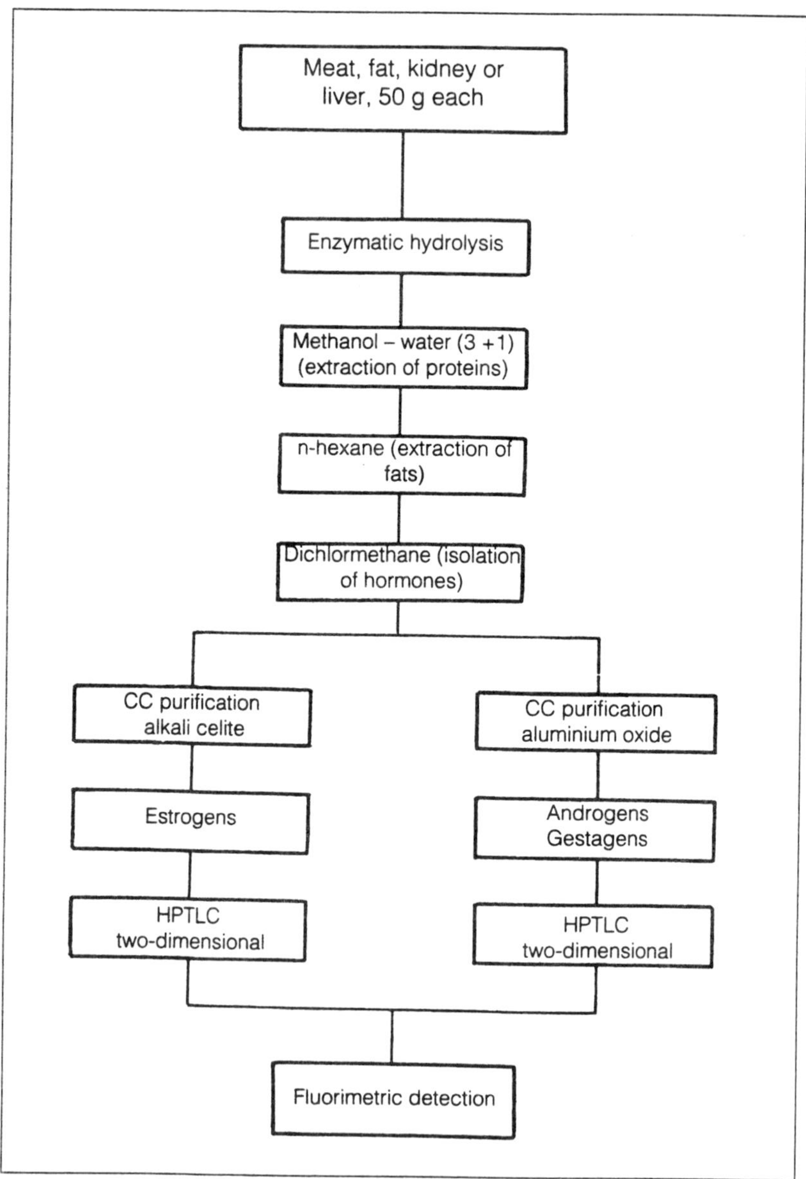

Fig. 8 Preparation of Meat and Meat Products to be Examined for Anabolic Residues according to [231]

Tab. 3 List of some Publications on TLC/HPTLC of Anabolics in Animals for Slaughter

Substances	Matrix	Clean up	Remarks and Literature
Estrone, Estradiol, Estriol, Hexestrol, Dienestrol, Urine, Diethylstilbestrol, Zeranol	meat, serum, urine	meat: cleaning aceto-nitrile extract of fat and acidy com-ponents; urine: hydrolysis and fract. extraction, methylation	TLC: suitable for 0.2-2 µg/chrom. zone; detection limit: 50-200 ng; HPTLC: 10-200ng/spot; corresponds to HPLC and GC; accordance with radio immuno assay [233]
Zeranol, Hexestrol, 17-ß-Estradiol, Ethinyl estradiol, Dienestradiol, Di-ethylstilbestrol	meat-, liver-, kidney tissue	cleaning by phenolate formation and dansylation; Sephadex LH 20	detection of 50 ppb Zeranol and 20 ppb of the other estrogens. Above 200 ppb the re-covery rate is 60%. Reproducibility: 8-19% [234]
Estradiol, Estriol, Estrone	meat	Merckogel 6000 column prechro-matographic derivatization	recovery rate at approx. 50%; detection of 1 µg zeranol or 10 µg estradiol and estrone, resp. per kg meat [235]
Trenbolon	liver, kidney	silica gel 60-column	determination limit = 5 ppb (meat); recovery rate 63%±6.5% [236]
Diethylstilbestrol	faeces, urea	cleaning by extracting; concen-trating zone	detection limit: 10 ng/ml or 30 ng/g, resp. [232]
Estrogens	--	--	TLC-screening test [237]
Trenbolon	meat	extraction, redisolve	determination limit: 250 pg; progesterone and testosterone do not interfer [238]

Tab. 3

Substances	Matrix	Clean up	Remarks and Literature
Diethylstilbestrol	meat, urine	extraction, glucuronide cracking, dansylation	detection limit: approx. 20 ppb [239]
Diethylstilbestrol, Estradiol, Estriol, Estrone	meat, liver, kidney of beef, veal and pig	Sep-Pak C-18 cartridge	recovery rate. e.g. diethylstilbestrol: 69–71 %; estrone: 76-85%, time requirement for each analysis: 90 min [240]
Diethylstilbestrol	milk, meat, faeces, urine	liquid/liquid extraction, silica gel column, hydrolysis of eluate and dansylation	detection limit: <0.5 µg/kg [241]
Zeranol	–	–	[242]
Trenbolon	pure substance	no clean up	semi-quantitative determination: detection limit: 1 ng [243]

6.4.4 Antibiotics and Chemotherapeutics

In the modern animal breeding the intensive feeding techniques have succeeded for economical reasons [244]. For avoiding and treating infection diseases, antibiotics and chemotherapeutics are employed. Antibiotics are substances, originating from live cells or their metabolic products, which damage or kill pathogens. Chemotherapeutics have the same effect. But they are produced synthetically. Table 4 lists corresponding examinations in animals for slaughter.

6.4.5 Mycotoxins

Mycotoxins are a world-wide danger for contamination of food and feed stuffs. Their detection must be successful and reliable in the ppb and ppt range. Most absorption photometric methods fail in this range and therefore, fluorimetric measurements are increasingly employed as screening methods [239,256–258,368]. Fluorimetric measurements have the following advantages:

– An increased selectivity as compared to absorption measurements because the excitation wavelength as well as the emission range show many substance specific characteristics (compare e.g. Figure 7). Many substances absorb radiation but fewer emit fluorescent light.

– The measurement sensitivity is 100 to 1000 times higher than for absorption measurements. Underground interferences are hardly present because the stationary phase normally does not fluoresce at the selected excitation wavelength. Only a fraction of the test solution is applied. The chromatograms are no longer overloaded, the zone formation improves and the amount of interfering substances is significantly decreased.

Increased reliability when identifying fungus toxins can be achieved by treating the chromatogram with reagents. Often treating the stationary phase with acids or ammonia in gas form is sufficient to move the absorption maximum and the fluorescent range characteristically caused by a change of the pH-value.

Table 5 lists several publications regarding identification and (semi-)quantitative determination of the most important mycotoxins. Not mentioned are those publications dealing with detection and determination of aflatoxin B_1, B_2, G_1, and G_2 in peanut extracts (e.g., [259–269]). The determination limit is in the ppb-range [270], i.e. at 25 to 50 pg substance per chromatogram zone [239]; and with HPTLC-layers even lower [271]. The reproducibility at

Tab. 4   List of some Publications on TLC/HPTLC of Antibiotics and Sulfonamides in Meat and Meat Products

Substances	Matrix	Clean up	Remarks and Literature
A. Antibiotics			
Chloramphenicol	trout	not mentioned	From 80 examined samples 11 contained residues between 0.8 and 17 µg/kg [244]
Chloramphenicol	meat, eggs	liquid/liquid extraction	recovery rate: 100%; determination limit: 10 ppb [37]
Chloramphenicol	kidney, meat	phosphate buffer; extraction, RP-18-column	determination limit <5C ng [67]
Chloramphenicol	skim milk	Chromosorb 102, cation exchanger, reduction, diazotization, liquid/liquid extraction	screening method for milk: determination limit: 4 ppb [246a]
Tetracycline, Neomycin, Ampicillin, Streptomycin etc.	milk	liquid/liquid extraction	bioautographical detection; determination limit: 1ppm [245]
Chloramphenicol	milk, eggs, etc.	–	survey on screening methods [245]
Cuprimyxin	milk, udder tissue, plasma, urine, gall-bladder	mostly no clean up required	Cuprimyxin decomposes in water spontaneously to myxin; recovery rate approx. 80%; reproducibility: VK = ±7% (c= 0.5µg/ml); detection limit: 20-50 ng/g biologic sample [247]
Gramicidin	fermentation batches	–	[248]
Tetracycline, Streptomycin, Chloramphenicol, etc.	pure substances	no clean up	derivatization, characterization with 11 reagents [249]

Tab. 4

Substances	Matrix	Clean up	Remarks and Literature
B. Sulfonamides			
Sulfadimidine	pork and veal, innards	–	from 32 meat samples 3 contained residues between 9 and 200 µg/kg. For innards the corresponding values were between 140 µg/kg and 500 µg/kg [250]
Sulfachinoxaline, Sulfamethazine, Sulfadimethaxin, Sulfathiazole	turkey, duck, pork	liquid/liquid extraction	the mean recovery rate for all tests was 95%. For repeatability VK =± 7.7% and for reproducibility VK = ± 10.5 were reported. [251]
Sulfamethazine	pork	liquid/liquid extraction	screening method; reproducibility VK = <6% (concentration approx. 0.1 ppm) [252]
Sulfadiazine	milk, eggs, liver, fat, skin, kidneys, plasma	adjustment of pH-value, no special clean up	recovery rates vary for the individual samples: 54-91%; reproducibility VK = ±4.2 – 15%, depending on matrix [253]
25 Sulfonamides	muscle and kidney tissue, serum	liquid/liquid extraction with adjusted pH-value	screening method; detection limit: 10–50 ppb, for serum samples 10 ng/ml [254]
20 Sulfonamides	pork, calf kidneys	extraction	detection at 0.1 ppm (10 times more sensitive than microbiological detection [370,371]
20 Sulfonamides	pure substances	no clean up	methodology [250,255]

Tab. 5 List of some Publications on Identification and (Semi)Quantitative Determination of Mycotoxins by TLC/HPTLC

Substances	Matrix	Clean up	Remarks and Literature
Aflatoxin B_1, B_2, B_{2a}	cotton seeds	silica gel column	Aflatoxin B_{2a} detection: 17 ppb [285]
Aflatoxin B_1, G_1, B_2, G_2	nutmeg	silica gel column	recovery rate: 92-108%; variation coefficient: VK = ±5.2% [286]
Aflatoxin B_1, G_1, B_2, G_2	–	simple clean up	the recovery rate is better and the analysis time shorter than for other methods [287]
Aflatoxin B_1, G_1	–	–	0.2 ng range: laser-fluorescence-scanning [288]
Aflatoxins B_1, G_1, B_2, G_2, M_1 and M_2, Ochratoxin A, Sterigmatocystine, Penicillic acid, Patulin	nuts, pistachios, apple sauce, milk, cheese	Sep-Pak-silica gel cartridge	recovery rate: 70-75% or, after fat separation 75-80%, resp. [289]
Aflatoxin B_1	grain	silica gel column	method is quicker and less costly than other methods; recovery rate: 86-93%; VK = ±18.6% [290]
Aflatoxin B_1, G_1, B_2, G_2	buckwheat, grain products, peanuts, raw or roasted, almonds	defattening with Frigen 113 F	Instead of $CHCl_3$, CH_2Cl_2 is used as extraction agent [267,291]
Aflatoxin B_1, G_1, B_2, G_2	grain	–	rapid screening method [292]
Aflatoxins, Ochratoxin A, Zearalenone, Vomitoxin	bran	RP-column chromatography	recovery rate: 80% (ppb range), determination limit aflatoxins: <0.5 ng/g, ochratoxin A 10 ng/g [293]
Aflatoxin B_1, G_1, B_2, G_2	wine	extraction	[294]

Tab. 5

Substances	Matrix	Clean up	Remarks and Literature
Aflatoxin B_1,G_1,B_2,G_2	fungus suspension	liquid/liquid extraction ; extrelut	method variation coefficient, the extrelute clean up is much more effective (VK =±1.0–1.8%) than liquid/liquid extraction (VK = ±5.8–11%) [295]
Aflatoxin B_1,G_1,B_2,G_2	culture medium	no clean up	reproducibility: VK = ±11%, better than HPLC examinations [296]
Aflatoxin B_1,G_1 and 9 other mycotoxins	culture medium	TLC as clean up	derivatization; reproducibility B_1 VK = ±1.6; G_1: VK = ±2.9% [361]
Aflatoxins	meat	–	[297]
Aflatoxins	meat	–	[298]
Aflatoxin B_1,M_1	liver, kidney	silica gel column	[299]
Aflatoxin M_1	milk, powdered milk	Sep-Pak C_{18} cartridge	less solvent consumption, high sample throughput, detection limit 3-5 ng/kg [323]
Citrinin	fungus extracts	–	screening method [300]
Citrinin, Patulin, Sterigmatocystine Zearalenone and 8 other mycotoxins	animal food	membrane clean up	[301]
Citrinin, Griseofulvin, Penicillicacid, Zearalenon, etc.	various food	–	systematic TLC-analysis [302]
Citrinin, Patulin, Ochrat oxin and 5 other mycotoxins	–	–	[303]
Citrin	–	–	]304]

Tab. 5

Substances	Matrix	Clean up	Remarks and Literature
HT-2-Toxin	rice, corn, wheat, oat, rye, peas	Sep-Pak cartridge	recovery rate: 88% [305]
Trichothecene (=HT-2-Toxin)	fermentation batches 57 wheat samples	activated carbon-Al_2O_3-column	detection limit: 0.1-0.2 µg/chrom. zone [81, 306]
Mycophenolic acid	mushrooms	–	determination limit: 100 ng/chrom. zone; relative standard deviation VK = ±9-13% [307]
Ochratoxin	rice, coconut, oatmeal, peas	defattening; first TLC-development for clean up	recovery rate: 88-110%, reproducibility: ±7.9%; detection limit: 6 ng/spot [308]
Ochratoxin, Zearalenone, Aflatoxin	peanut butter	silica gel column	screening method [309]
Ochratoxin A,B,C	not mentioned	–	determination limit: 0.5 ng/spot after exposure to NH_3vapor [310]
Ochratoxin A,B,C, Sterigmatocystin, Zearalenon, etc.	olive oil, peanut oil, soybean oil	TLC-development	screening method [311]
Ochratoxin, Zearalenone, penicillic acid	–	–	screening method [312]
Penicillium roque-forti-toxin	cheese	silica gel column	recovery rate: 30-50%, detection limit: 0.2 ppm [313]
Penicillic acid	culture media	no clean up	recovery rate approx. 85% [314]
Penicillic acid	culture medium of *P. cyclopium* Westling	no special clean up	determination limit: approx. 50 µg/ml [315]
Penicillic acid	mouldy grain	no clean up	NH_3 treatment: blue flourescent zones [316]
Patulin	unfermented grape juice, wine	silica gel column	detection limit: 80 ng/spot [184]

Tab. 5

Substances	Matrix	Clean up	Remarks and Literature
Patulin	apple juice, cider	Dowex 50 W-H$^+$ column, silica gel column; 2,4-DNP-impregnated celite column	recovery rate: 85% ±14.2%; determination limit: 50 ppb [317]
Patulin	fruit juices	liquid/liquid extrac.	recovery rate:97-119% [318]
Patulin	fruit juices, nectars	Extrelut column	recovery rate: 70-80%; detection limit: 20 ppb [319]
Patulin	fruit juice, fruit puree, fruits, vegetables	Florisil column	detection limit 5 µg/l juice or kg fruit, resp.; control by derivatization [319a]
Patulin	apple juice	Extrelut column	screening method; recovery rate: 97 ± 4 for 50 ppb; detection limit: 20 ppb [320]
Patulin	apple juice	no clean up	TLC better by genuine samples; otherwise GC-MS [321]
Patulin	apple juice	silica gel column	detection limit: 0.025-0.1 ppm [322]
Patulin	tomatoes, pears, apples, cucumbers, plums	Celite 545 column	determination limit: 100 µg/kg; recovery rate: 82-90% [107]
Patulin	fruit and vegetable products	silica gel column	determination limit: 20 µg/kg [324]
Patulin	grain	preparative TLC	detection limit: 40 µg/kg [325]
Patulin	mouldy bread and backed goods	no clean up	applicable as screening method [326]
Sterigmatocystin	food	–	[327]
Sterigmatocystin	coffee beans	no clean up	brick-red fluorescence; screening method [328]
Sterigmatocystin	grain	–	[329]

500 pg is given as VK = ±3.0 to 4.5% [271]. With that HPTLC corresponds in regard of accuracy, reproducibility, sensitivity, and recovery rate to the HPLC [272]. One laboratory assistant is able to complete at least 20 samples a day [271,273].

Determinations of aflatoxin M S51 T in milk and milk products show similar results [274–282]. The detection limit is significantly lower (3–5 ppt for milk [323] and 30 ppt for cheese [283]). This toxin can also clearly be detected in ice cream or sorbet under screening conditions.

6.4.6 Preservatives and Antioxydants

According to the German Zusatzstoff-Zulassungs-VO (German regulation on food additives) sorbic-, benzoic-, formic-, and propionic acid, and PHB-ester as well as their sodium-, potassium- and calcium salts are allowed for use as preservatives. Furthermore, diphenyl, o-phenylphenol and thiabendazole may be applied for surface treatment as long as those substances are declared and certain maximum quantities are not exceeded. Consequently, analysis procedures must be available for detection and determination of such low quantities. Table 6 contains a list of corresponding publications.

6.4.7 Alkaloids

Alkaloids are nitrogen containing plant bases usually bound onto organic acids and having a physiological effect. They can be found as hot, pungent or bitter substances in spices or as constituents in "luxury food" (coffee, tea, etc.). They are subject to state food control. The content of capsaicinoids in bell pepper is determined for quality reasons. For tobaccos of the so-called nicotine-free species (content <0.1%) and species with low nicotine content (<0.7%) are being processed, requiring at least determination of these limiting values.

Also, the nicotine content in chewing tobacco and snuff (Fig. 9) is interesting. At present, there are no legal regulations. Screening methods are not only used for determination of limits in residue analysis (PAH's, pesticides, etc.) or for control of allowed as well as forbidden additives (e.g. coloring agents, preservatives) but also for detection and determination of genuine active substances and constituents in food and "luxury food" made from plant and animal sources. Table 7 gives a short overview of the TLC/HPTLC-publications in this field.

Tab. 6 List of some Publications on Analysis of Preservatives and Antioxidants in Food by TLC/HPTLC

Substances	Matrix	Clean up	Remarks and Literature
Benzoic acid, sorbic acid	yogurt, apricot jam, herring fillet, meat salad, mayonnaise,	steam distillation, alkaline re-extraction	recovery rate: sorbic acid 96-110% benzoic acid 87-100% [330]
Benzoic acid, sorbic acid	white- and red wine,	no clean up	quantitative determination: VK = ±0.95-2.83% [331]
Benzoic acid, sorbic acid	gooseberries, strawberries, peaches, apricots, etc.	liquid/liquid extraction with pH-adjustment	derivatization at the starting point, recovery rate 80-105% [331a]
Sorbic acid, benzoic acid PHB-ester	aqueous solution	no clean up	detection limit:benzoic acid 0.2µg sorbic acid 0.02 µg, PHB-ester 0.05µg/chromatogram zone [332]
Benzoic acid, PHB-ester	orange juice, pineapple jam, salad sauce, edible oil, pepper salami	diethylether. extraction no further clean up	The result of the in- situ measurement corresponds to the colorimetry, HPLC and GC. The separation is not optimal, but photometric selectivity of evaluation is given [333]
Sorbic-, benzoic-, salicylic acid, PHB-ester	food	selective extraction in fatty samples	polyamide-TLC, detection limit for sorbic acid, p-hydroxy benzoic acid and ester, salicylic acid: 0.5 µg [334]
PHB-ester	beer, fruit juices, cosmetics	no clean up	recovery rate: 100-108%; reproducibility: VK = ±2.3-2.5%, the results are comparable to those of GC and HPLC [335]
PHB-ester	food	steam distillation	multiple development [369]
PHB-ester	–	no clean up	[336]
PHB-ester	body lotion	no clean up	evaluation of peak areas and heights [337]

Tab. 6

Substances	Matrix	Clean up	Remarks and Literature
Butylated hydroxy toluene (BHT), butylated hydroxy anisole (BHA) etc.	artificial food	for BHT possible distillation with subsequent diethyl ether extraction.	recovery rate: BHT: 60-83%, BHA: 70-76% [338]
Butylated hydroxytoluene (BHT); butylated hydroxyanisole	fat from bouillon, sauce, marzipan, nougat	methylation; TLC-clean up	reproducibility of octyl gallate determination: VK = ±4% [339]
Octyl- and dodecylgallate			
Butylated hydroxytoluene (BHT); butylated hydroxianisole (BHA) octyl- and dodecylgallate, PHB-ester, diphenyl, o-Phenylphenol	potato chips, margarine, fat for frying, chewing gum	TAS-method, except for gallates	visual detection limit: 0.2 µg to 0.4 µg (molybdato phosphoric acid) [341]
o-Phenylphenol	citrus fruits	dansylation	reproducibility: VK = ±4-5%; detection limit: 5 ng/zone [342]
Thiabendazole	citrus fruits	dichlormethane extract, no clean up	measurements of fluorescence are more preferable than absorption measurements, reproducibility VK = ±2.67% [342a,343]
Thiabendazole	citrus fruits, juice, oil	liquid/liquid extraction	working range: 0.2-6 ppm, recovery rate: 95-99% [344]
Thiabendazole	ethanolic solutions	no clean up	reproducibility: VK = ±1.4% (n=20 [345]
Hexachlorophene	talcum powder, sprays, lotions, cremes, soaps	no clean up	recovery rate: 95.8% ±2.1%; reproducibility: VK = ±2.85% [346]
Chloramine T	ice cream, whipped cream	derivatization; hyflo-supercel-extraction	not allowed as preservative in food [347]

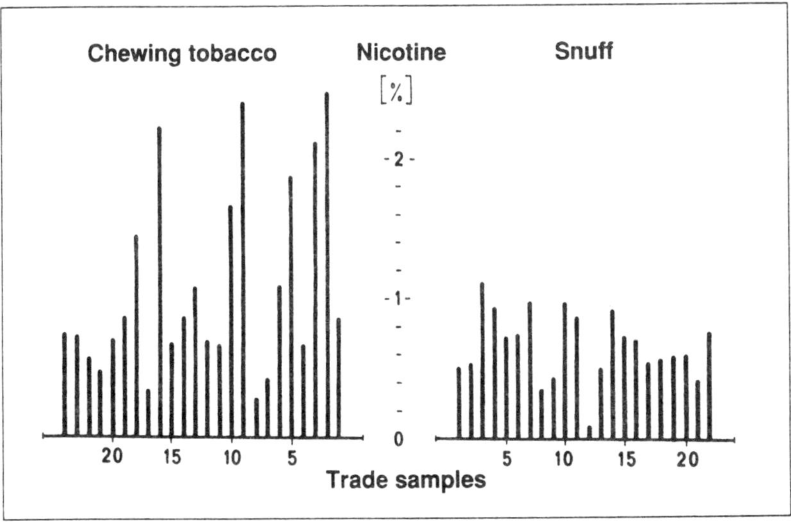

Fig. 9 Result of Two Screening Tests Regarding Nicotine Content of Snuff and Chewing Tobacco Products Commercially Available

Tab. 7 List of Some Publications on Thin-Layer Chromatographic Characterization of Alkaloids in Food and Luxury Food

Substances	Matrix	Clean up	Remarks and Literature
Caffeine	coffee, tea, caffeine containing food	Extrelut	[348]
Caffeine	cola, roasted coffee	evtl. preceding H_2O extraction (roasted coffee)	no figures regarding recovery rate, reproducibility, detection limit [239]
Caffeine	beverages	–	reproducibility: cola: VK = ±4.75% [349]
Caffeine	beverages	no clean up	suitable as screening method, low interference susceptibility [349a]
Caffeine	coffee powder, caffeine-free coffee, caffeine-containing lemonade	Al_2O_3- and Extrelut columns	recovery rate: 93-99%, reproducibility: VK = ±3.2% [350]
Caffeine	roasted coffee beans 5 different sorts	no clean up	the caffeine content was determind using the relation $\sqrt{F} = k \cdot \log.a$. VK approx. ± 10% (n = 5). Comparison: Kjeldahl method. The time and work effort is lower for TLC [351]
Caffeine	roasted coffee, extract coffee, caffeine-free coffee	liquid/liquid extraction	low manual effort, reproducibility VK = ±2% [352]
Caffeine	roasted coffee, coffee-extract, caffeine containing beverages	no clean up	detection limit: 10 ng/chromatogram zone [239]

Tab. 7

Substances	Matrix	Clean up	Remarks and Literature
Caffeine, Trigonelline	raw- and roasted coffee, coffee powder	DEAE-cellulose-ion exchanger in acetate form	reproducibility:caffeine VK =±3.9% trigonelline: VK = ±2.8% [354]
Caffeine, Theobromine, Theophylline	cocoa seed peel, defattened cocoa powder	Soxhlet extraction	alkaloid content was determined by the relation $\sqrt{F} = k \times \log.a$ [355]
Caffeine, Theophylline	blood serum	–	working with internal standard, reproducibility VK = ±5%; good correlation to Emit method (r = 0.985) [356]
Theobromine	chocolate, cocoa products, cuverture	no clean up	reproducibility: VK = 2.7 %; recovery rate: 93–98 % [357]
Theobromine	cocoa powder	cellulose ion exchanger DEAE in acetate form for adsorption of browning agents	standard error 1.5 - 2.1± [358]
Chinin	bitter lemon, tonic water	no clean up	no figures regarding recovery rate reproducibility, detection limit [359]
Chinin	tonic water, fruit juice, lemonade	no clean up	screening method, reproducibility at 250 ng/zone: VK = ±2.6% to 3.4% recovery rate: 99.7% [360]
Capsaicin	chilli fruits	no clean up	recovery rate: 97-104%; reproducibility: VK = ±2% [362]

Tab. 7

Substances	Matrix	Clean up	Remarks and Literature
Capsaicinoids	bell pepper	evtl. TLC clean up	no figures regarding recovery rate reproducibility, detection limit [363]
Capsaicinoids	chilli fruits	no clean up	detection limit: 30 ng Capsaicinoid, reproducibility: VK = ±4.19% for internal standard ±2.79% [94]
Dihydrocapsaicin	bell pepper	liquid/liquid extraction; protein precipitation	no figures regarding recovery rate determination limit and reproducibility [364]
Nicotine, Anabasine, Nicotyrine	snuff	no clean up	detection limit: nicotine: 10 ng, nicotyrine: 20 ng; anabasine: 40-50 ng, recovery rate: nicotine 01–104%; total error: VK = ±1.33% [365]
Nicotine, Anabasine, Nicotyrine, Nornicotine	chewing tobacco	SPE-extraction, C-18 or CN phases, resp.	detection limit: nicotine 10 ng, nicotyrine 20–30 ng, anabasine and nornicotine 30–40 ng; recovery rate: nicotine 98-99%, anabasine 96–101%; nornicotine 97–98%, reproducibility: VK = ±1.5% (N=10, data pair) [366]
Nicotine, Anabasine, Nicotyrine	cigarettes, chewing tobacco, snuff	no clean up	recovery rate: 97-99%, detection limit: nicotine 30-45 ng, anabasine 40-50 ng; reproducibility VK = ±1.25% (data pair) [367]

6.5 Summary and Discussion

In each laboratory that runs identity and purity tests on food, all three chromatographic methods, gas-, high pressure liquid- and thin-layer chromatography, are used. All methods are suitable for special applications and should be employed there where they perform best. The TLC, for example, is best for separation of many individual samples, especially for screening tests requiring the decision if the required content of active substance had been reached, exceeded or is inadequate.

Such determinations of limiting values allow a certain degree of "sorting" and with that reduction of work quantity. Furthermore, TLC is a very cost-efficient analysis method. Kelker [33] published corresponding figures for a medium-sized analytical industrial laboratory in which yearly approximately 160,000 gas chromatograms, 130,000 (HP)LC chromatograms and 35,000 thin-layer chromatograms were taken by 70 gas chromatographs and 25 employees, 28 HPLC units with 18 employees and 4 TLC places with 5 employees. If one chromatogram is taken as one counting unit each, no matter whether calibration or analysis run, this means that annually 6,400 GC-, 7,200 LC-, and 7,000 TLC-results had been produced for each coworker. Including investment cost for a system, the GC costed 3.5 mill. DM, the HPLC 1.4 mill. DM and the TLC maximal 0.15 mill DM. This corresponds to about DM 550.00 for one GC-analysis, slightly less than DM 200.00 for a HPLC-analysis and maximal DM 25.00 for a TLC-analysis [337]. This is also due to the low operating cost of the TLC in the daily routine operation. More advantages of TLC as screening method have been compiled in the beginning of chapter 6.1, preceding the quantitative evaluation (6.2) and automatization (6.3) of the method.

The second part of this article lists the most important publications from 7 topics on TLC separation and quantitative evaluation. Starting with polycyclic aromatic hydrocarbons as well as insecticides, fungicides and herbicides which can contaminate food through water, soil or air and then pass on into the human body.

For the meat and met product sector, anabolics and antibiotics have been listed and shown that the TLC as screening method can be used equivalent to other chromatographic methods. The Analogy is valid for the mycotoxins and preservatives, which play a larger role in food of vegetable origin. They had therefore been chosen as an example to demonstrate the application of TLC/HPTLC as a control method.

The survey on publications in the field of alkaloids as present in spices and "luxury food" shows that the TLC/HPTLC not only gives valuable information on determination of contaminations but also on detection of genuine constituents. Amino acids and biogenic amines, food-grade acids, vitamins and sugar can be treated similarly.

In the future, employment of TLC as screening method will increasingly cover the ppb- and ppt-range. Those efforts are supported by the possibilities of using selective fluorimetric determination methods with pre-and post-chromatographic derivatization. With that, an optimization of clean up methods is self-evident.

REFERENCES

1. JÄNCHEN, D.: 1. Intern. Symp. Instrum. HPTLC, Bad DürhnPlm IfC-Bad Dürkheim 1980

2. JORK, H.: Fresenius Z.Anal. Chem. **318**, 177 (1984)

3. JORK, H., FUNK, W., FISCHER, W., WIMMER, H.: Dünnschicht- Chromatographie Reagentien und Detektion, Bd. 1, Verlag Chemie Weinheim 1987

4. STAHL, E.: Dünnschicht-Chromatographie, ein Laboratoriumshandbuch 2. Aufl., Springer, Berlin-Heidelberg-New York, 1967

5. WALLHÄUSER, K. H. in E. STAHL [4]

6. JORK, H.: GIT-Supplement 3 'Chromatographie' 1986, 79—87

7. JORK, H., WIMMER, H.: Quantitative Auswertung von Dünnschicht-Chromatogrammen Arbeitsblatt 1/1—7 und 1—8. GIT-Verlag Darmstadt 1986

8. FRERES, D., VALDEBOUZE, P.: J. Chromatogr. **87**, 300 (1973)

9. THOMAS, A. H., NEWLAND, P., SHARMA, N. R.: J. Chromatogr. **291**, 219 (1984)

10. BOGAERTS, R., DE VOS, D., DEGROOOT, J.-M.: Fleischwirtschaft **64**, 208 (1984)

11. EASTERBROOK, S. M., HERSEY, J.A.: J. Chromatogr. **121**, 390 (1976)

12. KLINE, R. M., GOLAB, T: J. Chromatogr. **18**, 409 (1965)

13. MAEHR, H., SCHAFFNER, C. P.: J. Chromatogr. **30**, 572 (1967)

14. WAGMAN, G. H., BAILEY, J.V.: J. Chromatogr. **41**, 263 (1969)

15. BEGUE, W. J., KLINE, R. M.: J. AOAC **56**, 20 (1973)

16. MARTINEZ, E. E., SHIMODA, W.: J. AOAC **66**, 1506 (1983)

17. PASCAL, C., GAILLARD, C., MOREAU, M.-O.: J.AOAC **62**, 976 (1979)

18. NICOLAUS, B. J. R., CORONELLI, C., BINAGHI, A.: Experientia **17**, 473 (1961)

19. FOUDA, H. G., LYNCH, M.J.: J. Agric. Food Chem. **26**, 463 (1978)

20. HEIL, K, PETER, F., CIELESZKY, V.: J. Agric. Food Chem. **32**, 997 (1984)

21. BICKEL, H., GAEUMANN, E. et al.: Helv. Chim. Acta **45**, 1396 (1962)

22. OKUYAMA, D., OKABE, M., FUKAGAWA, Y., ISHIKURA, T: J. Chromatogr. **291**, 464 (1984)

23. VON ARX, E., FAUPEL, M: J. Chromatogr. **154**, 68 (1978)

24. FAUPEL, M., FELIX, H. R., VON ARX, E.: J. Chromatogr. **193**. 511 (1980)

25. BRODASKY, T. F., LEWIS, C., EBLE, T. E.: J. Chromatogr. **123**, 33 (1976)

26. ASZALOS, A., DAVIS, S., FROST, D.: J. Chromatogr. **37**, 487 (1968)

27. BARBIERS, A. R., NEFF, A.W.: J. AOAC **59**, 849 (1976)

28. BRODASKY, T. F.: Anal. Chem. **35**, 343 (1963)

29. ZUIDWEG, M. H. J., OOSTENDORP. J.G., BOS, C.J.K.: J. Chromatogr. **42**. 552 (1969)

30. SHAFER, H., VANDENHEUVEL, W.J.A., ORMOND, R., KUEHL, F.A., WOLF, F.J.: J. Chromatogr. **52**, 111 (1970)

31. THOMAS, A. H., HOLLOWAY, J.: J. Chromatogr. **161**, 417 (1978)

32. IGLÓY, M., MIZSEI, A.: J. Chromatogr. **28**, 456 (1967)

33. VERPOORTE, R., KOS-KUYCK, E., TJIN-TSOI, A., RUIGROK, C. L. M., DE JONG, G., BAERHEIM SVENDSEN. A: Planta medica **48**, 283 (1983)

34. VALANJU, N. N., BADEN, M. M., VALANJU, S. M., MULLIGAN, D., VERMA, S. K.: J. Chromatogr. **81**, 170 (1973)

35. KLUG, E., TOFFEL, P.: Fresenius Z. Anal. Chem. **294**,46 (1979)

36. VERPOORTE, R., RUIGROK, C. L. M., BAERHEIM SVENDSEN, A.: Planta medica **46**, 149 (1982)

37. KARKOCHA, 1: Rocz. Paustw. Zakt. Hig. **29**, 193 (1978); **30**, 43 (1979)

38. ZIEMANN, K.: Staatsexamensarbeit, Universität des Saarlandes, Saarbrücken 1985

39. TOTH, L., PIORR, W.: Z. Lebensm. Unters. Forsch **133**, 322 (1967)

40. STAHL, E.: Arch. Pharmaz. **293**, 531 (1960)

41. GEIKE, F.: J. Chromatogr. **44**, 95 (1969)

42. SALO, T., SALMINEN, K., FISKARI, K.: Z. Lebensm. Unters. Forsch. **117**, 369 (1962)

43. ENGST, R., SCHNAAK, W.: Pharmaz. Z´halle **106**, 382 (1967)

44. AMBRUS, A., HARGITAIN, E. KAROLY, G., FÜLÖP, A., LANTOS, J.: J. AOAC **64**, 743 (1981)

45. PETERSON, C.A., EDGINGTON, L.V.: J. Agric. Food Chem. **17**, 898 (1969)

46. HOMMANS, A. L., FUCHS, A.: J. Chromatogr. **51**, 327 (1970)

47. REUSSER, F.: J. Biol. Chem. **242**, 243 (1967)

48. WOLTERS, B.: Planta Medica **17**, 42 (1969)

49. DURACKOVA, Z., BETINA, V, NEMEC, P.: J. Chromatogr **116**, 155 (1976)

50. LINNELL, J. C., HUSSEIN, H. A.-A., MATTHEWS, D. M.: J. Clin. Pathol. **23**, 820 (1970)

51. ONO, T., KAWASAKI, M.: Bitamin **30**, 280 (1964); ref. C. A. **62**, 1957 (1965)

52. KREUZIG, F.: Fresenius Z. Anal. Chem. **255**, 126 (1971)

53. SALEM, A. R. . FOSTER, M . A.: Biochem. et Biophys. Acta **252**, 597 (1971)

54. REUSSER, P.: J. Biol. Chem. **242**, 345 (1967)

55. KUROIWA, Y, HASHIMOTO, H.: J. Inst. Brew. **67**, 347, 352 (1961)

56. WEIGERT, E., SCHORN, P.J.. Tribuna Farmaceutica **30**, 48 (1962)

57. SCHORN, P. J. in E. STAHL [4]

58. TELEK, L., MARTIN, F.W., RUBERTÉ, R.M.: J. Agr. Food Chem. **22**. 332 (1974)

59. BAUR, CHR. GROSCH, W., WIESER, H., JUGEL, H.: Zeitschr. Lebensm. Unters. Forsch. **164**, 171 (1977)

60. HILLER, K., LINZER, B., PFEIFER, S.: Pharmazie **21**, 182 (1966)

61. HILLER, K., LINZER, B.: Pharmazie **22**, 321 (1967)

62. VAGUJFALVI, D., HELD, GY., TÉTÉNYI, P.: Arch. Pharm. **299**, 812 (1966)

63. HARASZTI, E., VETTER, J.: Botanikai Kozlemenyck **67**, 125 (1980)

64. YOON, K., WROLSTAD, R.: J. Agric. Food Chem. **32**, 691 (1984)

65. WINKLER, W.: Kolloid-Z. **177**, 63 (1961)

66. SHERMA, J., KOVALCHICK, A.J., MACK, R.: J.AOAC **61**, 616 (1978)

67. JOHANNES, B., KORFER, K. H., SCHAD, J., ULBRICH, I.: Arch. Lebensm. Hyg. **34**, 1 (1983)

68. SMOCZKIEWICZ, M. A., NITSCHKE, D., STAWINSKI, T M.: Mikrochim. Acta **2**, 597 (1977)

69. JORK, H.: Fresenius Z. Anal. Chem. **221**, 17 (1966)

70. JORK, H., J. Chromatogr. **33**, 297-306 (1968)

71. JORK, H.: Habilitationsschrift, Universität des Saarlandes, Saarbrücken 1969

72. JORK, H., KRAUS, LJ. in F. KORTE: Methodicum Chimicum Bd. 1, Kap. 2.4, G. Thieme, Stuttgart 1973

73. COURT, W. E. in SHELLARD, E. J.: Quantitative Paper- and Thin-Layer Chromatography, Akademic Press, London-New York, 1968

74. ISSAQ, H. J., BARR, E.W., ZIELINSKI JR., W.L.: J. Chrom. Sci. **13**, 597 (1975)

75. STROJNY, N., DE SILVA, A. F.: J. Chrom. Sci. **13**, 583 (1975)

76. ISSAQ. H. J., SCHROER, J.A., BARR, W.: Chem. Instrum. **8**, 51 (1977)

77. HARDEGGER, B. D.: Dissertation, ETH Zurich 6029 (1977)

78. RENNER, B., GERSTNER, E.: Z. Naturforsch. **33C**, 340 (1978)

79. DE JONG, J., VON MIEUWERK, H. J., SCHOLTEN, A. H. M. T., BRINKMAN, U. A. T, FREI, R.W.: J. Chromatogr. **166**, 233 (1978)

80. ISSAQ, H. J.: J. Liquid Chromatogr. **6**, 1213 (1983)

81. Eppley, R. M., Trucksess, M.W., Nesheim, S., Thorpe, C.W.. Wood, G.E. Pohland, A. E.: J. AOAC **67**, 43 (1984)

82. Brumley, W. C., Truckless, M. W., Adler, S. N., et al.: J. Agric. Food Chem. **33**, 326 (1985)

83. Vitek, R. K., Seul, C.J., Baier, M., Lau, E.: Americ. Labor. Febr. 1974

84. Falk, H., Krummen, K.: J. Chromatogr. **103**, 279 (1975)

85. Jork, H.: J. Pharmac. Belgique (Brüssel) 213 (1963)

86. Jork, H.: 111. Intern. Sympos. Chromatogr. Brüssel 1964, Symp. Bd. 1966, 295–303

87. Fairbain, J. W. in E. J. Shellard: Quantitative Paper- and Thin-Layer Chromatography. Academic Press. London-New York. 1968

88. Frei, R. W., Zürcher, H., Pataki, G.: J. Chromatogr. **45**, 284 (1969)

89. Frei, R. W., Lawrence, J. F., Belliveau, P. E.: Fresenius Z. Anal. Chem. **254**, 271 (1971)

90. Radecka, C., Wilson, W. L.: J. Chromatogr. **57**, 297 (1971)

91. Seher, A., Wortberg, A.: Erdöl+Kohle-Erdgas-Petrochemie, vereinigt mit Brennstoff-Chemie **27**, 428 (1974)

92. Herold, G.: Dissertation, Marburg 1975

93. Conway, R. L., Hood, F.: J. Chromatogr. **129**, 415 (1976)

94. Glasl, H., Ihrig, M.: Pharmaz. Ztg. **129**, 609 (1984)

95. Baltes, w.: Z. Lebensm. Unters.-Forsch. **145**, 34 (1971)

96. Hubl W., Buchner, M.: Z. Chem. **14**, 443 (1974)

97. Bethke, H., Santi, W., Frei, R.W.: J. Chromatogr. Sci. **12**, 392 (1974)

98. Wintersteiger, R., Gübitz, G.: Sci. Pharm. **45**, 18 (1977)

99. Haegele, E.-O., Müller, L., Lehmann, P.: Fresenius Z. Anal. Chem. **311**, 455 (1982)

100. Boice, R., Seidman, M, Southworth, B. C.: J. Pharm. Sci. **58**, 1527 (1969)

101. Messerschmidt, W.: J. Chromatogr. **39**, 90 (1969)

102. Koof, H. P., Noronha, R.V.: Fresenius Z. Anal. Chem. **250**, 124 (1970)

103. Shroff, A. P., Shaw, C.J.: J. Chrom. Sci. **10**, 509 (1972)

104. Bell, S.: J. AOAC **58**, 717 (1975)

105. Amin, M., Sepp, W.: J. Chromatogr. **118**, 225 (1976)

106. Friedrich, H., Wiedemeyer, H.: Planta medica **30**, 223 (1976)

107. Polzhofer, K.: Z. Lebensm. Unters. Forsch. **163**, 183, 272 (1977)

108. Möller, H., Steinbach, D.: Pharmaz. Ztg. **124**, 478 (1979)

109. Amin, M., Hassenbach, M.: Analyst **104**, 407 (1979)

110. THOMA, K., MÖLLER, H., STEINBACH, D.: Pharmaz. Ztg. **125**, 2073 (1980)

111. SZEKELY, G.: Mitt. Gebiete Lebensm. Hyg. **73**, 155 (1982)

112. SHERMA, J., ZORN, S.: Intern. Labor. **1982**, Juli/August S. 68

113. STREWE, K., NEUMANN, H.-J.: Umwelt und Technik **1982**, 47; Dissertation Hannover 1979

114. BERGNER-LANG, B., KACHELE, M., STENGEL, E.: Dtsch. Lebensm. Rundschau **79**, 400 (1983)

115. JOST, W., HAUCK, H.E., EISENBEISS, F.: J. Chromatogr. **256**, 182 (1983)

116. PITZ, B.: Staatsexamensarbeit, Universitat des Saarlandes, Saarbrücken 1985

117. BRASS, M., MILDAU, G.. JORK, H.: GIT-Supplement 3 'Chromatographie' **1983**, S. 35

118. JORK, H., KANY, E.: GDCh-Fortbildungskurs No 65, Saarbrücken 1980

119. FREI, R. W., LAWRENCE, J. F.: J. AOAC **65**, 1259 (1972)

120. SIBINSKI, R.: Diplomarbeit, FH Gießen, FR Umwelt- und Hygienetechnik 1980

121. RÜDT, U., HERBOLZHEIMER, D.: Lebensmittelchemie und gerichtliche Chemie **30**. 73 (1976) sowie Jahresbericht der Chem. Landesuntersuchungsanstalt, Stuttgart 1976

122. EICH, E., GEISSLER, H., MUTSCHLER, E., SCHUNACK, W.: Arzneim. Forsch. **19**, 1895 (1969)

123. WINTERSTEIGER, R., GÜBITZ, G., HARTINGER, A.: Microchim. Acta (Wien) **1979**, 235

124. FREYTAG, W. E.: Planta medica **40**, 278 (1980)

125. WINKLER, W., TRÖGER-JANSEN, J.: Privatmitteilung und Ingenieurarbeit der FH Aachen, FB 3, 1974

126. KREUZIG, F.: Fresenius Z. Anal. Chem. **282**, 447 (1976)

127. KERLER, R.: Diplomarbeit FH Gießen, FB Techn. Gesundheitswesen 1981

128. JORK, H., WEISS, M.: GIT Fachz. Labor **23**, 385 (1979)

129. LUITJENS, K.-D., FUNK, W., RAWER, P.: J. HRC & CC **4**, 136 (1981)

130. GÜBITZ, G., WENDELIN, W.: Analyt. Chem. **51**, 1690 (1979)

131. KLAUS, R.: Chromatographia **20**, 235 (1985)

132. FREI, R.W., MALLET, V., THIEBAUD, M.: J. Environm. Anal. Chem. **1**, 141 (1971)

133. KLAUS, R.: J. Chromatogr. **40**, 235 (1969)

134. KLAUS, R.: J. Chromatogr. **62**, 99 (1971)

135. FREI, R. W., ZÜRCHER, H.: Mikrochim. Acta **1971**, 209

136. RIPPHAHN, J., HALPAAP, H.: J. Chromatogr. **112**, 81 (1975)

137. EBEL, S., HEROLD, G.: Arch. Pharmaz. **308**, 839 (1975)

138. EBEL, S., GLASER, E., ROST, D.: J. HRC & CC **1**, 250 (1978)

139. ABE, F., SAMEJIMA, K.: Analyt. Biochem. **67**, 298 (1975)

140. EBEL, S., HEROLD, G.: Chromatographia **8**, 35 (1975)

141. JORK, H., KANY, E.: GDCh-Fortbildungskurs No 302, Saarbrücken 1986

142. EBEL, S., HEROLD, G., HOCKE, J.: Chromatographia **8**, 573 (1975)

143. HAGIWARA, T., SHIGEOKA, S., UEHARA, S., MIYATAKE, N., AKIYAMA, K.: J. HRC & CC **7**, 161 (1984)

144. LOCKWOOD, G. B., BRAIN, K.R., TURNER, T.D.: J. Chromatogr. **95**, 250 (1974)

145. FENIMORE, D. C., DAVIS, C. M., MEYER, C. J.: Clin. Chem. **24**, 1386 (1978)

146. REH, E., JORK, H.: Fresenius Z. Anal. Chem. **318**,264 (1984)

147. ROST, D.: Dissertation, Marburg 1979

148. JÄNCHEN, D. E.: Proc. 2nd Intern. Symp. Instrumental HPTLC, Interlaken, IfC, Bad Dürkheim 1982

149. BURGER, K.: Fresenius Z. Anal. Chem. **318**, 228 (1984)

150. BURGER, K.: Diskussionsvortrag, Chromatographie-Tagung E. Merck, Darmstadt 1982

151. BURGER, K.: GIT Supplement 4 "Chromatographie", 1984, S. 293

152. JÄNCHEN, D.: 3. Int. Symp. Instrum. HPTLC, Würzburg, IfC, Bad Dürkheim, 1985

153. PERRY, J. A.: US patent 3,864,250

154. PERRY, J. A., HAAG, K.W., GLUNZ, L.J.: J. Chromatogr. Sci. **11**, 447 (1973)

155. PERRY, J. A., JUPILLE, T. H., GLUNZ, L. J.: Ind. Res. **16**, 55 (1974)

156. PERRY, J. A., GLUNZ, L.J.: J. AOAC **57**, 832 (1974)

157. JUPILLE, T. H., PERRY, J.A.: J. Chromatogr. Sci. **13**, 163 (1975)

158. PERRY, J. A., JUPILLE, T. H., CURTICE, A.: Separation Sci. **10**, 571 (1975)

159. Camag: Firmenschrift AMD-85-D

160. KREUZIG, F.: Z. Anal. Chem **282**, 457 (1976)

161. ZEISS, C.: Prototyp des Chromatogramm-Spektralphotometers, Privatmitteilung 1967; Achema Frankfurt 1968

162. KYNAST, G.: Fresenius Z. Anal. Chem. **250**, 105 (1970)

163. KYNAST, G.: Fresenius Z. Anal. Chem. **251**, 161 (1970)

164. KUSSMAUL, H.: Dissertation, Marburg 1972

165. EBEL, S., GEITZ, E., HOCKE, H., und KAAL, M.: Kontakte (Merck) 1982 (1), S. 39–44

166. JORK, H.: Informationstag "Dünnschicht-Chromatographie 1986" Brüssel-Overisje

167. BALTES, W.: Fresenius Z. Anal. Chem. **296**, 365 (1979)

168. GÜNTHER, H. O.: Fresenius Z. Anal. Chem. **290**, 389 (1978)

169. HEZEL, U.: Mitteilungsbl. Lebensmittelchem. u. gerichtl. Chemie **30**, 217 (1976); GIT-Fachz. Labor **22**, 595 (1978)

170. THIER, H. P.: Angew. Chem. **86**, 244 (1974)

171. TOMINGAS, R., und BROCKHAUS, A.: Staub-Reinhaltung der Luft Nr. 12 (1973)

172. BUTLER, H.T., CODDENS, M.W., KHATIB, S., POOLE, C.F.: J. Chromatogr. Sci. **23**, 200 (1985)

173. SEIFERT, B., STEINBACH, 1: Fresenius Z. Anal. Chem. **287**, 264 (1977)

174. KRAFT, J., HARTUNG, A., LIES, K.-H., SCHULZE, J.: J. HRC & CC **5**, 489 (1982); 2. Intern. Symposium Instrum. HPTLC, Interlaken, IfC, Bad Dürkheim, 1982

175. TOMINGAS, R., GROVER, Y P.: Fresenius Z. Anal. Chem. **313**, 414 (1982); **315**, 515 (1983)

176. BROWN, K. K POOLE, C. F.: J. HRC & CC **7**, 520 (1984)

177. KASHANI, D. T.: GIT Labor-Medizin **2**, 128 (1979)

178. SCHÖSSNER, H., FALKENBERG, W., ALTHAUS, H.: Z. Wasser-Abwasser-Forsch. **16**, 132 (1983)

179. Deutsche Norm: DIN 38409, Teil 13, Juni 1981, Beuth-Verlag, Berlin

180. WEIL, L., HAUCK, E. H.: GIT Fachz. Lab. **24**, 538 (1980)

180.a WEIL, L., GRIMMER, G., HELLMANN, H. et al.: Z. Wasser, Abwasser Forsch. **13**, 108 (1980)

181. HELLMANN, H.: Fresenius Z. Anal. Chem. **287**, 148 (1977); **295**, 388 (1979)

182. WOIDICH, H., PFANNHAUSER, W., BLAICHER, G., TIEFENBACHER, K: Lebensmittelchemie u. gerichtliche Chemie **30**, 141 (1976); Chromatographia **10**, 140 (1977)

183. HERRMANN, R.: Catena **5**, 165 (1978)

184. ALTMEYER, B., EICHHORN, K.W., PLAPP, R.: Z. Lebensm. Unters. Forsch. **175**, 172 (1982)

185. SEHER, A., WORTBERG, B.: Erdöl und Kohle-Erdgas-Petrolchemie **27**, 428 (1974)

186. GERTZ, Ch.: Z. Lebensm. Unters. Forsch. **167**, 233 (1978)

187. GERTZ, Ch.: Z. Lebensm. Unters. Forsch. **173**, 208 (1981); Lebensmittelchemie u. gerichtliche Chemie **35**, 99 (1981)

188. SHERMA, J.: Whatman TLC Techn. Series, Vol. 1, Clifton (N. J.) (1981)

189. POOLE, C. F., BUTLER, H. T., CODDENS, M. E., KHATIB, S., VANDERVENNET, R.: J. Chromatogr. **302**, 149 (1984)

190. RÜSSEL, H.: Rückstandsanalytik von Wirkstoffen in tierischen Produkten, G. Thieme, Stuttgart-New York, 1986

190a. SHERMA, J.: J. Liquid. Chromatogr. **5**, 1013 (1982)

191. MacNEIL, J. D., FREI, R. W.: J. Chrom. Sci. **13**, 279 (1975)

192. FREAR, D. E. H. J.: Pesticide Index, College Science Publishers, State College, PA, USA

193. Caissie, G. E., Mallet, V. N.: J. Chromatogr. **117**, 129 (1976)

194. Jork, H., Kany, E. in: Jork, H., Wimmer, H. 17]

195. Frei, R. W., Mallet, V. N.: Intern. J. Environ. Anal. Chem. **1**, 99 (1971)

196. Bidleman, T. F., Nowlan, B., Frei, R.W.: Anal. Chim. Acta **60**, 13 (1972)

197. Frei, R. W., Lawrence, J. F., Belliveau, P. E.: Z. Anal. Chem. **254**, 271 (1971)

198. Frei, R. W., Mallet, V., Pothier, C.: J. Chromatogr. **59**, 135 (1971)

199. Frei, R. W., Lawrence, J. F.: J. Chromatogr. **67**, 87 (1972)

200. Lawrence, J. F., Frei, R. W.: Analyt. Chem. **44**, 2046 (1972); Intern. J. Environ. Anal. Chemie **1**, 317 (1972)

201. Lawrence, J. F., LeGay, D. S., Frei, R. W.: J. Chromatogr. **66**, 295 (1972)

202. Frei, R. W., Lawrence, J. F.:, J. AOAC **55**, 1259 (1972)

203. Lawrence, J. F., Frei, R. W.: J. Chromatogr. **66**, 93 (1972)

204. Brun, G. L., Mallet, V.: J. Chromatogr. **80**, 117 (1973)

204.a Ragab, M T. H .: J . AOAC **50**, 1088 (1967)

205. Mallet, V. N., Zakrevsyk, J. G., Brun, G. L.: Bull. Soc. Chim. Fr. **1974**, 1755; J. Agric. Food. Chem. **23**, 754 (1975)

206. Frei, R. W., Lawrence, J. F., Legay, D. S.: Analyst **98**, 9 (1973)

207. Sherma, J., Koropchack, J.: Anal. Chim. Acta **91**, 259 (1977)

208. Sherma, J., Boymel, J. L.: J. Liquid Chrom. **6**, 1183 (1983)

209. Rheinstädtler, B.: Staatsexamensarbeit, Universität des Saarlandes. Saarbrücken 1980

210. Young-duck, A., Bergner, K.-G.: Dtsch. Lebensm. Rundschau **76**, 390 (1980); **77**,102(1981)

211. Jork, H., Roth, B. in: Jork, H., Wimmer, H. [7]

212. MacNeil, J. D., MacLellan, B. L., Frei, R. W.: J. AOAC **57**, 165 (1974)

213. Polzhofer, K.: Z. Lebensm. Unters. Forsch. **167**, 162 (1978)

214. Hauck, H. E., Amadori, E.: ACS Symp. Ser. **136**, 159 (1980)

215. Stivje, T.: Dtsch. Lebensm. Rundschau **76**, 234 (1980)

216. Goldhagen, H., Meier, W.: Mitt. Gebiete Lebens. Hyg. **71**, 248 (1980)

217. Puchwein, G., Schmidinger, G., Hahn, S., Krützen, D.: Z. Lebensm. Unters. Forsch. **169**, 339 (1979)

218. Ishizuka, M., Kondo, Y., Nagare, T., Takeuchi, Y.: J. Agric. Food Chem. **33**, 1070 (1985)

219. Frei, R. W., Freemann, C. D.: Microchim. Acta (Wien) **1968**,1214

220. Mallet, V., Surette, D. P.: J. Chromatogr. **95**, 243 (1974)

221. De Jong, J., van Nieuwkerk, H. J., Scholten, A. H. M. T., Brinkman, U. A. Th., Frei, R. W.: J. Chromatogr. **166**, 233 (1978)

222. FREI, R. W., NOUMURA, N. S.: Microchim. Acta (Wien) **1968**, 565

223. SHERMA, J., STELLMACHER, S.: J. Liquid Chromatogr. **8**, 2949 (1985)

224. POLZHOFER, K.: Z. Lebensm. Unters. Forsch. **163**, 109 (1977)

225. MacNEIL, J. D., HIKICHI, M.: J. Chromatogr. **101**,33 (1974)

226. MANDROU, B., BRUN, S., KINGKATE, A.: J. AOAC **60**, 699 (1977)

227. FRANCOEUR, Y., MALLET, V.: J. AOAC **60**, 1328 (1977)

228. MALLET, V., SURETTE, D., BRUN, G. L.: J. Chromatogr. **79**, 217 (1973)

229. SHERMA, J.: J. Chromatogr. **104**, 476 (1975)

230. BECKER, N.: GIT Supplement "Chromatographie" Juli 1981, S. 14

231. VERBEKE, R.: J. Chromatogr. **177**, 69 (1979)

232. VOGT, K.: Archiv für Lebensmittelhygiene **29**, 178 (1978)

233. JARC, H., RUTTNER, O., KROCZA, W.: J. Chromatogr. **134**, 351 (1977)

234. STAN, H.-J., HOHLS, F. W.: Z. Lebensm. Unters. Forsch. **166**, 287 (1978)

235. WORTBERG, B., WOLLER, R., CHULAMORAKOT, T.: J. Chromatogr. **156**, 205 (1978)

236. HOHLS, F. W., STAN, H.-J.: Z. Lebensm. Unters. Forsch. **167**, 252 (1978)

237. WALDSCHMIDT, M.: Archiv für Lebensmittelhygiene **23**, 76 (1972)

238. LAITEM, L.. GASPAR, P., BELLO, I.: J. Chromatogr. **147**, 538 (1978)

239. SCHMUTZ, H. R.: Labor-Praxis **6**, 38 (1982)

240. JARC, H.: Fleischwirtschaft **60**, 676 (1980)

241. GAUCH, R., LEUENBERGER, U., MÜLLER, U.: Mitt. Gebiete Lebensm. Hyg. **72**, 418 (1981)

242. GIRON, D., GROELL, P.: J. HRC & CC **1**, 67 (1978)

243. OEHRLE, K. L., VOGT, K., HOFFMANN, B.: J. Chromatogr. **114**, 244 (1975)

244. MALISCH, R.: Dtsch. Lebensm. Rundschau **82**, 222 (1986)

245. BOSSUYT, R., VAN RENTERGHEN, R., WAES, G.: J. Chromatogr. **124**, 37 (1976)

246. ALLEN, E. H.: J. AOAC **68**, 990 (1985)

246a. SCHWARTZ, D. P., McDONOUGH, F. E.: J. AOAC **67**, 563 (1984)

247. HAEFELFINGER, P., HAEFELINGER, K.: Fresenius Z. Anal. Chem. **302**, 121 (1980)

248. KREUZIG, F.: J. Chromatogr. **142**, 441 (1977)

249. LANGNER, H. J., TEUFEL, U.: Die Fleischwirtschaft **52**, 1610 (1972)

250. PAUNCZ, J. K.: J. HRC & CC **4**, 287 (1981)

251. THOMAS, M. H., EPSTEIN, R. L., ASHWORTH, R. B., MARKS, H.: J. AOAC **66**, 884 (1983)

252. THOMAS, M. H., SOROKA, K. E., SIMPSON, R. M., EPSTEIN, R. L.: J. Agric. Food Chem. **29**, 621 (1981)

253. SIGEL, C. W., WOOLEY JR., J. L., NICHOL, C. A.: J. Pharm. Sci. **64**, 973 (1975)

254. SCHLATTERER, B.: Z. Lebensm. Unters. Forsch. **176**, 20 (1983)

255. KNUPP, G., POLLMANN, H., JONAS, D.: Chromatographia **22**, 21 (1986)

256. BELJAARS, P. R.: Practical approach to governmental surveillance for Mycotoxins in Food and Foodstuffs to protect Human Health. Keuringsdienst van Waren, Maastricht, 1978

257. STAHR, H. M. in: Thin-Layer Chromatography (eds. TOUCHSTONE, J. C, ROGERS, D.) Chapter 14, J. Wiley, New York, 1980

258. HSIEH, D. Ph., FUKAYAMA, M. Y., RICE, D. W., WONG, J. J. in: Thin Layer Chromatography (eds. TOUCHSTONE, J. C., ROGERS, D.) Chapter 16, J. Wiley, New York, 1980

259. MILLER, N., PRETORIUS, H. E., TRINDER, D. W.: J. AOAC **68**, 136 (1985)

260. AYRES, J. L., SINNHUBER, R. 0.: J. AOAC **43**, 423 (1966)

261. PONS, W. A. JR., ROBERTSON, J. A., GOLDBLATT, L. A.: J. AOAC **43**, 665 (1966)

262. PETERSON, R. E., CIEGLER, A., HALL, H. N.: J. Chromatogr. **27**, 304 (1967)

263. BECKWITH, A. C., STOLOFF, L.: J. AOAC **51**, 602 (1968)

264. PONS, W. A. JR.: J. AOAC **51**, 913 (1968); **54**, 870 (1971)

265. SCHULLER, P. L., VERHÜLSDONK, C. A. H., PAULSCH, W. E.: Arzneim. Forsch. **20**, 1517 (1970)

266. NESHEIM, S.: J. AOAC **54**, 1444 (1971)

267. MÜLLER, H., SIEPE, V: Dtsch. Lebensm. Rundschau **74**, 133 (1978)

268. KLEMM, U.: GIT Supplement 'Chromatographie' **1982**, 9

269. JANICKI, J., SZEBIOTKO, K., CHELKOWSKI, J. et al.: Acta Alimentaria Polonica 1. 2 0 7 (1975)

270. EL-NAWAWY, F. A.: Wien. tierärztl. Mschr. **61**, 66 (1974)

271. RIPPHAHN, J., HALPAAP, H.: J. Chromatogr. **112**, 81 (1975)

272. TOSCH, D., WALTKING, A. E., SCHLESIER, J. F.: J. AOAC **67**, 337 (1984)

273. BELJAARS, P. R., FABRY, F. H. M., SICKOTT, M. M. A., PEETERS, M, J.: J. AOAC **55**, 775, 1310 (1972)

274. JUNG, M., HANSEN, E.: Food Cosmet. Toxicol. **12**, 131 (1974)

275. GAUCH, R., LEUENBERGER, U., BAUMGARTNER, E.: J. Chromatogr. **178**, 543 (1979)

276. BIONDI, P. A., GAVAZZI, L., FERRARI, G., MAFFEO, G., SECCHI, G.: J. HRC & CC **3**, 92 (1980)

277. STUBBLEFIELD, R. D., SHOTWELL, 0. L., SHANNON, G. M.: J. AOAC **55**, 762 (1972)

278. PONS, W. A. JR., CUCULLU, A. F., LEE, L. S.: J. AOAC **56**, 1431 (1973)

279. SCHULLER, P. L., VERHÜLSDONK, C. A. H., PAULSCH, W. E.: Pure and Applied Chem. **35**, 291 (1973)

280. STUBBLEFIELD, R. D., SHANNON, G. M., SHOTWELL, 0. L.: J. AOAC **56**, 1106 (1973)

281. GASIOROWSKA, U. W., STRZELECKI, E. L.: J. Chromatogr. **157**, 445 (1978)

282. Tuinstra, L. G. M. Th., Bronsgeest, J. M.: J. Chromatogr. **111**, 448 (1975)

283. Tripet, F., Riva, C., Vogel, J.: Mitt. Geb. Lebensm. Hyg. **72**, 367 (1981)

284. Wiseman, D. W., Marth, E.: Z. Lebensm. Unters. Forsch. **177**, 22 (1983)

285. Sekut, A. A., Dollear, F. G., Codifer, L. P. Jr.: J. Agric. Food Chem. **25**, 1314 (1977)

286. Beljaars, P. R., Schumans, J. C. H. M., Koken, P. J.: J. AOAC **58**, 263 (1975)

287. Kaminura, H., Nishijima, M., Yasuda, K., et al.: J. AOAC **68**, 458 (1985)

288. Berman, M. R., Zare, R. H.: Analyt. Chem. **47**, 1200 (1975)

289. Gertz, Ch., Böschemeyer, L.: Z. Lebensm. Unters. Forsch. **171**, 335 (1980)

290. Shannon, G. M., Shotwell, O. L., Kwolek, W. F.: J. AOAC **66**, 582 (1983)

291. Seitz, L. M., Mohr, H. E.: J. AOAC **59**, 106 (1976); Cereal Chemistry **51**, 487 (1974)

292. Spilmann, J. R.: J. AOAC **68**, 453 (1985)

293. Ehrlich, K., Lee, L.: J. AOAC **67**, 963 (1984)

294. Drawert, F., Barton, H.: Z. Lebensm. Unters. Forsch. **154**, 223 (1974)

295. Chalela, G., Schwantes, H. O., Funk, W.: Fresenius Z. Anal. Chem. **319**, 527, (1984)

296. Seiber, J. N., Hsieh, D. P. H.: J. AOAC **56**, 827 (1973)

297. Toth, L., Tauchmann, F., Leistner, L.: Die Fleischwirtschaft **50**, 1235 (1970)

298. Brown, N. I., Nesheim, S., Stack, M. E., Ware, G. L.: J. AOAC **56**, 1437 (1973)

299. Trucksess, M. W., Stoloff, L.: J. AOAC **62**, 1080 (1979)

300. Scott, P. M., Lawrence, J. W., van Walbeek, W.: Applied Microbiol. **20**, 839 (1970)

301. Roberts, B. A., Patterson, D. S. P.: J. AOAC **58**, 1178 (1975)

302. Durackova, Z., Betina, V., Nemec, P.: J. Chromatogr. **116**, 141 (1976)

303. Reimerdes, E.: 3. Intern. IUPAC Symp., Paris 1976

304. Damodaran, C., Ramadoss, C. S., Shanmugasundaram, E. R. B.: Anal. Biochem. **52**, 482 (1973)

305. Schmidt, R., Ziegenhagen, E., Dose, K.: Z. Lebensm. Unters. Forsch. **175**, 169 (1982)

306. Eppley, R. M.: J. AOAC **58**, 906 (1975)

307. Engel, G.: J. Chromatogr. **207**, 430 (1981)

308. Asensio, E., Sarmento, I., Dose, K.: Z. Anal. Chem. **311**, 511 (1982)

309. Eppley, R. M.: J. AOAC **51**, 67, 74 (1968)

310. Trenk, H. L., Chu, F. 5.: J. AOAC **54**, 1307 (1971)

311. Hagan, S. N., Tietjen, W. H.: J. AOAC **58**, 620 (1975)

312. Wilson, D. M., Tabor, W. H., Trucksess, M. W.: J. AOAC **59**, 125 (1976)

313. ENGEL, G., PROKOPEK, D.: Milchwissenschaft **34**, 272 (1979)

314. REIMERDES, E. H., ENGEL, G., BEHNERT, J.: J. Chromatogr. **110**, 361 (1975)

315. REIMERDES, E. H., ENGEL, G.: Milchwissenschaft **30**, 209 (1975)

316. CIEGLER, A., KURTZMAN, C. P.: J. Chromatogr. **51**, 511 (1970)

317. STINSON, E. E., HUHTANEN, CH. N., ZELL, T. E., SCHWARTZ, D. P., OSMAN, St. F.: J. Agric. Food Chem. **25**, 1220 (1977)

318. TANNER, H., ZANIER, C.: Schweiz. Z. f. Obst- u. Weinbau **112**, 656 (1976)

319. BERGNER-LANG, B., KÄCHELE, M., STENGEL, E.: Dtsch. Lebensm. Rundschau **79**, 400 (1983)

319a. KOCH, CH. E., THURM, V., PAUL, P.: Nahrung **23**, 125, 131 (1979)

320. LEUENBERGER, U., GAUCH, R., BAUMGARTNER, E.: J. Chromatogr. **161**, 303 (1978)

321. SCOTT, P. M., MILES, W. F., TOFT, P., DUBLE, J. G.: J. Agric. Food Chem. **20**, 450 (1972); **25**, 434 (1977)

322. EYRICH, W.: Chem. Mikrobiol. Technol. Lebensm. **4**, 17 (1975)

323. STEINER, W., BATTAGLIA, R.: Mitt. Gebiete Lebensm. Hyg. **74**, 140 (1983)

324. MEYER, R. A.: Die Nahrung **26**, 337 (1982)

325. POHLAND, A. E., ALLEN, R.: J. AOAC **53**, 686 (1970)

326. REISS, J.: Chromatographia **4**, 576 (1971); Naturwissenschaften **59**, 37 (1972); Zbl. Bakt. Abt. II **128**, 685 (1973)

327. SCHMIDT, R., NEUNHOFFER, K., DOSE, K.: Fresenius Z. Anal. Chem. **299**, 382 (1979)

328. PURCHASE, I. F. H., PRETORIUS, M.: J. AOAC **56**, 225 (1973)

329. ATHNASIOS, A. K., KUHN, G. 0.: J. AOAC **60**, 104 (1977)

330. SCHWEDT, G., SCHWADORF, K.: Dtsch. Lebensm. Rundschau **82**, 209 (1986)

331. RIOS, V. M.: Z Lebensm. Unters. Forsch. **147**, 331 (1972)

331.a ROBBIANI, R., BÜCHI, W.: Proc. Euro Food Chem. III **1985**, 216

332. DUDEN, R., FRICKER, A., CALVERLEY, R., PARK, K. H., RIOS, V. M.: Z. Lebensm. Unters. Forsch. **151**, 23 (1973)

333. GERTZ, CH., HILD, J.: Z. Lebensm. Unters. Forsch. **170**, 103 (1980)

334. WOIDICH, H., GNAUER, H., GALINOVSKY, E.: Z. Lebensm. Unters. Forsch. **133**, 317 (1967)

335. SHERMA, J., ZORN, St.: Intern. Laboratory **1982**, Juli/August, S. 68

336. NAGY, L.: Gyogyszereszet **30**, 127 (1986)

337. DE LA VIGNE, K.: Seifen-Öle-Fette-Wachse **111**, 275 (1985)

338. RÜDT, U., HERBOLZHEIMER, D.: Min. Geb. Lebensmittelchemie u. gerichtl. Chemie **30**, 73 (1976)

339. GROEBEL, W., WESSELS, A.: Dtsch. Lebensm. Rundschau **69**, 453 (1973)

340. WEI, R., SONG, F., GAO, H.: Chinese J. of Food & Ferm. Ind. **1985**, 32

341. LAUB, E., WOLLER, R.: Dtsch. Lebensm. Rundschau **72**, 276 (1976)

342. FREI-HÄUSLER, M., FREI, R. W., HUTZINGER, O.: J. Chromatogr. **79**, 209 (1973)

342a. OTTENEDER, H.: Mitt. Geb. Lebensmittelchem. u. gerichtl. Chem. **26**, 223 (1972)

343. OTTENEDER, H., HEZEL, U.: J. Chromatogr. **109**, 181 (1975)

344. NORMAN, S. M., FOUSE, D. C., CRAFT, C. C.: J. AOAC **55**, 1239 (1972)

345. EBEL, S., HEROLD, G.: Dtsch. Lebensm. Rundschau **70**, 133 (1974)

346. BELJAARS, P. R., RONDAGS, TH. M. M.: De Ware(N)-Chemicus **4**, 146 (1974)

347. BELJAARS, P. R., RONDAGS, TH. M. M.: J. AOAC **61**, 1415 (1978); De Ware(N)-Chemicus **8**, 11 (1978)

348. LEHMANN, G., HAUG, I., SCHLÖSSER, R.: Dtsch. Lebensm. Rundschau **71**, 61 (1975); Z. Lebensm. Unters. Forsch. **172**, 87 (1981)

349. SHERMA, J., MILLER, R.: Americ. Lab. **16**, 126 (1984)

349a. HARLOS, H.: Dtsch. Lebensm.-Ind. **20**, 412 (1973)

350. BELJAARS, P. R.: De Ware(n)-Chemicus **6**, 101 (1976); **7**, 169, 174 (1977)

351. WASHÜTTL, J., BANCHER, E., REDERER, P.: Z. Lebensm. Unters. Forsch. **143**, 253 (1970)

352. BALTES, W.: Z. Lebensm. Unters. Forsch. **145**, 34 (1971)

353. KELKER, H.: Nachr. Chem. Tech. Lab. **31**,786 (1983)

354. BALTES, W., PETERSEN, H., DEGNER, CH.: Z. Lebensm. Unters. Forsch. **152**, 145 (1973)

355. SENANAYAKE, U. M., WIJESEKERA, R. O. B.: J. Chromatogr. **32**, 75 (1968)

356. KAWAMOTO, H., YAMANE, T.: Japan. J. Clin. Pathology (Rinsho Byori) **33**, 217 (1985)

357. LAUB, E., ZIMMER, M.: Lebensmittelchemie u. gerichtl. Chemie **33**, 117 (1979)

358. BALTES, W., KLASMANN, M.: Chem. Microbiol. Techn. Lebensm. **1**, 195 (1972)

359. HEY, H.: Z. Lebensm. Unters. Forsch. **148**, 1 (1972)

360. BELJAARS, P. R., KOKEN, P. J.: J. AOAC **56**, 1284 (1973)

361. GRABARKIEWICZ-SZCZESNA, S., GOLINSKI, P. et al.: Nahrung **29**, 229 (1985)

362. RIOS, V. M., DUDEN, R.: Lebensm. Wiss. u. Technol. **4**, 97 (1971)

363. SUZUKI, T., KAWADA, T., IWAI, K.: J. Chromatogr. **198**, 217 (1980)

364. IWAI, K., SUZUKI, T., LEE, K.-R., KOBASHI, M., OKA, S.: Agric. Biol. Chem. **41**, 1877 (1977)

365. JORK, H., PITZ, B. in JORK, H., WIMMER, 11. [7]

366. JORK, H., JOCHUM, P. in JORK, H., WIMMER, H. [7]

367. MANG, P.: Staatsexamensarbeit, Univ. d. Saarlandes, Saarbrücken, 1979

368. SCOTT, P. M.: J. AOAC **68**, 242 (1985)

369. SCHMIDT, N.: GIT-Supplement 4 'Chromatographie' **1985**, 78

370. DORNHEIM, O.: Dissertation Berlin 1985

371. DORNHEIM, O., HILDEBRANDT: Fleischwirtschaft **66**, 1169 (1986)

High Performance Liquid Chromatography (HPLC)
Reflections on Application in Food Analysis

H. Engelhardt

7 High Performance Liquid Chromatography (HPLC)
Reflections on Application in Food Analysis

H. Engelhardt, Saarbrücken

7.1 Introduction

Within the last 20 years, the modern column liquid chromatography described with the acronym HPLC for High Performance (originally Pressure) Liquid Chromatography has developed to an analysis method that is an integral part of all modern laboratories [1–4] The application possibilities reach, supplementary to the gas chromatography, from separation of less volatile substances and substances that do not decompose on evaporation to polymer compounds [5]. The introduction of chemically bonded phases, especially of the reversed phases is largely responsible for the rapid development because samples can be applied directly from the aqueous solution into the chromatographic system. Introduction of columns packed with particles sized below 10 µm made it possible to perform highly efficient separations in short time, i.e. with high analysis speed. Columns with 10 000 theoretical plates at 25 cm length are now standard. Due to the pressure achievable by pumps (mostly 400 bar) the number of plates is limited to a maximum of approximately 200 000. This, however, is not a disadvantage because the selectivity in the liquid chromatography, contrary to the gas chromatography, can be varied within large limits and optimized by selection of the eluent. Capillary columns, used in gas chromatography with extremely high separation efficiencies, will hardly be practical for liquid chromatography because no detectors are available for detecting substances sensitively enough and in such extremely small volumes (1 nl) [6].

Despite many efforts the sensitive detection of the separated sample components in the eluate of the separation columns is still a problem which restricts the universal application of HPLC. It is no problem to detect substances which absorb light in the visible or ultraviolet spectral range or can be excited to fluorescence. With sufficiently high absorption coefficients detections is easily possible in the ng range with modern detectors. But especially in the food chemistry many important substances (e.g. fats, sugars, etc.) do not have sufficient absorption in a favorable wave length range and, therefore, they cannot be sensitively detected. For example, trace detection of contaminants also make great demands on the chromatographic system. In the following, possibilities will be shown to avoid these limitations, to enhance detection sensitivity, and to assure qualitative identification.

7.2 Enhancing the Detection Sensitivity by Application of Micro Bore Columns

During migration of the sample zones through the chromatographic column the band broadening causes a dilution of the substance zone, i.e. during the separation of the sample components the separated zones are always diluted by the eluent, thereby the sample concentration at the peak maximum decreases. This influences the detection sensitivity with concentration sensitive detectors, used exclusively for HPLC, negatively. The lower the band broadening in the column the higher the concentration at the peak maximum of the eluted substance zones. Columns packed with smaller particles always provide better detection sensitivities (lower band broadening corresponds to high number of theoretical plates and with that lower dilution). Because in separation columns packed with small particles the theoretical plates required for a special separation problem (given selectivity) are available in shorter separation columns. The separation can be done successfully with low analysis times [7]. Today, columns with 3–6 cm length, packed with 3 μm particles are used for high speed analysis and high detection sensitivity [8]. But modern equipment with small dead volumes (short connection tubes with inside diameter below 0.25 mm) and detectors with very short time constants are required for that.

The volume, in which the substance zone at the column exit is dissolved, can be decreased by using separation columns with diameters smaller than the usual 4–4.6 mm. The so-called micro bore columns [9] are very suitable for detection in the trace range, especially if the sample amount is limited and/or very small. Normally columns with inside diameters of 1–2 mm are used. With constant sample quantity the concentration at the peak maximum can be increased by the factor 4–16. This is shown in Figure 1 for columns with 4.2 and 1 mm inside diameter. Identical sample quantities had been applied to all columns [10]. Of course, with the same loading capacity, 16 times the sample quantity could be separated at the 4 mm column.

The advantages of the micro bore columns for trace analysis is complete only if a sensitive detector is available. The detector should have the same layer thickness used for normal separation columns even with the correspondingly small cell volumes. For micro bore columns with their low volume flow rates (50–100 μl/min at 1 mm column) different pumps as well as modified sample injectors (for maximal sample volumes of approx. 1 μl) are necessary. Substantial experience with HPLC is necessary to recognize and eliminate influences on the analytical result caused by the equipment. They have their advantages for chromatographic analysis with very limited sample amounts. The low volume flow rate additionally helps save eluents.

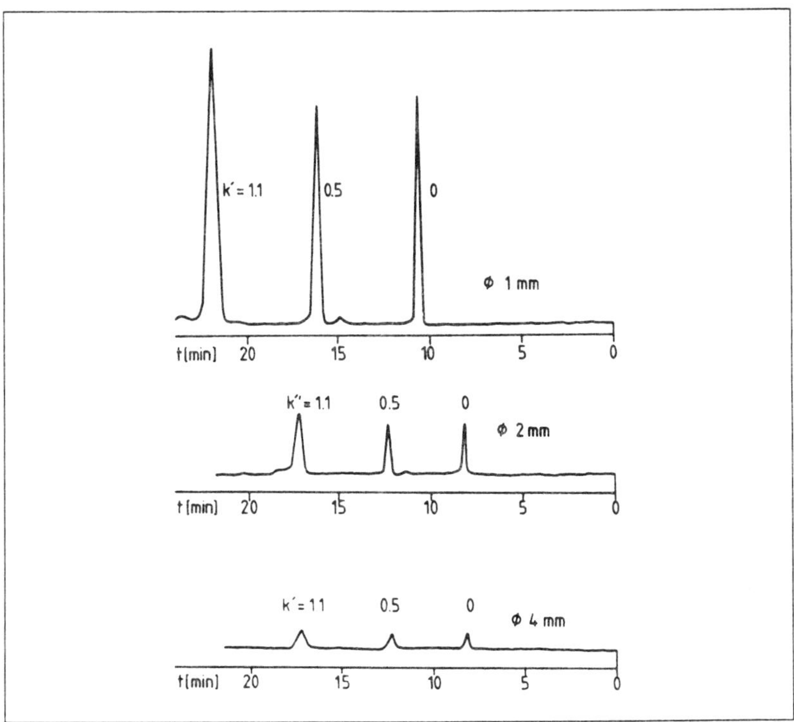

Fig. 1 Advantages of Separation Columns with Narrow Diameter and Identical Sample Quantity
Column dimension: 250 x 2 or 4 mm, resp. and 300 x 1 mm, packed with RP 18, dp 10 µm, Eluent: methanol
Samples: phenole (k'= 0), anthracene (k'= 0.5), chrysene (k'= 1.6)
Sample quantity: 1 µl each

7.3 Sample Preparation and Sample Concentration by Column Switching

Very often the detection of substances in complex matrix is preceded by sample preparations i.e. removal of disturbing matrix components in order to protect the HPLC separation columns. Classically, this was done by manifold distribution steps – but today, suitable stationary phases for sorptive pre-cleaning steps are available. This can be done separate from the chromatographic equipment with various cartridge systems. Two six-way switching valves allow the sample preparation to be carried out on-line, directly before the separation column. Figure 2 shows a corresponding layout.

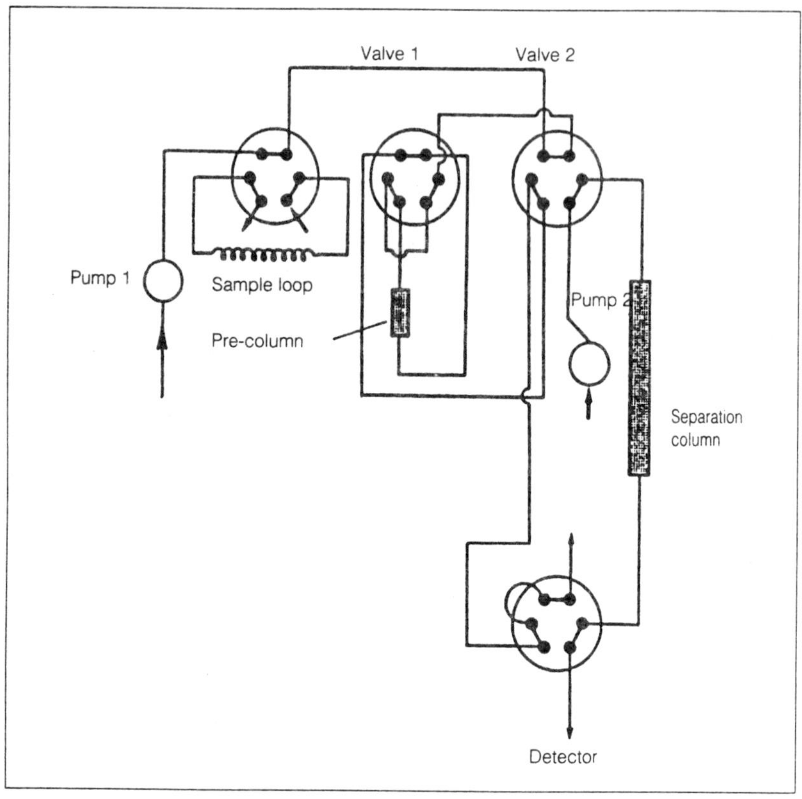

Fig. 2 Column Switching Layout for Sample Pre-Clean-up or Enrichment

The sample is usually separated in a short column. If the eluate of this pre-column contains the component to be determined it is connected with the main column in the second six-way switching valve. The cut with the interesting fraction is transferred into the main column (heart cut). The two columns are only connected during the transfer. During the separation in the main column (it is advantageous to have a second pump for that), the pre-column can be regenerated and another sample may be applied. The third valve, which connects the pre- and main column alternately with the detector, is not absolutely necessary. By selecting a suitable stationary phase for this pre-column and its combination with the main column, band broadening can be kept at a minimum for this column switching. The system is optimal if the sample to be determined are less retarded at the pre-column than at the main column. In this case the eluent must be changed after sample transfer from pre- to main column, i.e. its elution power must be strengthened. It is advantageous to combine a revered phase with short alkyl groups in the pre-column with a RP C_{18} separation column.

The same layout can be used for enrichment of substances which are present in the sample in low concentrations only. This can be done without any problems with relatively polar substances. Enrichment factors up to 1:100 can be achieved. Figure 3 shows the enrichment of caffeine from decaffeinated coffee and the reproducibility of the enrichment procedure [12]. Brewed coffee samples of 1.5 ml each (0.4 g coffee with 10 ml hot water) were enriched on a pre-column of 10 mm length. The caffeine peak in Figure 3 corresponds to a caffeine content of 0.05% in coffee. A simultaneous detection of chlorogenic acid seems to be possible. But if polar soluble substances in presence of non-polar sample components are to be detected, only low enrichment factors can be achieved. It is, however, possible to determine water-soluble vitamins (e.g. vitamin B_2) in beverages quantitatively [11].

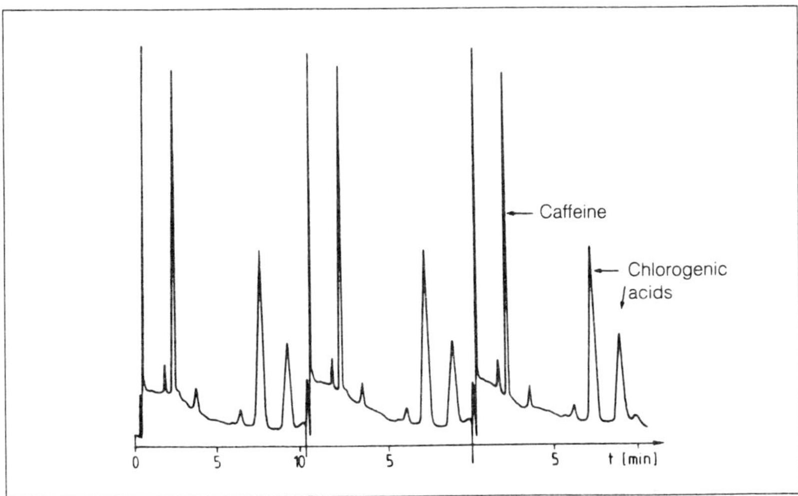

Fig. 3 Trace Enrichment: Caffeine Detection in Decaffeinated Coffee
Sample quantity: 1.5 ml 0.4% solution of powdered coffee in water
Pre-column: 10 x 4 mm packed with Hypersil ODS
Main column: 250 x 4 Lichrosorp RP 18
Eluents: enrichment: water; separation: acetonnitrile
0.015 m H_3PO_4 (10–90). Three times repeated to demonstrate the reproducibility.

It must be pointed out that the amount of time necessary for optimization of such enrichment and column switching detection methods is relatively high. The methods, therefore, seem to be suitable only for routine detections. This is especially true if the differences in the sorption strength between the substances to be determined and the matrix components are not very large. Slight fluctuations in the retention times due to high concentrations of the accompanying substance can lead to unreliable analytical results such as a peak loss or undesired components on the main column.

7.4   Coupling HPLC with UV/VIS Spectroscopy

The chromatographic result from HPLC is primarily quantitative. The detector indicates the concentration (g/ml) at the column exit. A qualitative allocation is possible only indirectly by means of the retention time and its congruency to corresponding comparison substances. For confirmation of the qualitative result the separation has to be repeated at least once in another separation system. Attempts have been made to couple HPLC with other physical-chemical analysis methods.

Only the direct coupling of liquid chromatography with UV/VIS spectroscopy has been accomplished. In principle, coupling with IR-spectroscopy is possible, but the choice of suitable chromatographic eluents, however, is quite limited and, due to the relatively low absorption coefficient in the IR range, relatively high sample concentrations are needed. The same is valid for coupling with NMR-spectroscopy. The direct coupling with mass spectrometry is possible with eluent splitting [13,14] or with micro bore columns, respectively, and most recently with thermospray ionisation techniques [15,16]. With the latter method positive and negative ions can be produced and evaluated. But high equipment costs restrict the use for routine applications. Furthermore, for the chemical ionization methods, applied here, practically only molecule peaks (molecular weight) and mass peaks due to addition of eluent components are obtained. Fragmentations are very seldom, thus information usually obtained by common mass spectrometry is not available, except if it is possible to fragmentize the molecular peak obtained in a second mass spectrometer (HPLC/MS/MS-coupling).

Today, a variety of diode-array-detectors for direct coupling of LC with UV/VIS spectroscopy are available. The units (assuming sufficient computer support) display the chromatogram (concentration at one wavelengths versus time) and the UV spectrum (extinction versus wavelength) simultaneously. Those so-called 3D plots are very impressive on color screens but their meaningfulness for the chromatograph is very limited (e.g. behind high absorption peaks other components could be hidden only recognizable if the 3D plot is mirrored). More meaningful are graphic diagrams in which points of identical absorption, like contour lines on maps, are connected with each other. Additional important information to peak purity is obtained by comparison of the spectra of the ascending and descending peak sides with the spectrum at the peak apex. Components with the same chromophore (same compound class) can easily be identified.

Figure 4 shows part of a chromatogram of an aqueous extract of green coffee [17]. The original printout lists, above the chromatogram, the UV spectra taken at different times. This allows the qualitative allocation of the peaks.

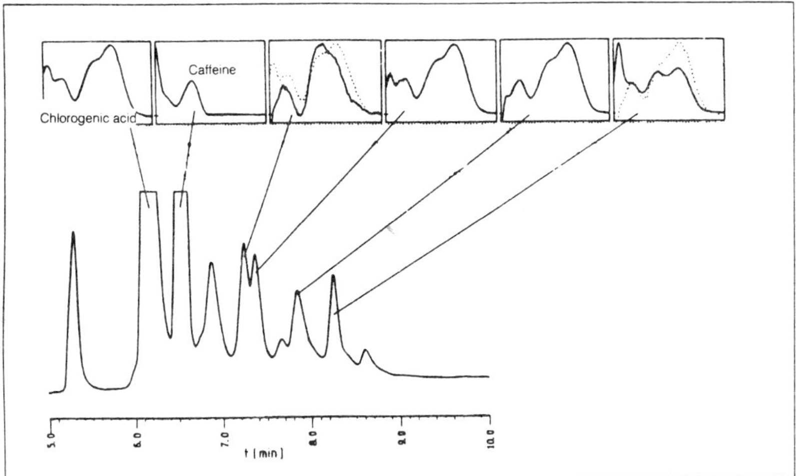

Fig. 4 Diode-Array-Detection
Part of a Chromatogram with UV Spectra of Individual Components
Chromatography of an aqueous extract of green coffee beans (Robusta). Separation on RP 18 with gradient elution; eluent A: 0.5% phosphoric acid; eluent B: methanol; program time 10 min.

The peak with the elution time of 6.52 min is identified undoubtedly as caffeine. The spectra taken at the ascending (6.47 min) and the descending (6.38 min) peak slope are identical. The peak does not contain any other component. By comparing the spectra, the different isomeric chlorogenic acids can immediately be identified as such. Co-elution of various components is also easily recognizable (e.g, far right spectrum window, spectra taken at 8.22. and 8.42 min, resp.). The basic data are stored in the computer, which is absolutely necessary, and can be processed further. For example, the quantitative detection of individual components can be done at their specific absorption maximum each. This improves the detection limit and includes a larger linear range of the calibration curve.

Furthermore, by comparing retention times and characteristics of absorption spectra (minima, maxima of spectra and their differentiations, integration over the normalized spectra range, etc.) with data from self-made spectra libraries, a definite qualitative identification of known components in mixtures are possible [17]. The calculation effort is not too large, but the chromatographic conditions must be kept constant, because position and strength of absorption in UV spectra may strongly be influenced by solvochrome effects. It seems possible to use these comparison methods for definite identification of substances with spectra consisting of a final absorption at low wavelength only. But, for those comparisons the sample concentrations may not be less than 10 ppm. Otherwise misinterpretations due to computer background noise are possible.

7.5 Improvement of Detection Sensitivity by Derivatization Reactions

Sample derivatization for introduction of chromophores in order to improve detection sensitivity can be done prior to or after the separation. In the first case, this is the classical method, there are no limitations regarding the derivatization reaction. It is favorable to optimize the separation system in a way that reagent and eventually formed by-products can be eluted prior to the

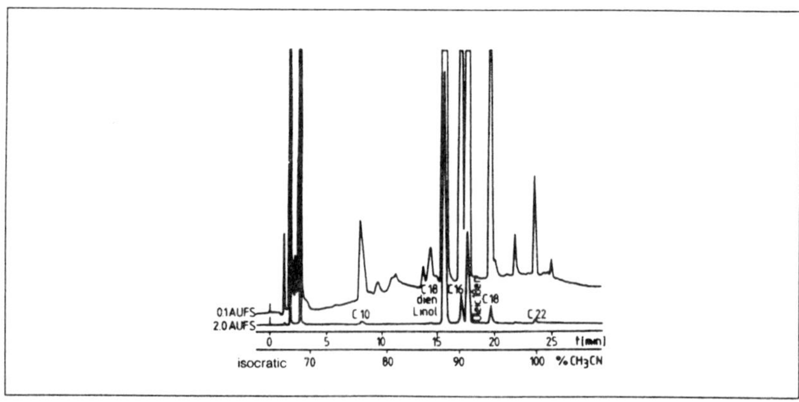

Fig. 5 Gradient Elution of Fatty Acid Phenacylester in Margarine
Gradient Elution at RP 18. Eluent A: 70% actonitrile in water, eluent B: acetonitrile, program time 20 min.

sample components to be determined and, therefore, do not disturb their detection. This is shown in Figure 5 on the example of gradient elution of phenacyl esters of fatty acids from a fat hydrolysate [18]. For derivatization prior to separation, there are some known problems regarding quantifications, and for samples with several functional groups the appearance of derivatives in variable frequency.

Even though the pre-column derivatization is a typical off-line method, additional equipment for HPLC units is now available for on-line pre-column derivatization directly before the separation column. Derivatization reagents for amino acids, i.e., o-Phthalaldehyde (OPA) and 9-Fluorenylmethyl chloroformate (FMOC) are commercially available. Advantages and disadvantages pre- and post-separation derivatization are listed in Table 1.

Tab. 1 Derivatization Before and After the Separation

	Before the separation	After the separation
Advantages	– Free choice of reaction conditions*) – Free reaction time selection – Standard HPLC equipment*) – Standard columns	– Selective detection of components in complex matrix – Continuous fully automatic method – Separation of not derivatized substances – Reagents and artefacts do not disturb – Reaction must not run quantitatively

Disadvantages	– Reaction must run definitely and quantitatively
	 – Disturbances by
	    – excessive reagent, artefact formation
	    – derivate variety by multifunctional compounds
	 – "Clean-up" step sometimes necessary
	 – Derivate stability should be identical
		– Complex costly instrumentation
		 – Dilution by reagent addition and band broadening in the reactor
		 – Reagent and reaction conditions must suitable for the eluent
		 – Reagent and reaction products must have different optical characteristics

*) not valid for automatic pre-column derivatization in the HPLC unit

Tab. 2 Comparison of Reactor Types for Reaction Detectors

	Packed reactors	Open tubes knitted	Open tubes with air segmentation
Advantages	– easy to produce – packed-bed can participate in all reactions (catalysis) – high temperature possible	– easy to produce – dispersion independent of flow – good heat exchanger – no problems at higher temperatures (teflon up to 120 °C, above that steel)	– low pressure drop – long reaction times possible – multi-step reactions easy – cheap reagent pumps (peristalic pumps)
Disadvantages	– danger of clogging – packed-bed stability against aggressive reagents – poor transfer in packed-bed – pressure drop adds to total pressure	– pressure drop adds to total pressure (2–3 times higher than for straight tubes)	– for phase separation demixing problematic – limited temperature range (vapor pressure) – problems with flow constancy (reaction time) – tube material (wetting) influences dispersion
Optimal construction	15 µm particles	0.25 mm i.d.	–
	Used in separation columns with 10–25 cm length, particle diameter 3–5 µm		
	20 µm particles	0.35 mm i.d.	1 mm i.d. segmentation frequency 2 sec^{-1}
	Used with packed columns, particle size 7–10 µm		
Optimal reaction times (RT)	5 min < RT < 1 min	0.35 mm i.d. 5 min < RT < 1 min 0.25 mm i.d. RT < 1 min	> 5 min

For post-column derivatization with the chemical reaction detector [19], the problems are different. The eluate and the reagent added must be stored during the entire reaction period, which must not run quantitatively and completely, without causing a disturbing dispersion of the sample zone. Excessive dispersion always leads to declining detection sensitivity. Eluent and reagent should, of course, not react with each other. The reaction products must differ from the reagent regarding absorption and/or fluorescence. The instrumentation for post-column derivatization is significantly larger. For reagent feed, a pump with constant flow, entirely pulseless and corrosion resistant is necessary. Reactors can be packed columns, open tubes with segmented flow or geometrically formed open tubes [20]. Table 2 lists advantages and disadvantages of the different reaction detection systems. For

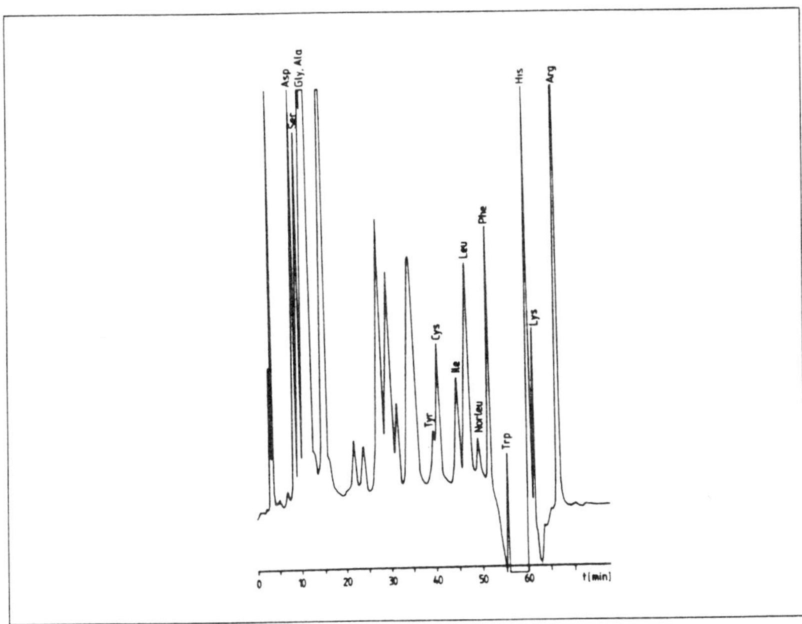

Fig. 6 Post-Column Derivatization; Gradient Elution of Amino Acids in Red Wine

Column: cation exchanger on diatomaceous earth; gradient: Lithium citrate buffer pH 2.0 to pH 7.5. post-column derivatization with o-Phthalaldehyd in 10 m knitted capillary with fluorescence detection, sample quantity 5 µl

very slow reactions (reaction time more than 5 min) segmented systems are advantageous, while for short reaction times (approx. 1 min) and aggressive reagents geometrically formed (knitted) reactors are preferable. Dispersion in those systems can be kept low allowing coupling to small-volume columns. Also with knitted reactors the dispersion is independent from the flow rate.

Application of reaction detectors for post-column derivatization does not influence the chromatographic system. Separation of amino acids can be achieved with ion exchange chromatography. Modern reactors (e.g., knitted capillary tubes) and OPA are used to detect amino acids in the pg range. Figure 6 shows the separation of amino acids in red wine by means of gradient elution on a cation exchanger with subsequent reaction detection with OPA [21].

Sugars can be separated in strong alkaline medium (pH > 10) on an anion exchanger [22]. The detection of reducing and non-reducing sugars is possible with thymol/sulphuric acid [23]. Even such high viscous and aggressive reagents may be used. With optimized systems sugar detection in the ng range is possible [24]. In Figure 7 the contents of various sugars in white wines of different origin are compared. Both wines are completely fermentated because the fructose and glucose content differs insignificantly from the

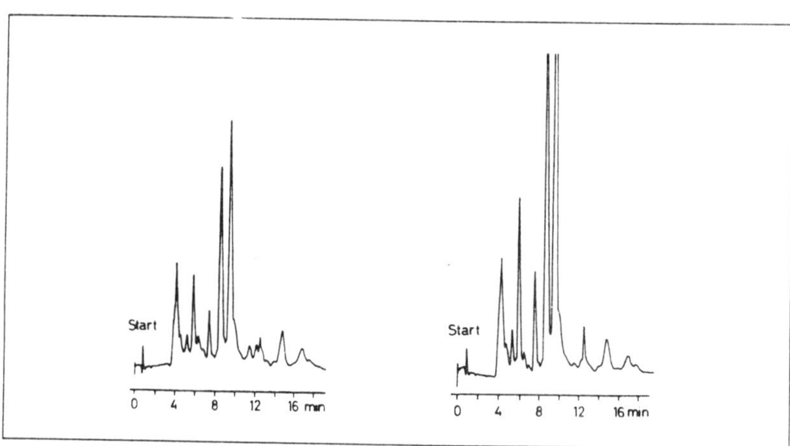

Fig. 7 Post-Column Derivatization; Detection of Sugar in Wine
Separation: anion exchanger (Dionex HPIC-AS6), Eluent: 0.15 ml NaOH. Post-column derivatization with thymol/sulphuric acid. Reactor: 10 m knitted capillary. Sample quantity: 20 μl each Soave (1982) and Entre Deux Mers (1978). Both main peaks are fructose and glucose.

content of other sugars. It seems possible to use the concentration ratio of the different sugars for characterization of the ripening degree of the grapes.

Another advantage is the selectivity of the chemical reaction that allows the identification of individual substances and groups of compounds, in addition to many non-reacting accompanying substances in complex matrices. For the above mentioned examples either the amino acids or the sugars in wine had been detected selectively. Pre-treatment of samples or separation did not take place. The advantages of this detection selectivity can be clearly seen on the example of residue analysis of carbamate pesticides in vegetables [25].

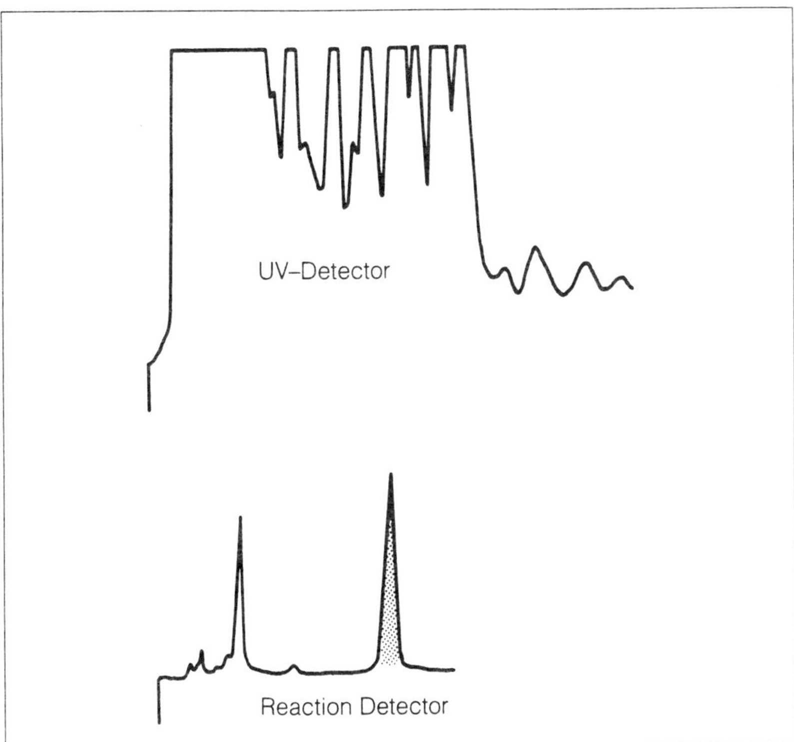

Fig. 8 Selective Detection in Complex Matrix by Post-Column Derivatization
Determination of carbamate residues (Curaterr) in Savoy cabbage. Two-step
reaction detector

For determination by gas chromatography an extensive clean-up procedure with heavy losses due to multi-step distribution steps of the plant extract is necessary. For HPLC with a reaction detector a simple extraction with methanol followed by centrifugation is sufficient. As shown in Figure 8 the detection is only possible with reaction detector. The chromatogram recorded by UV-detector does not allow any allocation and quantitative detection. In this case a multi-step reaction detector had been used for these detections. The carbamates in the chromatographic eluate had first been saponified with NaOH and the thereby liberated methylamine, after being treated with OPA, forms a fluorescent derivative that is then quantitatively determined.

The reaction detection systems, optimized individually for special applications, will surely find increased use in the next years, despite problems occurring with adaptation of derivatization reactions and chromatographic separations.

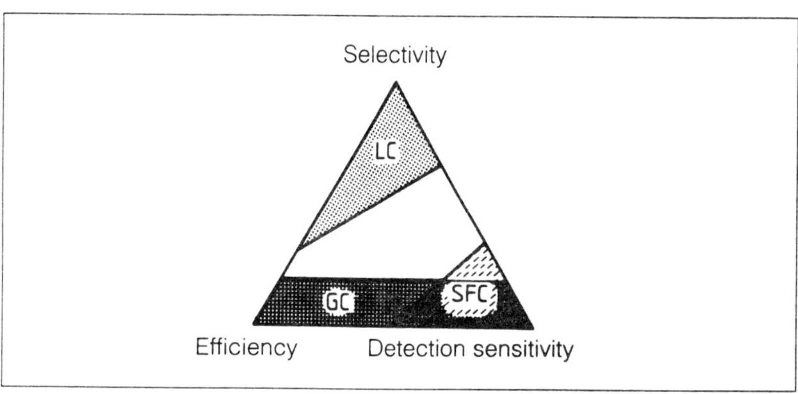

Fig. 9 Advantages of Chromatographic Separation Methods with Different Mobile Phases (Gas, Liquid, Fluid)

7.6 Outlook

The use of HPLC as analytical method is still rapidly growing. The limitation to the detection of substances with absorption in the visible or ultraviolet spectral range has not had any detrimental effect. With the chemical reaction detector one possibility to avoid this disadvantage is still open. Due to this limitation, however, the application of fluid (supercritical) gases as mobile phases, especially carbon dioxide has gained importance lately. The advantage is the gentle separation of less volatile compounds at lower temperatures than in the gas chromatography. Also for this application, the sensitive flame ionization detector is used which allows sensitive detection of components without chromophores. But also here, one is again limited to carbon dioxide, nitrous oxide, and xenon (1 kg ~ DM 6,000.00).

Because the physical-chemical characteristics of fluid media resemble the liquids more than the gases, the high efficiency of capillary gas chromatography cannot be achieved with fluid mobile phases. They will be similar to those of the liquid chromatographic columns due to the low diffusion coefficient. For non-polar substances with high boiling points, however, chromatography with fluid mobile phase has its advantages. Nothing is known about the selectivity of the separation systems with fluid phases and common stationary phases as used for gas chromatography and/or liquid chromatography. The elution strength of fluid media seems to increase with their density.

A further advantage may be the combination of extractions by means of hypercritical gases and chromatography from the same media. Figure 9 may clarify the advantages of the chromatographic methods with three different mobile phases, as seen today.

The gas chromatography cannot be reached regarding efficiency, analysis time and detection sensitivity with the flame ionization detector (FID), the electron capture detector (ECD), etc. Gas chromatography can be coupled with other physical-chemical analysis methods as mass spectroscopy, FTIR spectroscopy, etc., too. If carbon dioxide is used as mobile fluid phase the SFC (Super Critical Fluid Chromatography) surely can be treated as equivalent to the gas chromatography. The main advantage of liquid chromatography is its selectivity, which is not only given for the stationary phase, but which is predominately determined by the characteristics of the mobile phase. Regarding efficiency neither LC nor SFC can compete with the gas chromatography. Up to now relatively few data are available to allow reliable statements regarding the selectivity of SFC with capillary columns or packed columns. Equipment for chromatography with super critical CO_2 are available on the market by various manufacturers.

REFERENCES

1. KIRKLAND, J. J., SNYDER, L. R.: Introduction to Modern Liquid Chromatography, 2nd Edition, Wiley-Interscience, New York 1979

2. ENGELHARDT, H.: High Performance Liquid Chromatography, Springer-Verlag, Berlin 1979 (Deutsche Ausgabe 1977)

3. MEYER, V.: Praxis der Hochleistungs-Flüssigkeitschromatographie, Diesterweg-Salle, Frankfurt 1983

4. Practice of High Performance Liquid Chromatography, H. Engelhardt, ed., Springer-Verlag, Berlin 1986

5. GLÖCKNER, G.: Polymercharakterisierung durch Flussigkeitschromatographie, VEB Deutscher Verlag der Wissenschaften, Berlin 1980

6. GUIOCHON, G., COLIN, H. in P. Kucera, ed., Micro Column High Performance Liquid Chromatography, Elsevier, Amsterdam 1984

7. HALÁSZ, 1., GÖRLITZ, G.: Angew. Chem. **94**, 50 (1982)

8. ERNI, F.: J. Chromatogr. **282**, 371 (1983)

9. KUCERA, P., ed., Micro Column HPLC, Elsevier, Amsterdam 1984

10. DUHR, M.: Dissertation, Universität des Saarlandes, Saarbrücken 1985

11. JAUMANN, G., ENGELHARDT, H.: Chromatographia **20**, 615 (1985)

12. ENGELHARDT, H., JAUMANN, G., KÖNIG, TH.: ASIC, Paris 1986; Chemical Abstracts-Zitat: Colloq. Sci. Int. Cafe [C.R.l **1985**, 11th, 213

13. GUIOCHON, G., ARPINO, P. J.: J. Chromatogr. **271**, 13 (1983)

14. ARPINO, P. J.: J. Chromatogr. **323**, 3 (1985)

15. ALEXANDER, A. J., KEBARLE, P.: Anal. Chem. **58**, 471 (1986)

16. GÜNTHER, W.: Labo, Mai 1986, S. 94 ff.

17. KÖNIG, TH.: Dissertation, Universität des Saarlandes, Saarbrücken 1986

18. ENGELHARDT, H., ELGASS, H.: J. Chromatogr. **158**, 249 (1978)

19. LILLIG, B., ENGELHARDT, H.: in 1. S. Krull, ed., "Reaction Detection in Liquid Chromatography", Marcel Dekker, New York; 1986

20. ENGELHARDT, H., NEUE, U. D.: Chromatographia **15**, 91 (1982)

21. LILLIG, B.: Dissertation, Universität des Saarlandes, Saarbrücken 1984

22. ROCKLIN, R. D., POHL, A. C.: J. Liquid Chromatogr. **6**, 1577 (1983)

23. KAKÁC, B., VEJDELEK, Z. J.: Handbuch der photometrischen Analysen, Verlag Chemie, Weinheim 1974–83

24. OHS, P.: Dissertation Universität des Saarlandes, Saarbrücken 1986

25. ENGELHARDT, H., LILLIG, B.: Chromatographia **21**, 136 (1986)

Characterization of Changes during Processing and Storage by Chromatographic Determination of Tracer Substances

K. Eichner

8 Characterization of Changes during Processing and Storage by Chromatographic Determination of Tracer Substances

K.Eichner, Münster

8.1 Introduction

If thermal methods for preservation of food (e.g., drying, sterilization) are employed the non-enzymatic browning reaction (Maillard reaction) often causes quality decreasing nutritional and sensory changes [1–3]. Those changes can also happen during storage, especially in food with low water content.

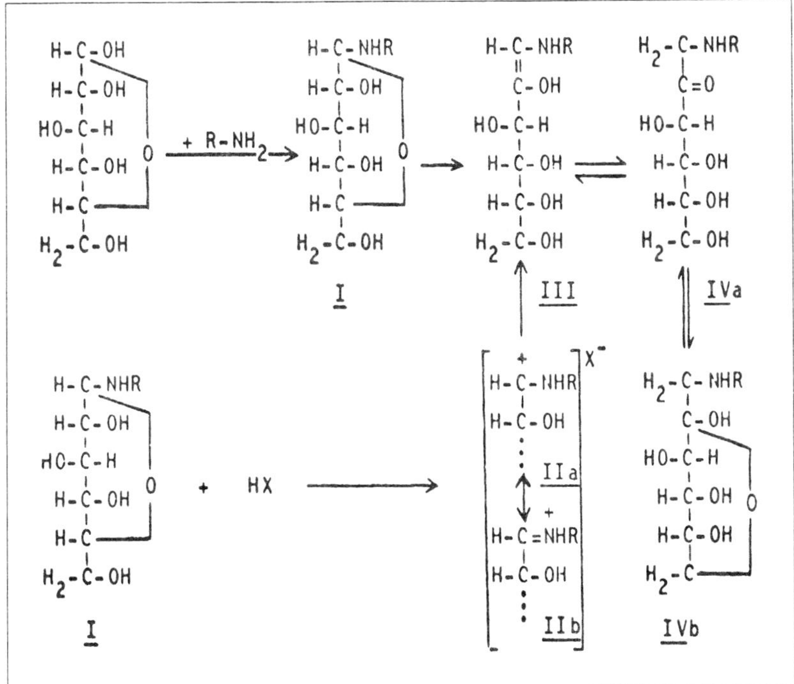

Fig. 1 Formation of Fructose-Amino Acids (IVa and IVb) from Glucose and Amino Acids by Amadori Rearrangement of Intermediate Glucosyl-Amino Acids (I). (R = CHR'- COOH)

Precursors of this reaction are, due to reactions between reducing sugars (aldoses) and amino acids, 1-N-amino acid-1-deoxyketoses (ketose-amino acids, "Amadori-compounds") (see Fig. 1); subsequently volatile flavor compounds and browning products are formed. These often quality decreasing reaction products develop later in the Maillard reaction. Therefore, it seems to make sense to use the Amadori-compounds, which are sensorically non-detectable precursors of the reaction, as indicators for early recognition of the Maillard reaction. Thermal stress should be detected as early as possible in order to control production methods in a way to allow development of only low amounts of these compounds [4–7]. These compounds, however, are not exclusively precursors of undesired flavor changes (e.g. at drying processes), but also precursors of desired aroma substances formed during broiling, baking, kiln-drying and roasting [8,9].

8.2 Separation and Detection of Amadori Compounds by Liquid Chromatography

Methods have been searched for that allow an efficient separation and specific detection of Amadori compounds as well as being simple and rapid analysis techniques.

Dried vegetable products containing sufficient amounts of reducing sugars and free amino acids were used for the investigations. For detection of Amadori compounds first the amino acid analysis was tested.

8.2.1   Amino Acid Analysis

Figure 2 shows the amino acid chromatogram of tomato powder obtained by a regular physiological separation program [10]. For analysis the dried product was cooled and extracted with chilled water. The extract was analyzed for amino acids directly after centrifugation.

The Amadori compounds (tagaturone-amino acids formed by reaction of amino acids with galacturonic acid, and fructose-amino acids), marked by letters in Figure 2, are indicated here together with unreacted amino acids. Both compound classes partly overlap each other. The area of peak C is proportional to the degree of thermal stress during the drying process. Under this aspect the amino acid chromatogram can be shortened correspondingly [4–7].

The Amadori compounds in peak C of the amino acid chromatogram have a different stability and effect on the following quality decreasing reactions. Therefore, a more efficient separation and detection system has been de-

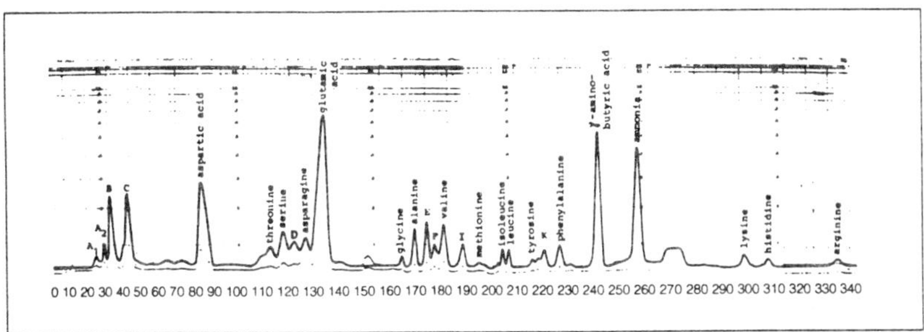

Fig. 2: Amino Acid Chromatogram of Thermally Treated Tomato Powder (time in min)

A_1 = Tagaturone-aspartic acid; A_2 = Fructose-aspartic acid
B = Tagaturone-glutamic acid; -asparagine, -serine, threonine;
C = Fructose-glutamic acid; -asparagine, -serine, threonine;
D = Tagaturone-γ-amino butyric acid
H = Fructose-γ-amino butyric acid
K = Fructose-lysine

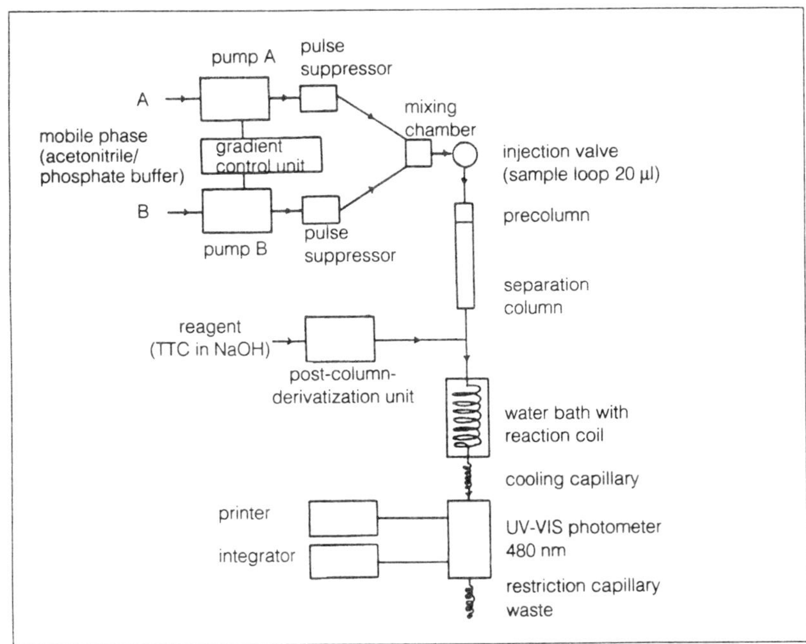

Fig. 3 HPLC-System with Post-Column Derivatization for Separation and Specific Detection of Amadori Compounds

veloped, aimed at separation and specific detection of all Amadori compounds occurring in dried vegetable products.

8.2.2 High Pressure Liquid Chromatography with Specific Post-Column Derivatization

Figure 3 shows a HPLC-system with post-column derivatization developed for separation and specific detection of Amadori compounds [11].

Separation is carried out on an anion exchange column on the basis of chemically modified silica gel (DEAE-Si 100, 3µm, Serva) with acetonitrile-phosphate buffer (70:30, v/v; 0.05 mol phosphate/l; pH 5.8) as mobile phase. For the detection an alkaline solution of triphenyl-tetrazolium-chloride (TTC) (3.5 g TTC/l 0.05 m NaOH) is added to the column eluate. In the follo-

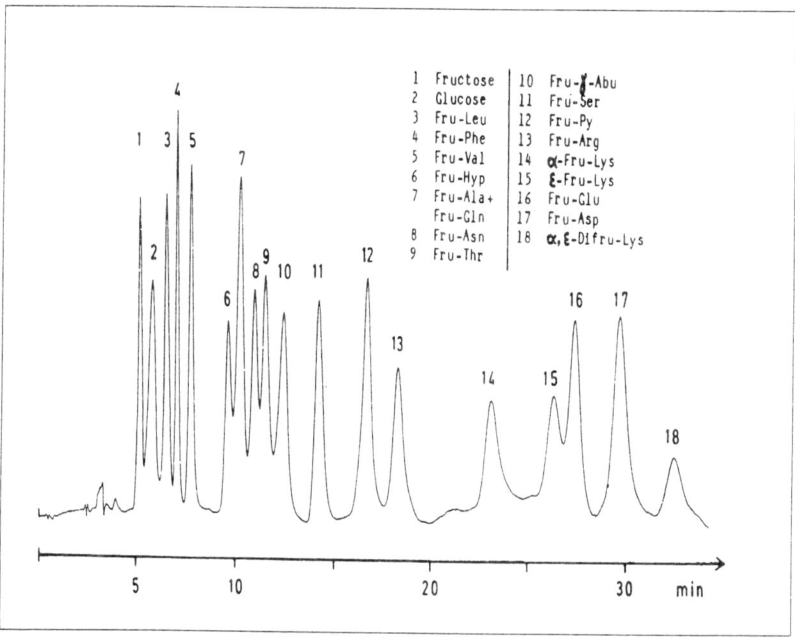

Fig. 4   HPLC-Chromatogram of a Standard Mixture of Amadori Compounds as well as Fructose and Glucose (Post-Column Derivatization with TTC)

(Fru-γ-Abu = Fructose-γ-amino butyric acid;
Fru-Py = Fructose-pyrrolidone-carboxylic acid)

wing reaction coil (85 °C, retention period 40 sec) the Amadori compounds reduce the TTC to the red 1,3,5-triphenyl-formazane which is then detected at 480 nm. Reducing sugars such as fructose and glucose react similarly.

Figure 4 shows the chromatogram of a standard mixture of Amadori compounds containing glucose and fructose. Contrary to the amino acid analysis, this separation method does not indicate free amino acids; therefore, the amino acids can not interfere the separation. Considering the elution sequence of the individual components it can be concluded that, in addition to the ion exchange, chromatographic distribution effects play a role here.

Application of HPLC is extremely time-saving as can be clearly seen by comparison with the amino acid chromatogram in Figure 2. Figure 5 shows a HPLC chromatogram of an aqueous extract of roller dried tomato powder.

With the separation system as described it is possible, contrary to amino acid analysis, to detect fructose-pyrrolidonecarboxylic acid (peak 12) (the cyclization product of frucose-glutamic acid or fructose-glutamine!) which indicates further thermal stress.

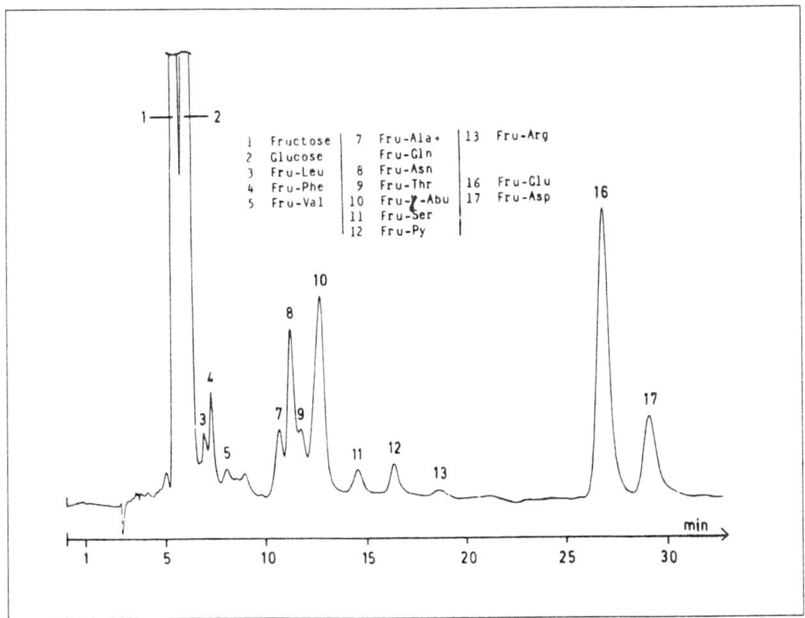

Fig. 5   HPLC-Chromatogram of Roller Dried Tomato Powder Extract (Post-Column Derivatization with TTC)

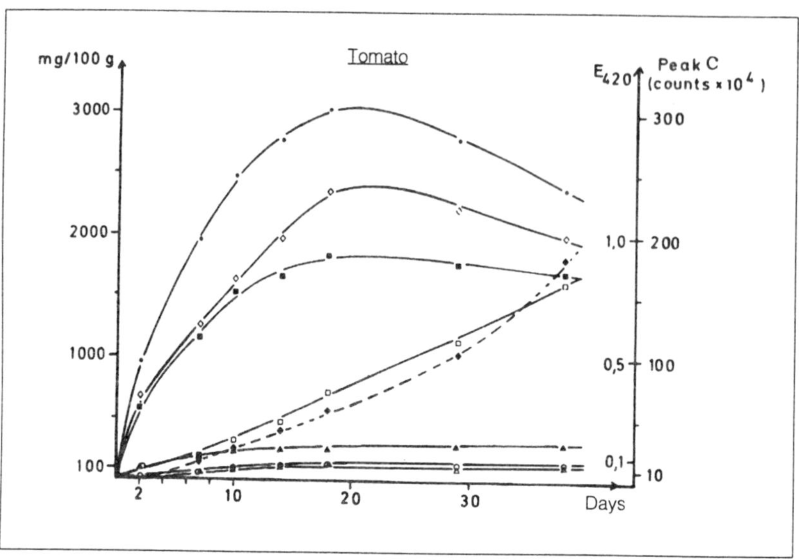

Fig. 6   Storage of Freeze-Dried Tomato Powder at 40 °C and a Water Activity (a_w) of 0.35

● = Fru-Glu, ■ = Fru-Gln + Fru-Ala, ▲ = Fru-Asn, ◆ = Fru-Py, ○ = Fru-Thr, △ = Fru-Ser, ◇ = peak C, indicated as area counts, □ = Browning (E_{420}); the extinction value refers to an extract of 0.5 g sample material in 25 ml water

Figure 6 shows a storage test with tomato powder [11]. The concentrations of fructose-glutamic acid (peak 16 in Figure 5) and fructose-glutamine (contained in peak 7, Figure 5) increase sharply at the beginning of storage and decrease in the course of time after reaching a maximum. On the other hand, the proportion of the relatively stable fructose-pyrrolidonecarboxylic acid increases slowly.

A characteristic similar to the one for fructose-glutamic acid can be recognized (Fig. 2) by the area of peak C in the amino acid chromatogram of tomato powder. This peak is a sum peak which contains mainly fructose-glutamic acid.

Figure 6 shows clearly that, for evaluating the course of Amadori compound formation due to thermal treatment or storage as criterion for a beginning Maillard reaction, the sum of the tracer substances fructose-glutamic acid, fructose-glutamine, and fructose-pyrrolidonecarboxylic acid must be considered.

The cyclization reactions leading to fructose-pyrrolidonecarboxylic acid are strongly promoted in acid foods such as tomatoes (pH-value approx. 4.2). At higher pH-values they are retarded and at pH 6 they are only of minor significance [11]. Table 1 lists the Amadori compounds detected in various dried vegetables [11].

Tab. 1 Occurrence of Amadori Compounds in Various Dried Vegetables (Concentrations in mg/100 g dry matter)

Amadori Compounds	Tomato Powder A	Tomato Powder B	Bell Pepper	Paprika	Asparagus	Cauliflower	Carrot	Celery
Fru-Leu + Fru-Ile	158	41	78	191	59	25	28	50
Fru-Phe	250	108	37	nn	33	15	20	32
Fru-Val	42	48	167	33	50	39	34	47
Fru-Ala + Fru-Gln	494	516	1599	302	1053	397	418	413
Fru-Asn	945	513	1073	783	1458	127	115	285
Fru-Thr	368	189	213	77	27	18	21	nn
Fru-γ-Abu	1462	1044	317	257	33	105	76	42
Fru-Ser	127	69	338	46	120	75	17	35
Fru-Py	594	2148	238	326	60	44	30	27
Fru-Arg	150	nn	288	80	nn	111	82	nn
Fru-Glu	3788	676	62	nn	32	89	52	265
Fru-Asp	948	659	155	62	64	55	34	70

(Fru-γ-Abu = Fructose-γ-amino butyric acid; Fru-Py = Fructose-pyrrolidonecarboxylic acid)

The tomato powder B listed in the second column of Table 1 is an overstored product, the fructose glutamic acid has been transformed primarily into fructose-pyrrolidonecarboxylic acid as compared to the tomato powder A listed in the first column.

Quantitative results regarding the degree of thermal damage of the dried products can only be obtained if the original molar amounts of the individual amino acids are known or calculated from the sum of reacted and unreacted amino acids detectable by HPLC and amino acid analysis. On this basis the molar reaction rates of the Amadori compounds can be calculated.

8.3 Separation of Amadori Compounds by Capillary Gas Chromatography [12]

8.3.1 Methodology

Prior to gas chromatographic separation the individual compounds concerned must be transferred into a volatile form. For compounds containing hydroxyl groups, the corresponding trimethylsilyl ethers are often formed.

Preparatively produced Amadori compounds as well as the corresponding sugars have been oximized with hydroxyl ammonium chloride in pyridine (30 min, 70 °C) followed immediately by the addition of N,O-bis-(trimethyl-silyl)-acetamide and trimethylchlorsilane leading to trimethylsilyl derivates (30 min, 70 °C). Due to the oximation prior to trimethylsilylation, formation of pyranosidic and furanosidic ring structures of the sugar moieties in their α- and β-anomeric forms could be avoided. In this way, only the syn- and anti-forms of the oximes were present.

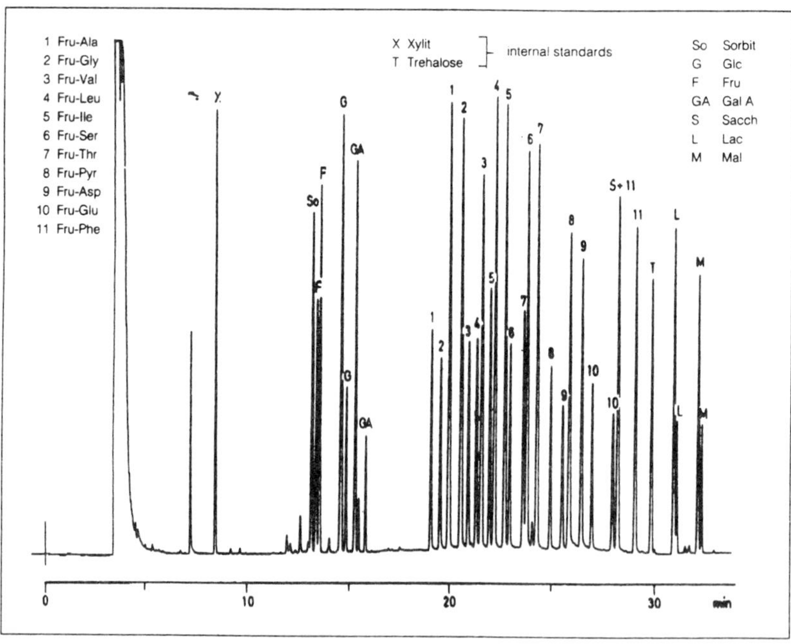

Fig. 7   Standard Capillary Gas Chromatogram of Oximized and Trimethyl-silylized Amadori Compounds and Sugars

The gaschromatographic separation was carried out on a 35 mm glass capillary (Duran, 0.3 i.d.), covered with methyl silicones (OV-101); temperature program: 120–280 °C with a heating rate of 4 °C/min.

Figure 7 shows a standard chromatogram of Amadori compounds and different sugars. Due to the fact that the retention times increase proportionally with the molecular weight of the derivated compounds, the Amadori compounds appear in the chromatogram in the range between the mono- and disaccharides.

Each compound appears characteristically as a double peak, corresponding to the syn- and anti-form of the oximes. The signal ratio has a specific value for each compound. Differences in the peak sizes are noticable. The first peak (syn-form) is always smaller than the second one (anti-form). Figure 7 shows further that the syn- and anti-forms of the oximized and silylized fructose-amino acids are spaced more widely apart than the ones for the corresponding sugar derivates.

8.3.2   Examination of Tomato Powder

Tomato powder contains glucose and fructose in excess compared to free amino acids. The Amadori compounds, therefore, were first bound on a cation exchanger and then, after washing out the sugars, eluted with 0.2 m trichloroacetic acid (TCA). The derivatization was carried out after extracting TCA with ether, evaporating the aqueous solution under vacuum, and freeze-drying. Figure 8 shows the corresponding gas chromatogram.

The disadvantage of the described method is that the fructose-pyrrolidonecarboxylic acid formed by cyclization of fructose-glutamine and fructose-glutamic acid is not bound on the cation exchange resin and, therefore, is washed out together with the sugars.

Furthermore, most of the isolated fructose-glutamic acid cyclizes under the conditions of derivatiza-tion as can be seen in the gas chromatographic analysis in Figure 8. Therefore, the sum of fructose-glutamic acid and fructose-pyrrolidonecarboxylic acid from the TCA fraction corresponds to the original content of fructose-glutamic acid in tomato powder.

Comparison with Figure 5 shows that most of the Amadori compounds already detected in roller dried tomato powder by HPLC appears in the gas chromatogram.

Heyns-rearrangement products (glucose- and mannose-amino acids, see [2,12]), expected from the reaction of fructose with amino acids, have no significance because of their low rate of formation and their low stability in food with low water content. Only a low amount of glucose-glutamic acid was detected in tomato powder (see Figure 8). Tagaturone-amino acids as reaction products with galacturonic acid, present in low amounts [10], could not be detected.

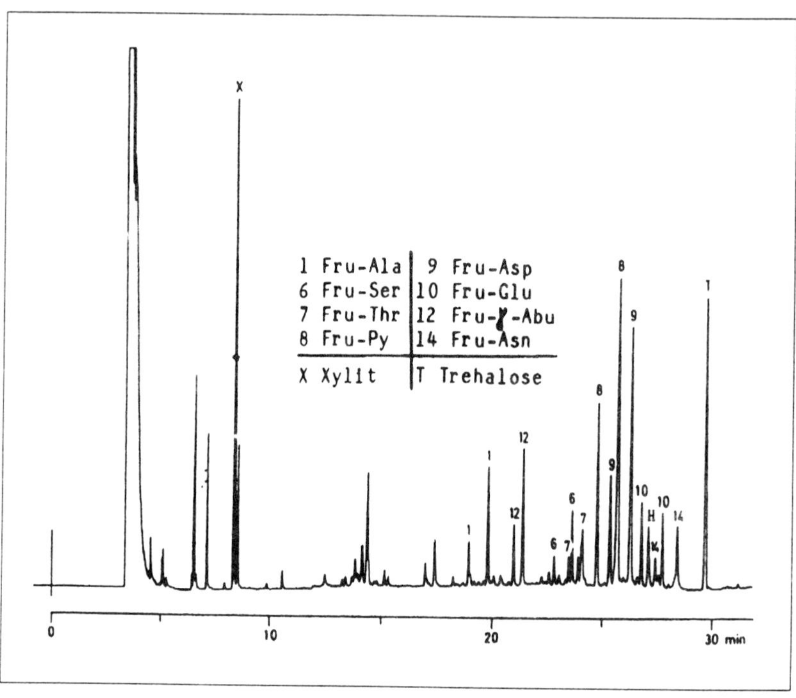

Fig. 8   Capillary Gas Chromatogram of Oximized and Trimethylsilylized Amadori Compounds Isolated from Spray-Dried Tomato Powder (H = Heyns-rearrangement product glucose-glutamic acid)

8.3.3 Examination of Malts and Beers

During malting the Maillard reaction produces the desired color and aroma. Depending on the degree of thermal stress, Amadori compounds are expected to be formed during this process.

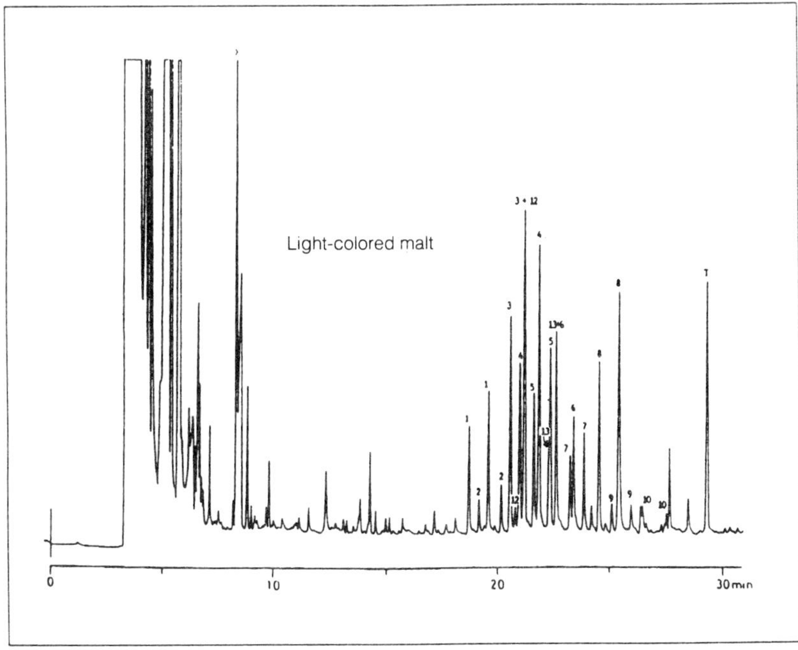

Fig. 9 Capillary Gas Chromatogram of Oximized and Trimethylsilylized Amadori Compounds Isolated from a Light-Colored Malt
Classification of the peaks: see Figure 7
Additional peaks: 12 Fru-γ-amino butyric acid, 13 Fru-Pro

Figure 9 shows a capillary gas chromatogram of Amadori compounds in a light-colored malt while Figure 10 is a corresponding chromatogram of a light-colored caramel malt which had been heat-treated for a longer time. Comparisons show that in the malt with higher thermal stress significantly higher concentrations of Amadori compounds could be detected.

For analysis of Amadori compounds present in the malt extract the sugar-free TCA eluate of the cation exchange resin (see chapter 8.3.2) was also used.

The TCA eluate also contains amino acids which do not interfere with the gas chromatographic separation (provided they form stabile derivates) because they appear before the fructose-amino acids in the chromatogram.

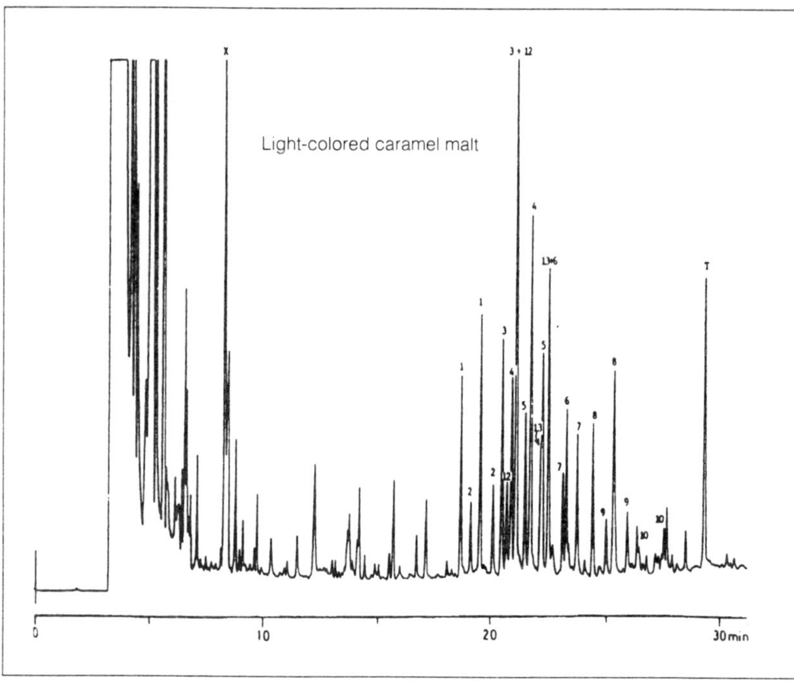

Fig. 10 Capillary Gas Chromatogram of Oximized and Trimethylsilylized Amadori Compounds Isolated from a Light-Colored Caramel Malt
Classification of the peaks: see Figures 7 and 9

As can be seen in Figure 11, darker malts contain higher concentrations of Amadori compounds than light-colored malts due to the more intense thermal treatment (temperatures above 100 °C). Furthermore, light and dark malts show characteristic patterns of Amadori compounds, each seems to be very significant for the (boldfaced line) group: fructose-alanine, fructose-valine and fructose-leucine. Noteworthy is that in dark malts the fructose-valine content does not increasing as much as the content of fructose derivates of the other neutral amino acids. This causes a reduction in the difference between fructose-alanine and fructose-valine, compared to the lighter malts, even though the fructose-leucine content is higher than the value for fructose-valine.

Beers produced from those malts again show typical Amadori compound patterns characterized by a relatively large decrease of fructose-valine

which is degraded considerably during the mashing and worting processes.

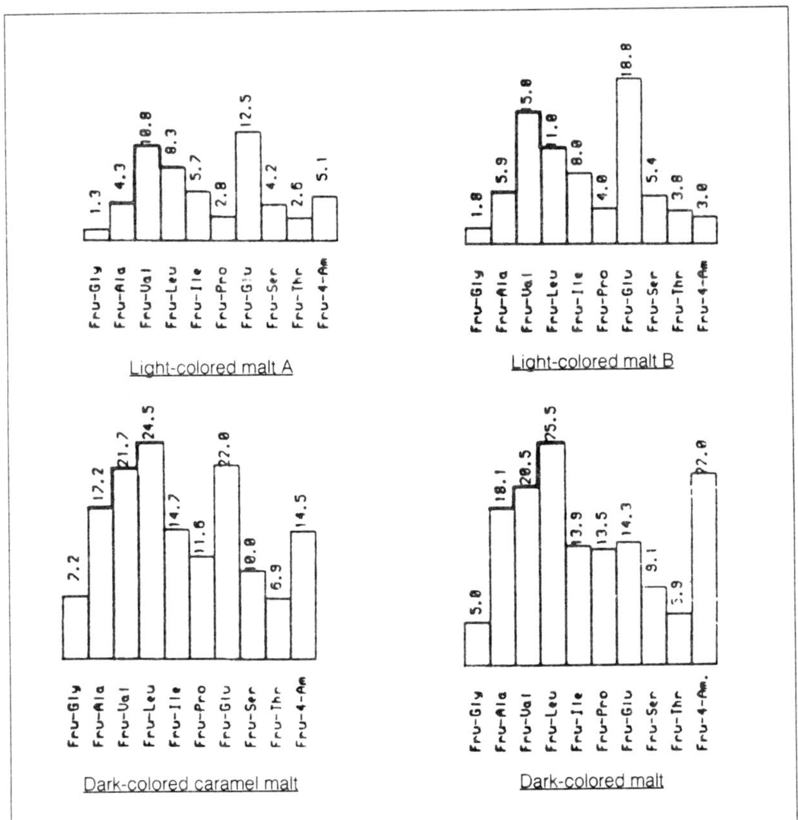

Fig. 11 Pattern of Amadori Compounds in Different Malts
(Concentration for Amadori compounds in mg/100 g malt;
Fru-4-Am = Fructose-γ-amino butyric acid)

On the other hand, in very dark colored malts, produced at 200 °C, no Amadori compounds can no longer be detected because they degrade quickly
under those conditions.

Gas chromatograms taken during examinations of "Altbier" (German top-
fermentation special beer) showed in some cases two additional peaks (A
and B) in the latter part, as can be seen in Figure 12.

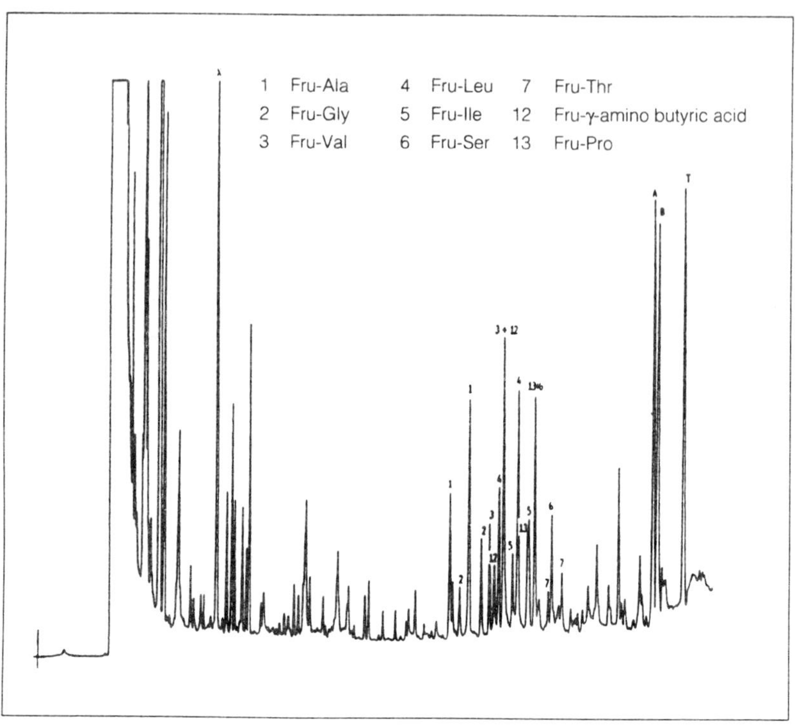

Fig. 12   Capillary Gas Chromatogram of Oximized and Trimethylsilylized Amadori Compounds Isolated from an "Altbier" with Added Ammonia Couleur
(Peak A = Deoxyfructosazine, 6-isomer; Peak B = Deoxyfructosazine, 5-isomer, see Figure 13)

A connection to the addition of caramel couleur was first suspected when both peaks were detected in a malt beverage, for which the use of caramel couleur was declared, according to legal regulations. Thereupon, caramel couleurs, in which glucose, fructose or saccharose were heated together with ammonia, were examined more closely. As reaction products in the lower molecular range deoxyfructosazines [13] could be detected. Those compounds come from a reaction between 3-deoxyhexosulose, a degradation product of Amadori compounds (1–3), and fructosamine (Amadori rearrangement product from reaction between glucose and ammonia) (Fig. 13). They correspond to the additional peaks A and B in Figure 12 (peak A = 6-isomer, peak B = 5-isomer). Common brewing couleurs contain those pyrazine derivates in the range of several percent, therefore, they are present in

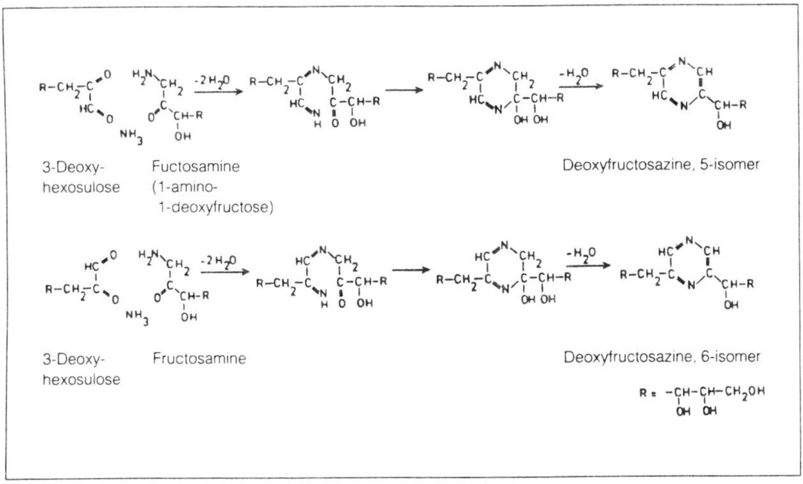

Fig. 13 Formation of Isomeric Deoxyfructosazines

detectable amounts (1–10 mg/100 ml beer) in beers colored with brewing couleurs. It is possible to use them for the detection of such ammonia couleurs in beers.

8.4 Analytical Determination of Degree of Roasting of Cocoa [14]

During fermentation of cocoa beans aroma precursors are formed, mainly consisting of reducing sugars, amino acids and peptides. The actual cocoa aroma is developed during the roasting process in which Maillard type reactions take place forming the characteristic aroma compounds [8,9].

Because the degree of roasting always correlates to certain aroma notes [15], the sensory evaluation will always determine the quality of a cocoa product. An analytical determination of the degree of roasting using defined tracer substances can serve to verify the sensory evaluation and would be advantageous for giving objective values independent of the sensory performance of the test panel. A linear correlation between the degree of roasting and the tracer substance concentrations would be desirable.

It was attempted, using rapid separation- and selective detection methods for specific roast constituents. to find concentration relationships relating to sensorically determined the degree of roasting.

HPLC with electro-chemical detection was employed, which should allow a selective detection of Maillard products and elimination of interferences by other cocoa constituents.

8.4.1 Methodology

8.4.1.1 Separations by HPLC

Figure 14 shows the HPLC-system with gradient control used for separation of roast-specific and other cocoa constituents as well as the compositions of the mobile phases A and B.

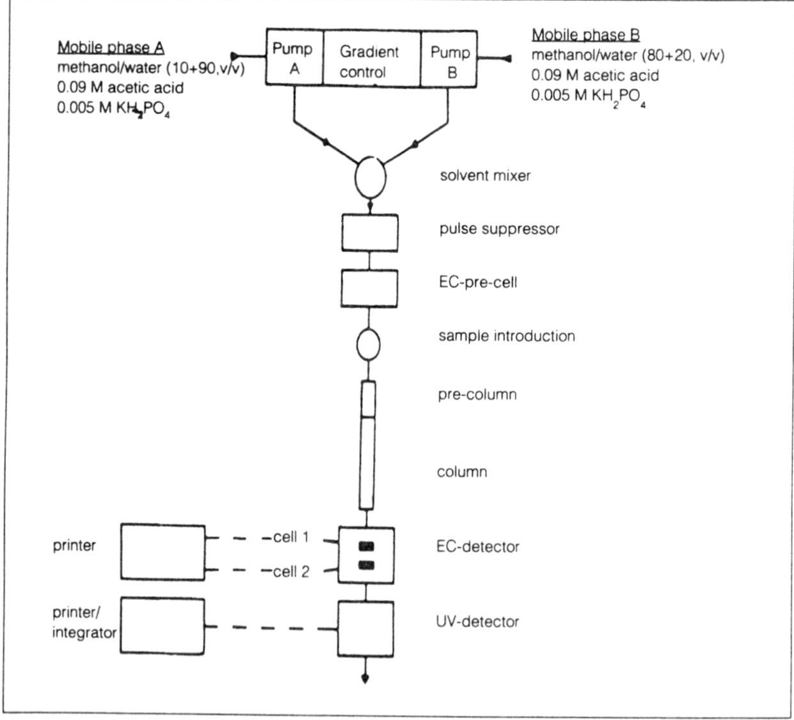

Fig. 14 HPLC System with Electro-chemical Detector for Separation and Detection of Cocoa Constituents

Lichrosorb RP 18 (5 µm, Merck) was used as the stationary phase. The detection was done using an electro-chemical detector (ESA, two cell unit Coulochem 5 100 A, Biotronic) or by means of an UV-VIS filter photometer (Interference filter 280 nm, Knauer).
Simultaneous application of both cells of the electro-chemical detector allows the detection at two different potentials during one analysis. For example, if a positive potential is applied to both cells, with pot. 1 lower than pot. 2, some components are detected at cell 2 only, while others, responding to a lower potential, are detected at cell 1. Furthermore, both potentials can be adjusted in a way that the test substance will be converted partly at cell 1 and completely at cell 2. In addition to the half-stage potential, the signal ratio of both cells can be used for specific detection of related pertinent compounds. Potentials with different charge signs are used for the redox method. A substance with a reversible redox reaction will be detected at both cells.

8.4.1.2 Extraction of Cocoa Beans

First, the fat is extracted from the peeled cocoa beans which are then extracted several times with methanol/water (70 + 30, v/v) under nitrogen.

8.4.1.3 Steam Distillation of Volatile Cocoa Constituents

20 g cocoa beans are disintegrated finely with addition of 20 g NaCl, 50 ml water is added, heated and loaded into a distillation apparatus according to Antonacopoulos. For analysis the first 10 ml of the steam distillate are used.

8.4.2 Tracer Substances for Determination of Degree of Roasting of Cocoa Beans

Figures 15 a and 15 b show HPLC chromatograms of extracts from unroasted and roasted cocoa beans.

The cocoa constituents theobromine and caffeine, present in larger amounts, are not indicated at a positive potential of the electro-chemical cell of 0.7 V while the polyhydroxyphenoles catechine and epicatechine, corresponding to the redox equilibrium as shown for epicatechine (Fig. 16), are indicated oxidatively (+ 0.7 V, EC-cell 1) as well as reductively (- 0.2 V, EC-cell 2). This behaviour is typical for polyhydroxyphenoles and additionally confirms the proper classification of this compound class.

The roast-specific substance appearing in peak 1 in Figure 15 b was proven electro-chemically active. It could be identified with reference to the lit-

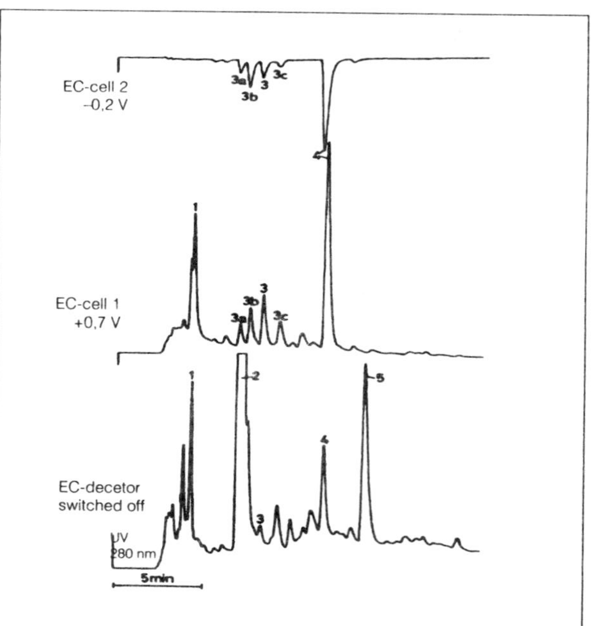

Fig. 15 a   HPLC Chromatograms of Unroasted Cocoa Extract, Variety Elfenbein

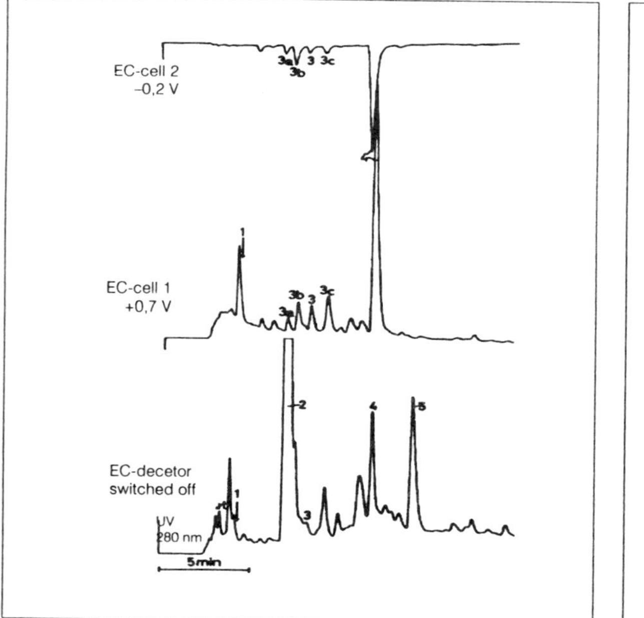

Fig. 15 b   HPLC Chromatograms of Roasted Cocoa Extract, Variety Elfenbein

(linear gradient from 14 % mobile phase B to 43 % mobile phase B in 15 min)

1 Roast-specific substance (2,3-dihydro-3-hydroxymaltol) **2** Theobromine **3** Catechine **3a, 3b, 3c** probably more polyhydroxyphenols **4** Epicatechine **5** Caffeine

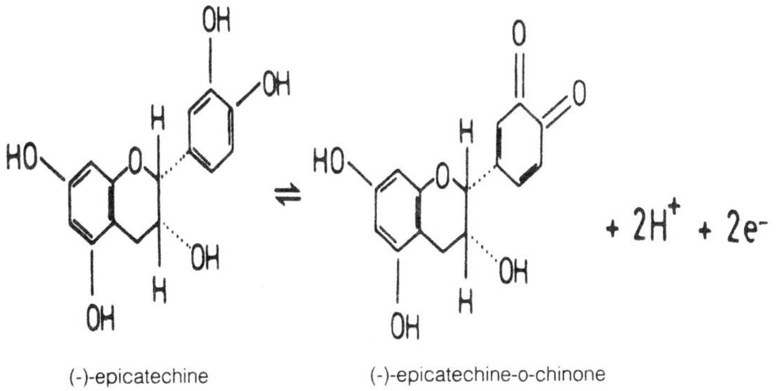

Fig. 16 Redox Equilibrium of Epicatechine and Epicatechine-o-chinone as Basis for Electro-Chemical Detection

erature [16,17,18] and compared to a corresponding reference substance as 2,3-dihydro-3.5-dihydroxy-6-methyl-4H-pyran-4-one (2,3-dihydro-3-hydroxy-maltol). This peak appears in the UV-chromatogram, too, but disappears if the compound is previously irreversibly oxidized at the electro-chemical cell.

Figure 17 shows the HPLC chromatograms of steam distillates of under-roasted and over-roasted cocoa. The detection was carried out in the UV with the electro-chemical detector switched off. Besides the increase of 2,3-dihydro-3-hydroxymaltol (peak 1) due to roasting, an increase of different furane-and pyrazine derivates [compare 19,20] can also be seen. Only the concentration of tetramethylpyrazine (peak 7), already present in unroasted cocoa, stays almost constant. The compounds shown in Figure 17, however, differ characteristically regarding their increase with time.

As shown in Figure 18 the concentration of 2,3-dihydro-3-hydroxymaltol for the cocoa types Elfenbein, Lagos and Ariba has a linear increase in the medium roasting range (between approx. 5 and 15 min roasting time at 130 °C), while for the conditions of overroasting a curve flattening can be observed.

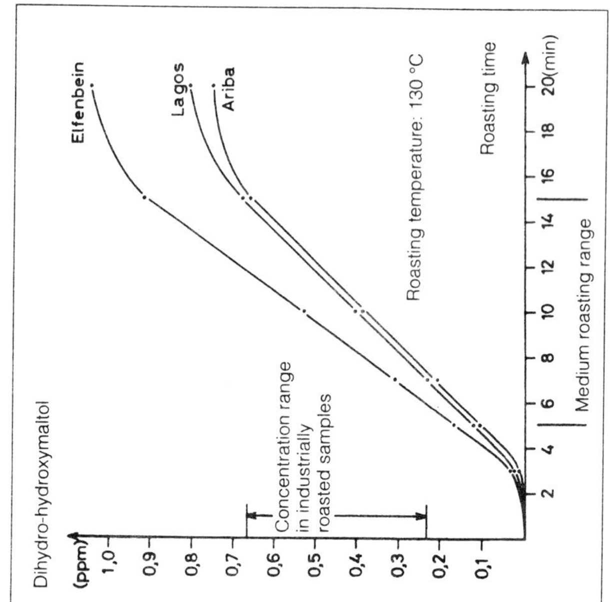

Fig. 18 Increase of 2,3-dihydro-3-hydroxymaltol Concentration in Relation to the Degree of Roasting of Cocoa Varieties Elfenbein, Lagos and Ariba (increasing roasting time at constant roasting temperature; amount of steam distillates in relation to total cocoa)

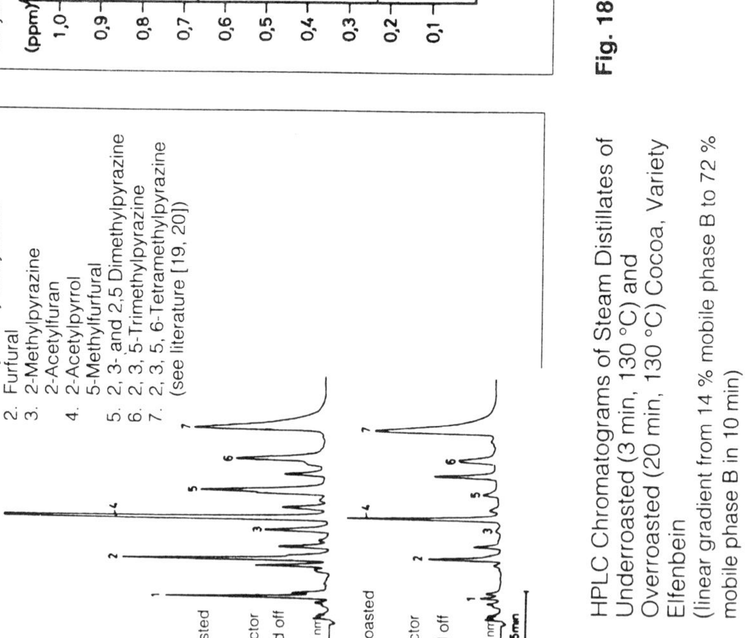

Fig. 17 HPLC Chromatograms of Steam Distillates of Underroasted (3 min, 130 °C) and Overroasted (20 min, 130 °C) Cocoa, Variety Elfenbein (linear gradient from 14 % mobile phase B to 72 % mobile phase B in 10 min)

In contrast, compounds contained in peak 2 to peak 6 increase significantly only with longer roasting times. This is very clear for dimethylpyrazines which do not change in the lower roasting range, but show a sharp increase at overroasting conditions, as Figure 19 demonstrates.

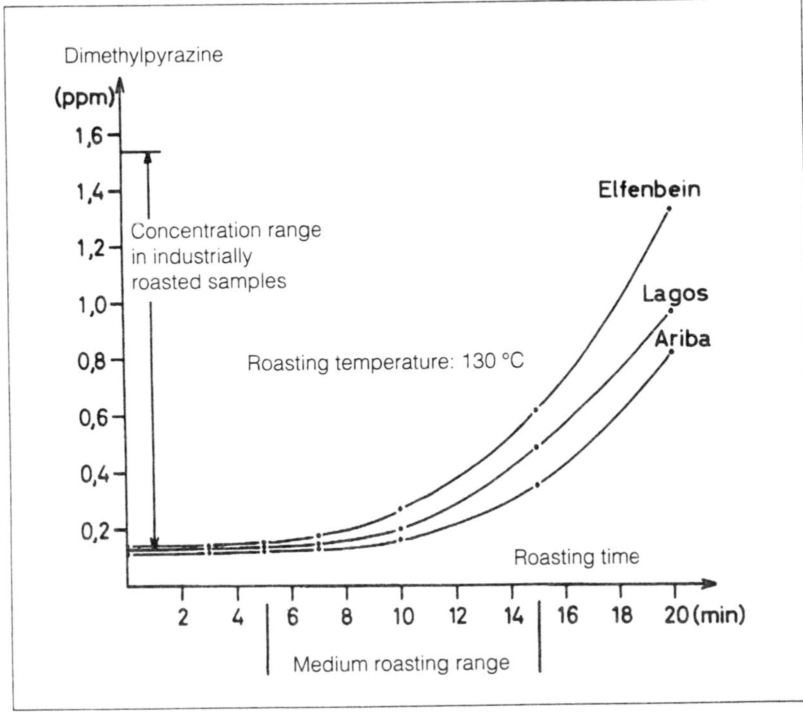

Fig. 19 Increase of Dimethylpyrazine Concentration in Relation to the Degree of Roasting of Cocoa Varieties Elfenbein, Lagos and Ariba (increasing roasting time at constant roasting temperature; amount of steam distillates in relation to total cocoa)

It is remarkable that industrially roasted samples can have relatively high dimethylpyrazine concentrations which are typical for the range of overroasting, as shown in Figure 19.

In conclusion, it can be stated, that 2,3-dihydro-3-hydroxymaltol seems to be a suitable tracer substance for evaluating the degree of roasting of cocoa beans in the normal roasting range. A large increase of dimethylpyrazines can indicate a beginning overroasting.

8.5 Concluding Remarks

Analytical determination of characteristical Maillard reaction products is suitable for the objective evaluation of chemical changes occurring during thermal processing such as drying, kilning or roasting.

Amadori compounds as tracer substances can be detected by HPLC or capillary gas chromatography and used for recognition of beginning Maillard reactions as well as for tracking the course of this reaction which leads to off-flavor compounds. In this way differences regarding the thermal stress of dried vegetable products can be detected prior to occurrence of undesired sensory changes. Furthermore, thermal processing steps can be controlled so that the tolerable amounts of those compounds are not exceeded.

Examples of such applications are dried vegetables and malts kilned under various conditions.

For evaluation of roasting processes reaction products from later stages of the Maillard reaction can be used. Taking the example of cocoa roasting, it could be shown that 2,3-dihydro-3-hydroxymaltol can be applied as tracer substance for evaluation of the degree of roasting in the normal roasting range, and certain pyrazine derivates can be used as tracer substances for the range of overroasting.

Parts of these investigations were supported by the BMWi within the scope of the industrial joint research for which we are very grateful.

REFERENCES

1. HODGE, J. E.: J. Agr. Food Chem. **1**, 928 (1953)

2. REYNOLDS, T. M.: Adv. Food Res. **12**,1 (1963);**14**,168 (1965)

3. BALTES, W.: Food Chem. **9**, 59 (1982)

4. EICHNER, K., WOU, W.: Chemie-Ingenieur-Technik **54**, 270 (1982) – Mikrofiche MS986/82

5. EICHNER, K.: Lebensmittelchem. Gerichtl. Chem. **36**, 101 (1982)

6. EICHNER, K., WOLF, W.: ACS Symposium Series **215**, 317 (1983) American Chemical Society, Washington D. C. 1983 (Hrsg.: G. R. Waller, M. S. Feather)

7. EICHNER, K., WOLF, W.: Lebensmittelchem. Gerichtl. Chem. **37**, 91 (1983)

8. BALTES, W.: Dtsch. Lebensm. Rdsch. **75**, 2 (1979)

9. BALTES, W.: Lebensmittelchem. Gerichtl. Chem. **34**, 39 (1980)

10. CINER-DORUK, M., EICHNER, K.: Z. Lebensm. Unters. Forsch. **168**, 9 (1979)

11. REUTTER, M., EICHNER, K.: Z. Lebensm. Unters. Forsch. **188**, 28 (1989)

12. WITTMANN, R., EICHNER, K.: Z. Lebensm. Unters. Forsch. **188**, 212 (1989)

13. TSUCHIDA, H., KOMOTO, M., KATO, H., FUJIMAKI, M.: Agr. Biol. Chem. **37**, (11), 2571 (1973)

14. SCHNEE, R., EICHNER, K.: Z. Lebensm. Unters. Forsch. **185**, 188 (1987)

15. MOHR, W.: Fette, Seifen, Anstrichm. **72**, 695 (1970)

16. BRUNNER, E.: Dissertation Universität München 1978

17. SEVERIN, TH., SEILMEIER,W.: Z. Lebensm. Unters. Forsch. **137**, 4 (1968)

18. LEDL, F., SCHNELL, W., SEVERIN, TH.: Z. Lebensm. Unters. Forsch. **160**, 367 (1976)

19. ZIEGLEDER, G.: Dtsch. Lebensm.-Rundsch.**78**, 77 (1982)

20. ZIEGLEDER, G., SANDMEIER, D.: Dtsch. Lebensm.-Rundsch. **79**, 343 (1983)

Rapid Sample Preparation for Instrumental Analysis

P. Schreier
O. Fröhlich

9 Rapid Sample Preparation for Instrumental Analysis

P. Schreier and O. Fröhlich, Würzburg

9.1   Introduction

Progress made during the last years in many fields of chemistry, pharmacy, biology, and medicine is closely connected with the application of instrumental analysis. The methods of instrumental analysis are important for industrial quality and food control as well as for environmental protection. The term "instrumental analysis" includes chromatographic and spectroscopic methods, as shown in Figure 1. As to chromatography (Fig. 1), the main emphasis lays on gas chromatography (GC) - listed here as capillary gas chromatography (high resolution gas chromatography, HRGC) [1]-, high performance liquid chromatography (HPLC) [2], and thin-layer chromatography (TLC), in particular in its HPTLC form [3]. Chromatography is an essential part of the instrumental analysis for separation of one or more components to be identified or determined.

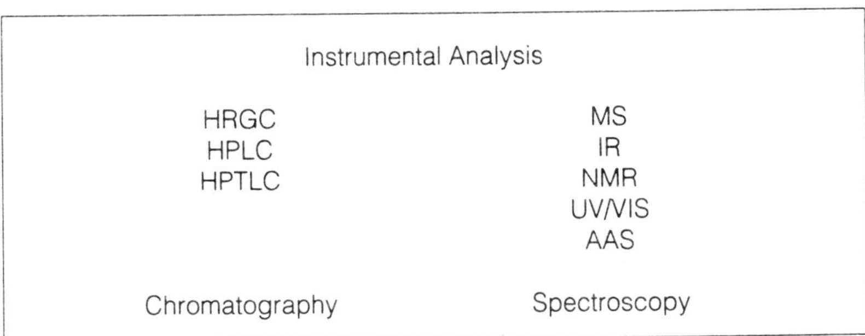

Fig. 1 Examples for Methods of Instrumental Analysis

For identification and quantification a number of methods is available, among them mass spectrometry (MS) [4], infra-red spectroscopy (IR) [5], NMR-techniques [6], spectral-photometric methods (UV/VIS) [7], and atomic absorption spectroscopy (AAS) [8], as listed in Figure 1. In this figure, HRGC and HPLC (on the left side) and MS- and IR-techniques (on the right side) are at the top of the list. This should symbolize a real top position as well as express the possibility to achieve on-line double information when combining a chromatographic with a spectroscopic method, as achieved in

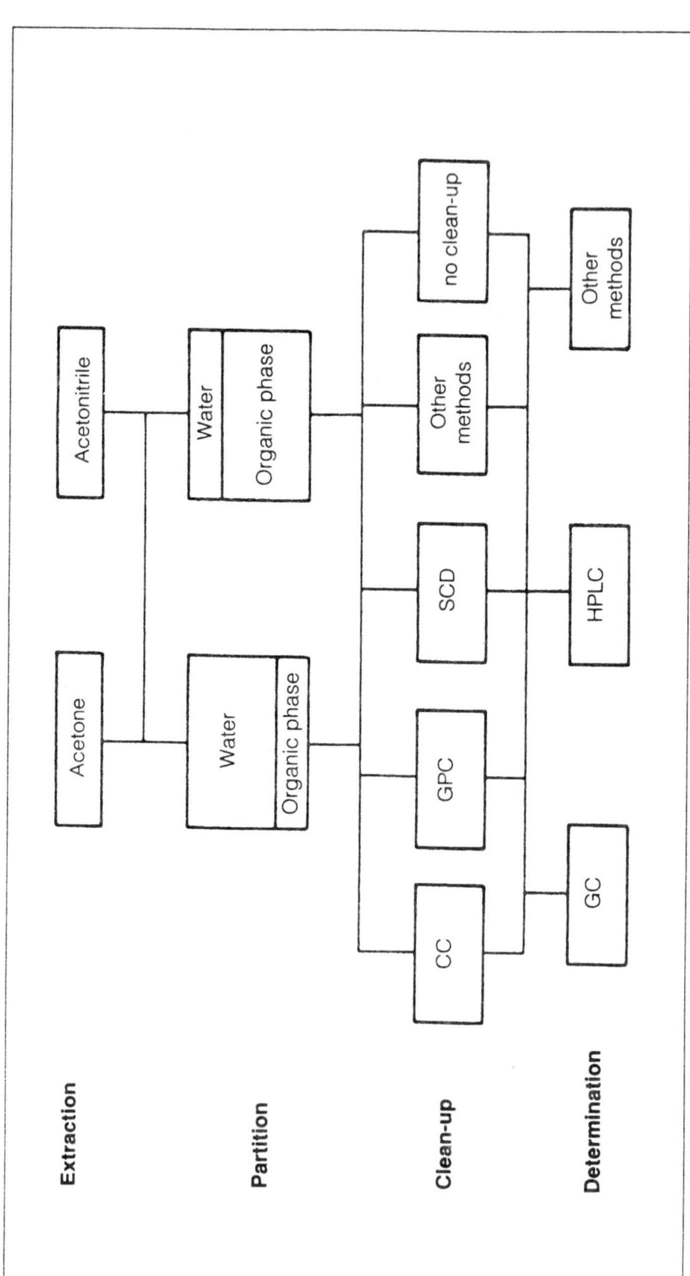

Fig. 2 Simplified Diagram of Working Steps for Preparation of Plant Material for Pesticide Analysis [13]
CC = Column chromatography, GPC = Gel permeation chromatography, SCD = Sweep-Co distillation, GC = Gas chromatography, HPLC = High pressure liquid chromatography

HRGC-MS- [9], HRGC-FTIR- [10] or HPLC-UV/VIS-, HPLC-MS- [11] and, in the future, HPLC-FTIR- [12] techniques.

However, direct chromatographic analysis of samples can only be performed in a few cases. Usually, the analysis starts with **isolation, clean-up** and - depending on the quantities to be determined - **concentration**. Sample preparation generally consists of these two or three steps. For each case sample preparation can be different. In the following, two extremes are described, i.e. on one hand a classical sample preparation technique with a series of laborious steps, and on the other hand the possibility to carry out an analysis without any sample preparation.

As to the first case, Figure 2 shows, in a simplified diagram, the steps for clean-up of plant material for analysis of pesticides [13]. It includes extraction with an organic solvent (acetone, acetonitrile), a distribution and concentration step, a series of different procedures for clean-up, and finally separation, identification and determination. For clean-up several possibilities are available. In Figure 2, the column chromatography (CC) [14], the gel permeation chromatography (GPC) [15] and the sweep-co-distillation (SDC) [16] are outlined as examples. For pesticide analysis the method according to Specht [17] proved good, in which the clean-up is carried out by GPC on a

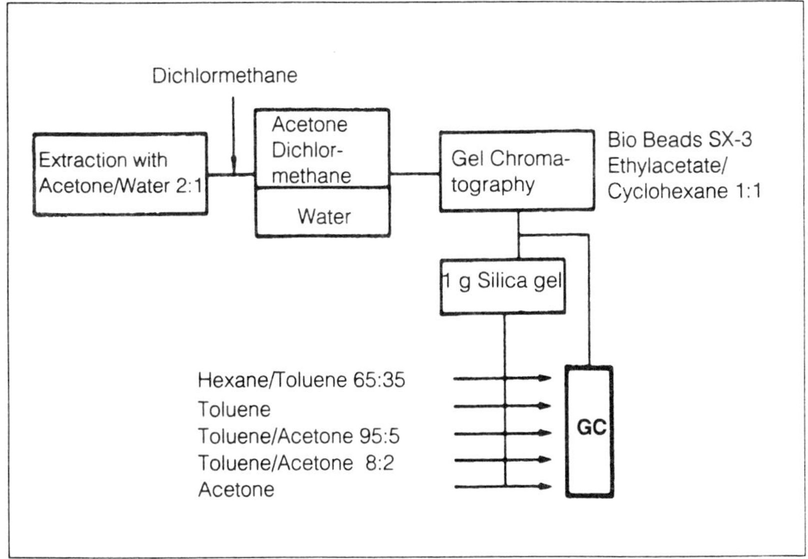

Fig. 3 Example of Universally Applicable Sample Preparation for Pesticide Analysis [17]

polystyrol gel (Bio Beads S-X3) (Fig. 3). GPC is followed by an additional clean-up using fractionation on a silica gel column, in the course of which the elution behaviour on the silica gel can be helpful for identification.

Figure 2 and 3 clearly show how laborious the pre-chromatographic steps are. Such procedure is time-consuming, costly and personnel intensive. For this reason it is obvious that, for some years now, efforts have been made to reduce this part of analysis work or, if possible, eliminate it completely.

Independent of special problems which for example can be solved by gas chromatographic headspace techniques [18] without sample preparation, the questions arises whether a procedure, as illustrated in Figure 4, can generally be realized. The answer is "yes, but with limitations". On one hand, the limitation is based on the fact that such a procedure can only be successful with correspondingly increased application of instrumental measuring techniques. As to the example shown in Figure 4, mass spectrometry can be improved by application of tandem-MS techniques [19] (Fig. 5). On the other hand, such a procedure is limited by the presence of multi-component mixtures [20]. If, however, clearly defined classes of uniform chemical structures should be analyzed from a complex mixture, an instrumental analysis without sample preparation can be realized. This has been verified successfully in environmental analysis, biochemistry, and toxicology [21].

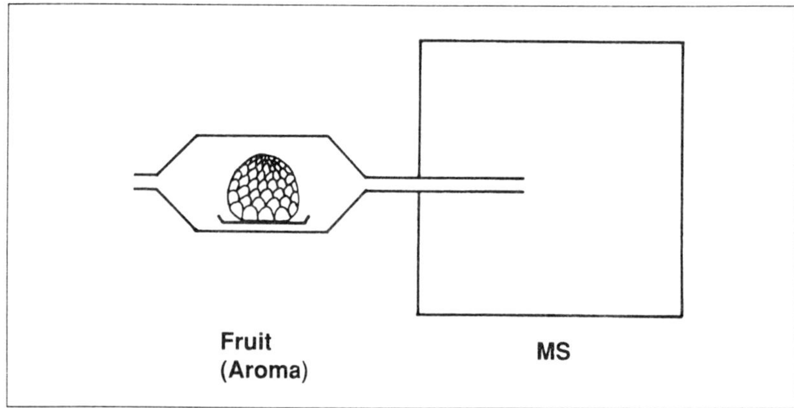

Fig. 4 Mass Spectrometric Aroma Analysis without Sample Preparation (Scheme) [19]

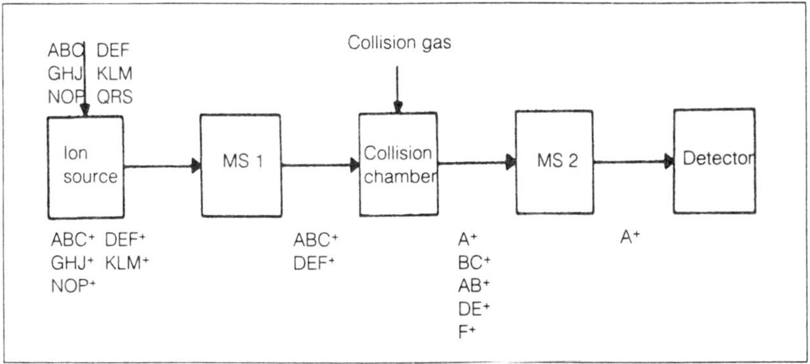

Fig. 5 Possibilities of Selection in a Mass Spectrometer [21]
The sample contains 6 individual components, five of them are ionized by the selected kind of ionization. Two of them have the same mass and are allowed to pass the first MS at the same time. In the collision chamber both disintegrate into an ion mixture from which the selective mass A⁺ is filtered by means of the second MS. A⁺ is then used for determination of ABC in the mixture.

9.2 Sorbent Extraction

During the last years, methods for sample preparation are used in instrumental analysis that represent a compromise between both extreme cases of the above-mentioned pre-chromatographic procedures. These are the methods of the sorbent extraction, for which the liquid/liquid technique and the liquid/solid technique have to be distinguished.

9.2.1 Liquid/Liquid Technique

Tab. 1: Sorbent Extraction, Liquid/Liquid Technique

Diatomaceous Earth	
Extrelut	Merck
ChemElut (ToxElut)	ict

As shown in Table 1, diatomaceous earth with wide pores is used as sorbent. It adsorbs the aqueous phase and distributes it over a large surface. As soon as a water-insoluble solvent drops slowly through the column it comes in contact with a thin-layer of the sample. Interactions at this phase cause retention of pigments, particle material and similar components of the sample while the compounds to be detected are transferred into the organic solvent. This method has been employed very successfully in forensic analysis but it is more and more used in the field of food chemistry [22]. Since the carrier material is stable between pH 1 and pH 13, the pH-value can be

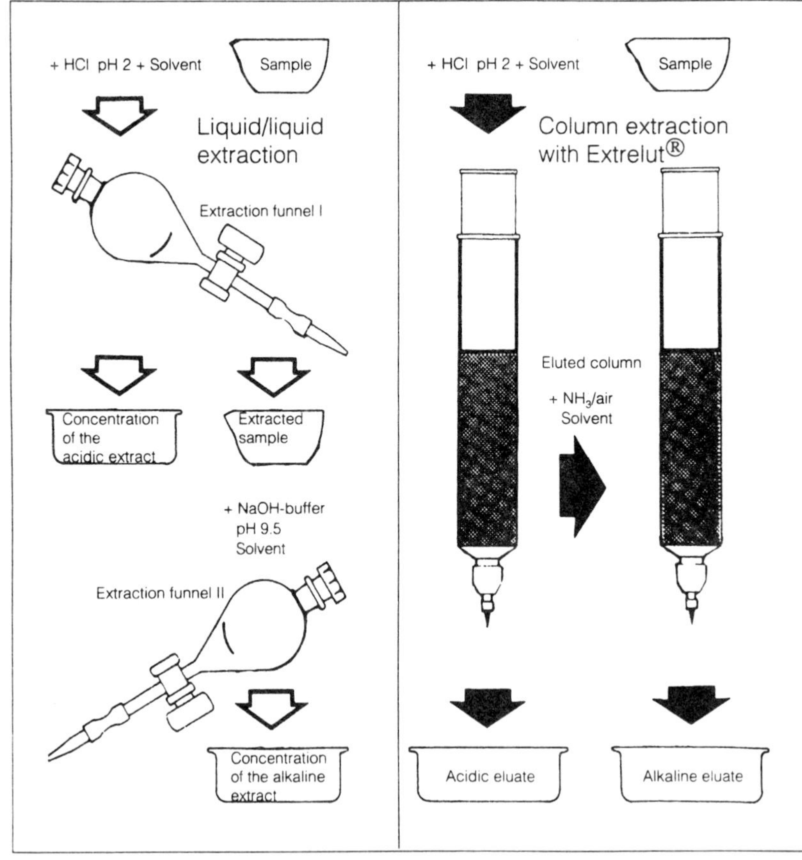

Fig. 6 Schematic Comparison between Conventional Extraction (left side) and Sorbent-Liquid/Liquid Extraction (right side) (modified according to [23])

adjusted to the pK-value of the substance to be extracted. Thus, extraction by means of sorbent liquid/liquid technique is possible with acidic as well as alkaline (and neutral) components (Fig. 6).

The advantages of this method compared to the conventional procedure are obvious: no formation of emulsions; saving of solvents, material and time (no drying process, automation easily possible); in many cases higher recovery values and yields of more pure eluates.

Tab. 2 Application of Sorbent-Liquid/Liquid Extraction for Sample Preparation for Determination of Ethlycarbamate in Hard Liquors [25]

Sample Preparation

- Mixing the carrier material of an Extrelut-column (Merck 11 737) with 10 g NaCl and filling the column

- Application of a 20 ml sample

- After 15 min, addition of 2 x 20 ml n-pentane, discarding the eluate

- Extraction with 40 ml ethyl acetate/CH_2Cl_2 (2:1)

- Again extraction with 20 ml of the solvent mixture

- Addition of 0.1 ml standard solution (5.0 mg undecanoic acid methylester in 100.0 ml ethanol) to the combined extracts and reduction to 2 ml under vacuum at 30-35 °C (rotating evaporator)

- GC/MS analysis (SIM-technique)

Table 2 shows an actual example for application of Extrelut in food analysis. At the end of 1985 in Canada, high concentrations (up to 6 mg/l) of ethyl carbamate, a multivalent, genotoxic cancerogen [24] had been determined in stone fruit brandy for the first time. In the meantime, investigations have been carried out all over the world to clarify the ethyl carbamate formation and to develop technological methods to reduce contents. A suitable analysis method is the basis for such studies. Table 2 shows the methodology, provided by Preu· [25], which had been adopted by other authors [26,27]. Table 3 shows a compilation of ethyl carbamate amounts in hard liquors recently determined by Christoph et al. [26]. The relatively high contents of this compound, particularly in stone fruit brandies, has been confirmed by these authors.

Tab. 3 Contents of Ethyl Carbamate in Hard Liquors [according to 26]

Product	Number of samples	Ethyl carbamate (mg/l)
Yugoslavian Slivowitz	20	1.2 – 7.0
Hungarian apricot spirit	20	0.3 – 1.5
Plum brandy	18	0.4 – 10.0
Kirsch	12	0.4 – 7.0
Pome fruit spirits	5	0.5 – 0.8
Syrian plum brandy	5	0.4 – 5.0
Scotch whisky	5	0.03 – 0.2
Apple jack	4	0.1 – 0.3
Pear brandy	4	0.1 – 0.3
Bourbon whisky	3	0.12 – 0.3
Canadian whisky	3	0.03
Irish whisky	3	0.1
Raspberry spirit	2	0.1
Calvados	2	0.1 – 0.3
Brandy	2	0.05 – 0.07
Cognac	1	0.6
Armagnac	1	0.24
Rum; gin; grappa; gentian brandy; grain brandy; vodka	1 each	0.4

9.2.2 Liquid/Solid Technique

For application of the liquid/solid technique as rapid sample preparation for instrumental analysis, a series of modified silica gels are available on the market. Their basic structures are listed in Figure 7. The modifications range from silica gel to ethyl-, octyl-, octadecyl-, cyclohexyl-, and phenyl-substituted silica gels to cyanopropyl-, diol-, aminopropyl-, and propylethylene diamine structures. Furthermore, a series of ion exchange functions on silica gels has to be mentioned.

The treatment of liquid samples with solid carriers in columns or cartridges from which subsequently organic compounds are directly dissolved in a solvent offers several advantages, compared to direct extraction by solvents: (i) contaminations in the solvent can be removed by pre-treatment; (ii) the amount of carrier material can be much smaller than the sample amount; (iii) tailing of interfering main components, e.g., from biological samples, can be avoided; (iv) the techniques can be carried out easily and quickly. Therefore, in particular, extractions with lipophile materials (cf. the reversed phase carrier materials (RPM), listed in Figure 7) are gaining more and more

<u>Structures</u>

<u>Non-polar</u>

Ethyl	$-\overset{\mid}{\underset{\mid}{Si}}-C_2H_5$
Octyl	$-\overset{\mid}{\underset{\mid}{Si}}-C_8H_{17}$
Octadecyl	$-\overset{\mid}{\underset{\mid}{Si}}-C_{18}H_{37}$
Cyclohexyl	$-\overset{\mid}{\underset{\mid}{Si}}-cyclC_6H_{13}$
Phenyl	$-\overset{\mid}{\underset{\mid}{Si}}-Phe$

<u>Polar</u>

Silica	$-\overset{\mid}{\underset{\mid}{Si}}-OH$
Cyanopropyl	$-\overset{\mid}{\underset{\mid}{Si}}-CH_2CH_2CH_2CN$
Diol	$-\overset{\mid}{\underset{\mid}{Si}}-CH_2CH_2CH_2OCH_2CH-CH_2OH$ with OH
Aminopropyl	$-\overset{\mid}{\underset{\mid}{Si}}-CH_2CH_2CH_2NH_2$
N-Propylethylenediamine	$-\overset{\mid}{\underset{\mid}{Si}}-CH_2CH_2CH_2NHCH_2CH_2NH_2$

<u>Ion Exchange</u>

Benzenesulfonylpropyl	$-\overset{\mid}{\underset{\mid}{Si}}-CH_2CH_2CH_2-Phe-SO_3^{\ominus}$
Carboxymethyl	$-\overset{\mid}{\underset{\mid}{Si}}-CH_2COO^{\ominus}$
Diethylaminopropyl	$-\overset{\mid}{\underset{\mid}{Si}}-CH_2CH_2CH_2\overset{\oplus}{N}H(CH_2CH_3)_2$
Sulfonylpropyl	$-\overset{\mid}{\underset{\mid}{Si}}-CH_2CH_2CH_2-SO_3^{\ominus}$
Trimethylaminopropyl	$-\overset{\mid}{\underset{\mid}{Si}}-CH_2CH_2CH_2\overset{\oplus}{N}(CH_3)_3$

Fig. 7 Sorbent Extraction: Basic Structure of Silica Gels Used for Liquid/ Solid Technique

importance for rapid sample preparation for instrumental analysis. Such products are available under the trade names outlined in Table 4.

Tab. 4 Trade Names of Well-Known Reversed Phase Carrier Materials

Bakerbond	Baker
BondElut	Analytichem
PrepSep	Kontron
Sep-Pak	Waters

How is the liquid/solid extraction employed in practice? The diagram shown in Figure 8 explains the procedure for the problem of separation of a component (∗) from contaminations (■) [28]. Figure 8 shows C_{18}-modified silica gel as example for the solid carrier, the numbered circles represent the different solvents. For "a" hexane is used for conditioning, for "b" methanol/water is employed. For "c" the sample application is carried out with a 20 min. drying period, and for "d" the interesting component (∗) is eluted with hexane.

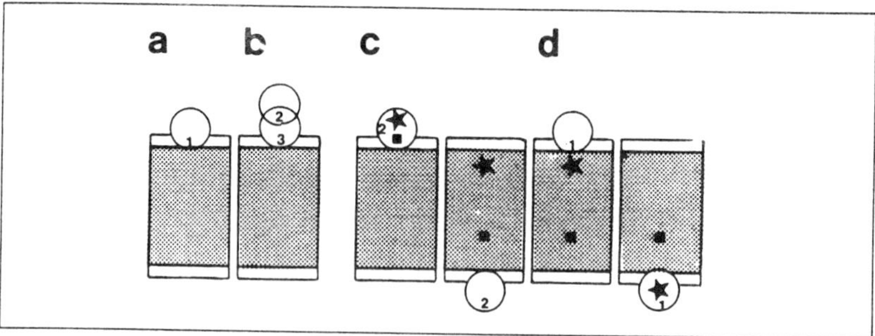

Fig. 8 Diagram of an Application of Liquid/Solid Extraction for Separation of One Component (∗) from Contaminations (■) on Modified C_{18}-Silica Gel (modified according to [28])
The numbered circles represent different solvents;
1 = hexane, 2/3 = metha-nol/water
a = conditioning with hexane b = conditioning with methanol/water
c = sample application, drying period 20 min d = hexane-elution

Figure 9 shows an extended scheme concerning a more complicated problem, i.e. the fractionation and separation of wine constituents (■ anthocyanins; ● organic acids; ▲ carbohydrates), according to [29]. In this case, the combination of a cyclohexyl-(I) with a strong alkaline ion exchange silica gel (II) is used. Again, the numbered circles represent different solvents. For "a" the conditioning is carried out with methanol/water, in "b" the wine sample is applied. As can be seen from Figure 9, first of all, the anthocyanin portion is retained in "b", which can then be eluted in "c" by means of methanolic HCl. The mixture of organic acids and carbohydrates is separated on the anion exchange column in such way that the carbohydrates are in the eluate while the acid portion is retained ("b"-II in Fig. 9). Elution of the acids can be performed in "d" by use of HCl.

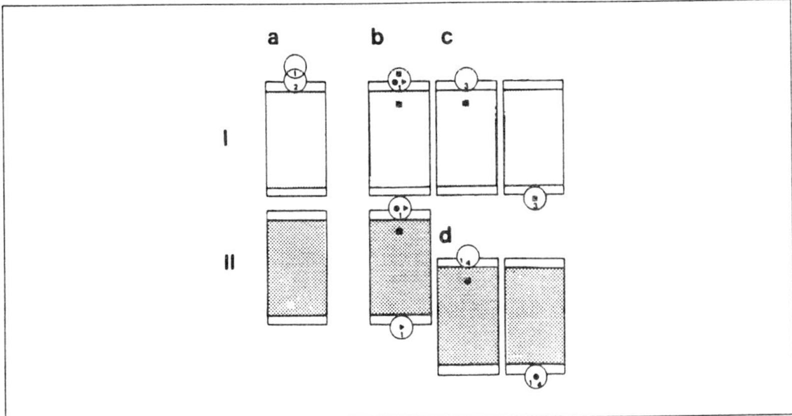

Fig. 9 Diagram of an Application of Liquid/Solid Extraction with Cyclohexyl-(I) and with Strong Alkaline Ion Exchange Modified Silicagel (II) for Fractional Separation of Anthocyanins (■), Organic Acids (●) and Carbohydrates (▲) from Wine (modified according to [29])

The numbered circles represent different solvents; 1/2 = methanol/water, 3 = methanolic HCl, 4 = HCl a = conditioning with methanol/water b = sample application c = elution with methanolic HCl d = elution with HCl

9.3 Practical Examples

Some practical examples now follow the previous explanations concerning the procedures of liquid/solid extraction. Very simple is the detection of, e.g., diethylene glycol ether in aroma concentrates. As shown in Table 5, it is done by application of an aroma concentrate onto Sep-Pak Florisil (magnesium silicate with about 15% MgO and 85% SiO_2), subsequent elution with CH_2Cl_2 and analysis of the sample. Some time ago, the official food control has been confronted with the presence of diethylene glycol ethers.

Tab. 5 Rapid Sample Preparation for Determination of Diethylene Glycol Ethers in Aroma Concentrates

– Application of 0.5 ml aroma concentrate onto Sep-Pak (Florisil)
– Eluation with 5 ml CH_2Cl_2
– HRGC and HRGC-MS analysis

For food analysis, a list of working instructions for liquid/solid extraction, as shown in Figure 10 [30], has been provided by a manufacturer of carrier materials. In each case, it deals with the fast sample preparation for analysis of aflatoxins in cocoa beans, lecithin in chocolate, vitamin D_3 in fish liver oil (on silica gel each), organic acids in wine (on a strong alkaline ion exchanger), fungicides in oranges (on a diol-phase) as well as aflatoxins in milk (on C_{18}-RPM). For all these examples, HPLC analysis is the method of choice.

Systematic investigations on suitability of different RP-material for rapid sample preparation in food analysis have been carried out, among others, for the xanthine derivatives, the examination of organic contaminations in water as well as the analysis of aroma compounds. Some years ago, Reid and Good [31] were able to prove the suitability of liquid/solid extraction, in particular on C_{18}-RPM, for rapid sample preparation for caffeine determination.

As can be seen in Table 6, recovery values of 93% and 98%, respectively, have been determined; the more favorable result was obtained by elution with $CHCl_3$. For quantification in coffee, tea and cola, HPLC has been employed (Fig. 11).

Tab. 6 Recovery Values of Xanthines in 5 ml Water Samples Using Different RPM[1]) [31]

Sorbent	Recovery (%)		
	Theobromine	Theophylline	Caffeine
C_{18}	95	98	93
C_8	66	81	86
Phenyl	94	95	93
Benzenesulfonic acid	18	22	32
COOH	24	22	57
CN	6	5	8
C_{18}[2])	93	94	98

1) 2 x 500 µl methanol as eluent - 2) 2 x 500 µl $CHCl_3$ as eluent

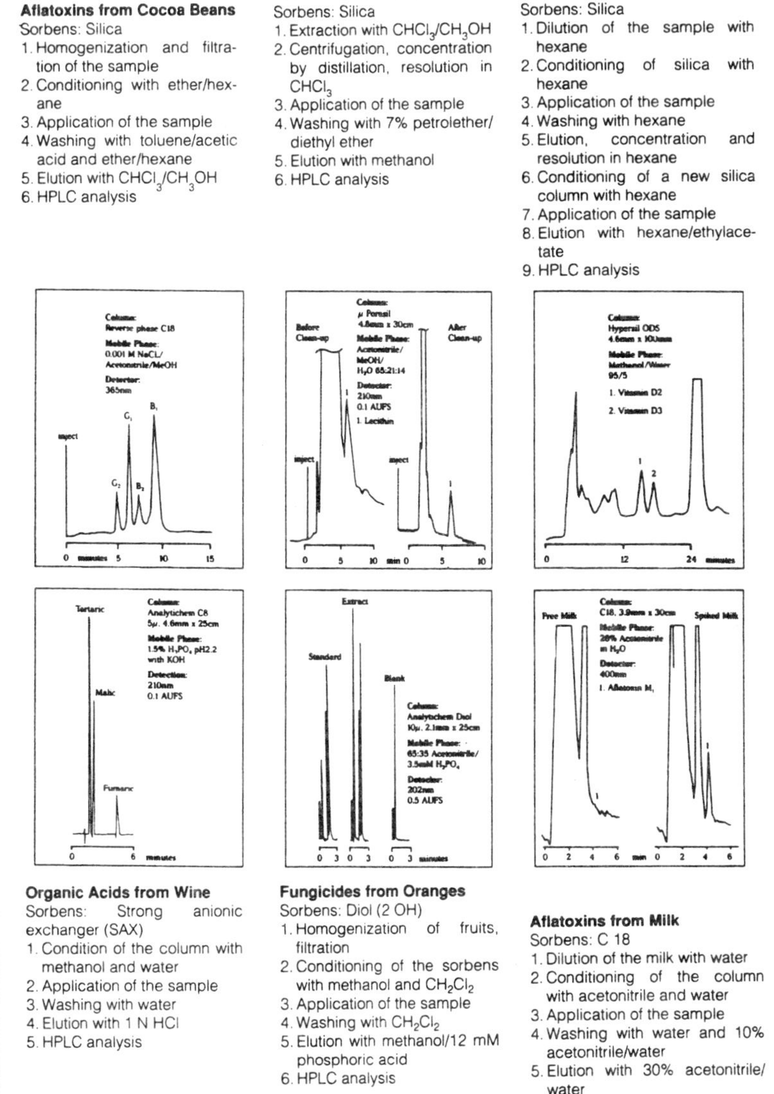

Fig. 10 Analysis Instructions (Analytichem [30]) for Application of Liquid/Solid Extraction for Rapid Sample Preparation in Food Analysis

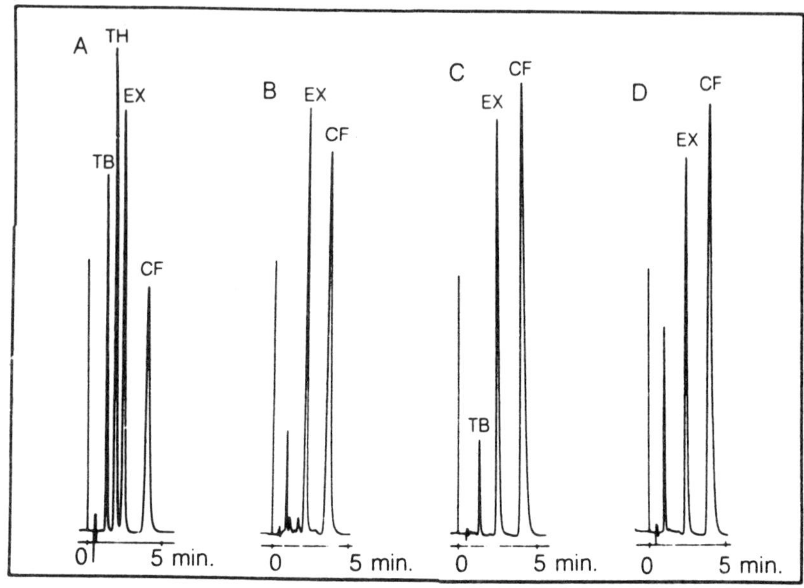

Fig. 11   HPLC-Analysis (Sepralyte 10 mm C$_{18}$; 2.1 mm x 25 cm) of Beverage Extracts after Liquid/Solid Extraction on C$_{18}$-BondElut [31]

A = standard solution with 25 µg theobromine (TB) and 50 µg theophylline (TH) and caffeine (CF) each as well as 2-hydroxyethyl theophylline (EX; external standard); B = Extract from 250 µl coffee brew; C = Extract from 500 µl tea brew; D = 1 ml extract from a cola beverage

Tab. 7 Model Studies on Recovery Values of Phenol Using Different RPM (**without** NaCl-addition) [1]) [32]

Sorbent	Recovery (%)
C$_4$	10
C$_8$	10
C$_{18}$	20
CN	10
Diol	10
CX	37
XE 347	10

1) 100 ml samples with 1 ppm phenol. Methanol elution.

Extended investigations regarding the suitability of liquid/solid extraction as rapid sample preparation for detection of organic'contaminations in waste waters have been carried out by Chladek and Marano [32]. In systematic tests of different RP carrier materials, first of all, these authors were able show for the first time that the recovery values, e.g., for the determination of phenol were low (Table 7). However, after addition of NaCl it was possible to obtain a sufficient result with a cyclohexyl phase, as can be seen from the data outlined in Table 8.

Tab. 8 Model Studies on Recovery of Phenol Using Different Cyclohexyl (CX)-RPM (**with** NaCl-addition) [1] [32]

NaCl (g/v, %)	Recovery (%)
0	37
5	55
10	72
15	83
20	92
25	95

[1] 100 ml samples with 1 ppm phenol. Methanol elution.

In further systematic studies, the authors [32] demonstrated that the obtained results also depend on the pre-treatment of the sorbent, e.g., addition of methanol.

Tab. 9 Model Studies on Recovery Values of Organic Contaminations in Water - Influence of Methanol Addition[1] [32)

% Methanol	Naphtalene	Dimethyl-phtalate	Diethyl-phtalate	Di-n-butyl-phtalate	Butylbenzyl phtalate	Bis(2-ethyl-hexyl)phtalate
			Recovery (%)			
0	90	105	100	90	60	8
10	70	50	70	82	65	15
20	80	17	80	90	73	28
30	70	3	30	90	84	23
40	46	1	10	87	96	68
50	8	0	0	60	60	65

[1] C_{18}-RP; solvent, acetonitrile

As can be seen from Table 9, the recovery values for naphtalene and various phtalates decreased with increasing methanol addition. On the other hand, methanol addition was required to recover satisfactory yields of, e.g., bis(2-ethyl-hexyl)phtalate. Table 10 finally shows the importance of the influence of the eluent for the recovery values, even though the differences were not so serious as for the previous case (Table 9).

Tab. 10 Model Studies on Recovery Values of Organic Compounds in Water – Influence of the Eluent[1]) [32]

| Solvent | Recovery (%) | | | | |
	Naphtalene	Dimethyl-phtalate	Diethyl-phtalate	Di-n-butyl-phtalate	Bis(2-ethyl-hexyl)phtalate
CH_2Cl_2	113	104	98	98	7
Acetonitrile	80	85	90	93	4
Ethylacetate	99	103	96	98	5
Hexane	71	23	48	64	10
Tetrahydrofuran	103	104	91	93	6

1) $C_{18}{}^{RP}$

Regarding the rapid sample preparation for aroma analysis, recently our group carried out a model study for checking liquid/solid extraction with different RP carrier materials. The principle of the method is shown in Figure 12. Two model solutions (I,II), each with seven aroma substances from different classes in different concentration ranges (I. approx. 0.2 mg per aroma substance; II: approx. 3 mg per aroma substance), were used.

The diagrams in Figure 13 represent the percentages of recovery obtained by conventional extraction (pentane-dichloromethane, 2:1) with separation funnel (S) as well as by liquid/solid extraction with the RP-carrier materials C S118 T-, C S18 T-, cyclohexyl (CX) and phenyl(Phe)-silica gel for the individual aroma substances (**1**= dodecane; **2**= hexylacetate; **3**= ethyl-3-hydroxybutanoate; **4**= 1-octanol; **5**= 2-methylbenzaldehyde; **6**= gerianol; **7**= 2-phenylethanol), in accordance with the conditions listed in Figure 12. At first glance, the graph shows for liquid/solid extraction that the relatively polar compound 3 could not be recovered. Compared to **3**, the most favorable result was obtained with C_{18}-RPM.

Figure 14 shows the quantitative dependence of the results using the examples of S and C_{18}-RPM. The dark-colored bars in the diagram represent the

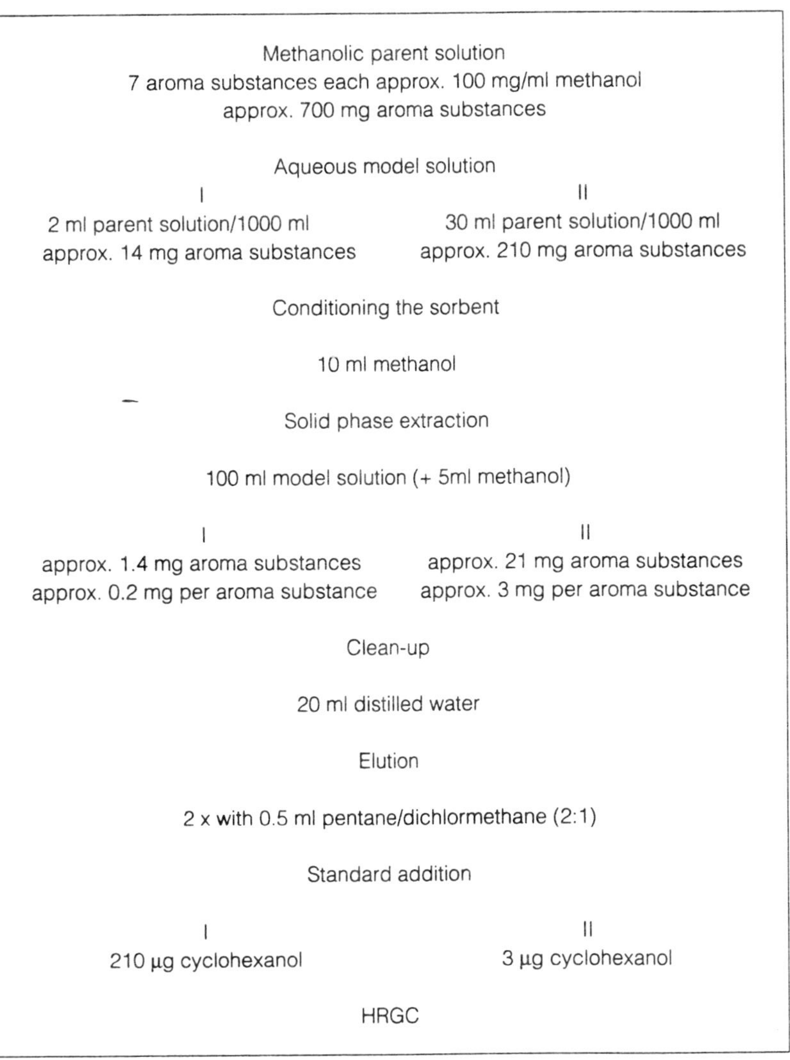

Fig. 12 Model Study for Reviewing Sorbent Liquid/Solid Extraction as Method for Rapid Sample Preparation for Analysis of Aroma Substances

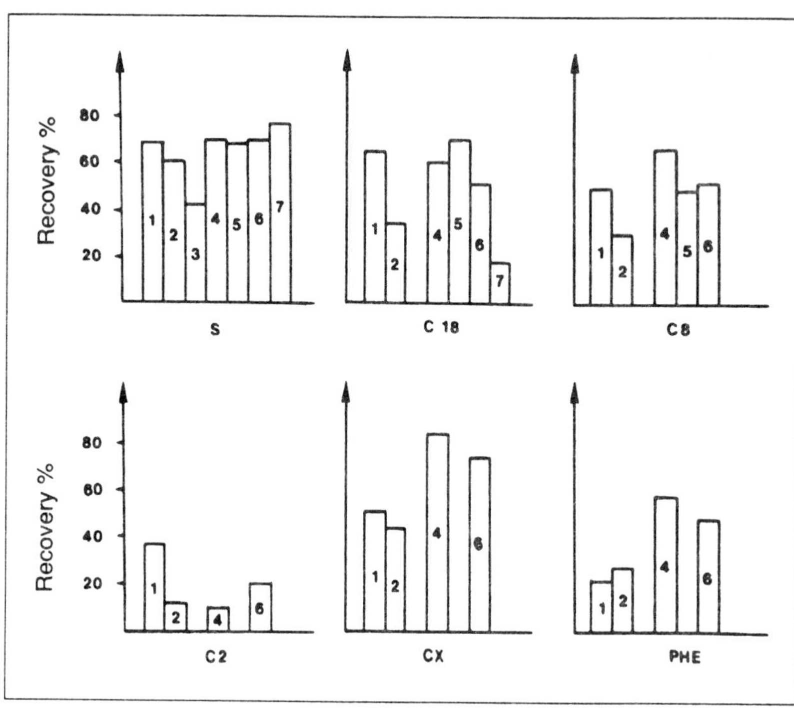

Fig. 13   Recovery Values in Percent for Aroma Substances
(**1** = dodecane; **2** = hexylacetate; **3** = 3-hydroxybutyric acid ethyl-ester; **4** = 1-octanol; **5** = 2-methylbenzaldehyde; **6** = gerianol; **7** = 2-phenylethanol) with separation funnel (S) as well as with liquid/solid extraction, in accordance with the conditions listed in Figure 12 (model solution I), with C_{18}-, C_8-, C_2-, CX- and Phe-RP-carrier materials

percentage of recovery values of the model solution (I), the light-colored bars those obtained with model solution (II). Except for 2-phenylethanol (No. **7**), in all cases higher recovery values were obtained with higher original aroma substance concentrations in tests carried out with C_{18}-RPM. Also in this test series, ethyl-3-hydroxybutanoate could not be recovered by liquid/solid extraction.

Potential improvement of recovery values of organic compounds observed after addition of NaCl and methanol in liquid/solid extraction (cf. Tables 7 - 10) had been described in the literature. Therefore, we adopted this technique in our study and used methanol as well as NaCl addition for the subsequent tests. Figure 15 shows the percentage of recovery values for the

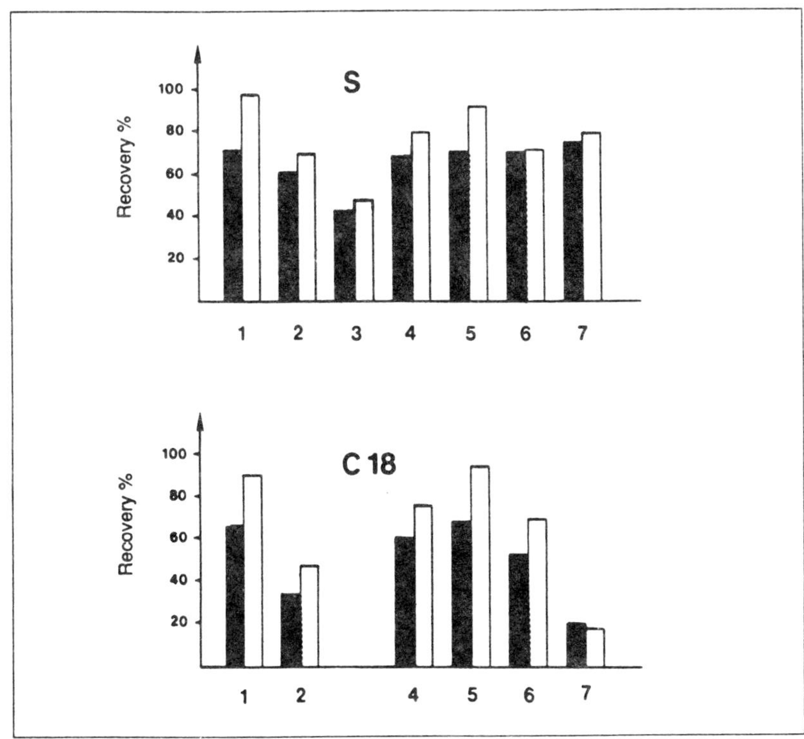

Fig. 14 Recovery Rate in Percent for Aroma Substances
(Numbers and terms see Figure 13) with classic extraction (pent-ane-
dichlormethane, 2:1) in separation funnel (S) as well as for application of
liquid/solid extraction with C_{18}-RPM under the conditions, listed in Figure 12 (dark colored bars: model solution I; light colored bars: model solution II).

above-mentioned aroma substances, determined on C_{18}-RPM after methanol addition (5%) (dark-colored bars) and NaCl saturation (light-colored
bars) or after double extraction - methanol- and NaCl addition - (striped
bars). From this diagram the importance of the NaCl addition can clearly be
seen: ethyl-3-hydroxybutanoate could now be recovered in satisfactory
yields; the percentage of recovery of 2-phenylethanol could be increased
significantly. As shown in Figure 15, double extraction was not necessary.

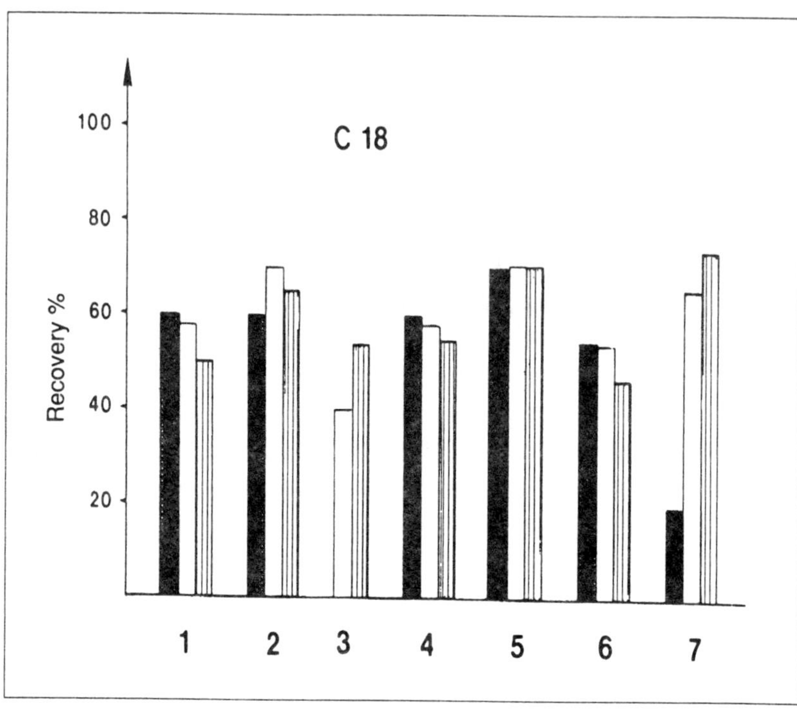

Fig. 15　Recovery Values in Percent for Aroma Substances
(Numbers and terms see Figure 13) for liquid/solid extraction with C_{18}–RPM (mode/solution I, cf Figure 11). Dark-colored bars: extraction with addition of methanol; light-colored bars: extraction after NaCl-saturation; striped bars: double extraction: 1. saturation with NaCl, elution and 2. addition of methanol, second elution.

A comparison with the results of the conventional preparation (S), outlined in Figure 13, reveals for the rapid method of the liquid/solid extraction after NaCl-addition an almost corresponding result. The above-mentioned tests carried out in a series of model examinations, which could only partly be described here, have been transferred to some analyses of natural aroma substances. These results are discussed elsewhere [33].

9.4 Summary

From the described examples it is obvious that the qualitative as well as the quantitative composition of the sample, the pre-treatment of the solid carrier as well as the eluent can influence the result of the liquid/solid extraction. The Use of model experiments for control purposes is recommended; within short time the conditions for sample preparation can be optimized. Of course, this can be easier realized if less components have to be analyzed. For complex multi-component mixtures this task becomes more difficult, as one has to develop, first of all, the most suitable preparation conditions for each group of components. But even for such a complex matter as analysis of aroma substances, application is possible if one is critical enough to check the obtained results by means of model studies. If this is done, the methodology of the sorbent liquid/solid extraction can be an useful alternative not only for analytical but also for preparative purposes due to simple handling, solvent saving, short-time requirement and automatization possibilities [34].

Acknowledgement: We thank Mrs D. Feineis for the skillful performance of the model studies of aroma analysis.

REFERENCES

1. SCHULTE, E.: Praxis der Kapillar-Gas-Chromatographie, Springer, Berlin, Heidelberg, New York, 1983

2. ENGELHARDT, H.: Hochdruck-Flüssigkeits-Chromatographie, Springer, Berlin, Heidelberg, New York, 1977

3. KAISER, R. E. (Hrsg.): Proceedings of the third international symposium on instrumental high performance thin-layer chromatography, Inst. für Chromatographie, Bad Dürkheim, 1985

4. MC LAFFERTY, F. W.: Interpretation of mass spectra, Univ. Sci. Books, Mill Valley, 1980

5. GÜNZLER, H. und BÖCK, H.: IR-Spektroskopie, Chemie, Weinheim, 1975

6. GÜNTHER, H.: NMR-Spektroskopie, Thieme, Stuttgart, 1983

7. GAUGLITZ, G.: Praktische Spektroskopie, Attempto, Tübmgen, 1983

8. WELZ, B.: Atomabsorptionsspektrometrie, Chemie, Weinheim, 1983

9. FRIGERIO, A. und MILON, H. (Hrsg.): Chromatography and mass spectrometry in nutrition science and food safety. Elsevier, Amsterdam, 1984

10. SCHREIER, P. und IDSTEIN, H.: Z. Lebensm. Unters.-Forsch. **181**, 183 (1985)

11. VESTAL, M. L.: Eur. Spec. News **63**, 22 (1985)

12. TAYLOR, L. T.: J. Chromatogr. Sci. **23**, 265 (1985)

13. Thier, H. P. und Frehse, H.: Rückstandsanalytik von Pflanzenschutzmitteln, Thieme, Stuttgart, New York, 1986

14. Mills, P. A., Onley, J. H. und Gaither, R. A.: J. Assoc. Off. Anal. Chem. **46**, 186 (1963)

15. Johnson, L. D., Waltz, R. H., Ussary, J. P. und Kaiser, F. E.: J. Assoc. Off. Anal Chem. **59**, 174 (1976)

16. Eichner, M.: Z. Lebensm. Unters.-Forsch. **167**, 245 (1978)

17. Specht, W. und Tillkes, M.: Z. Anal. Chem. **322**, 443 (1985)

18. Nitz, S. und Jülich, E.: in Analysis of volatiles (P. Schreier. Hrsg.), 151, W. de Gruyter. Berlin, New York, 1984

19. McLafferty, F. W. (Hrsg.): Tandem Mass Spectrometry, J. Wiley & Sons, New York, 1983

20. Labows, J. N. und Shushan, B.: Amer. Lab. **15**, 56 (1983)

21. Schubert, R.: Labor Praxis **1986**, 191

22. E. Merck AG (Hrsg.): Application of Extrelut—Literatursammlung, E. Merck, Darmstadt, 1985

23. E. Merck AG (Hrsg.): Extrelut. Neues Verfahren zur Extraktion lipophiler Stoffe. Firmenschrift Diagnostica Merck, Darmstadt, 1985

24. Schmähl, D., Port, R. und Wahrendorf, J.: Int. J. Cancer, **19**, 7 (1977)

25. Preuss, A.: Dtsch. Lebensm. Rdsch., in Vorbereitung

26. Christoph, N., Schmitt, A. und Hildenbrand, K.: Alkohol-Industrie **15**, 347 (1986)

27. Hübschmann, H. J. und Matter, L.: in Vorb.

28. Blevius, D. D., Burke, M. F., Good, T. J., Harris, P. A., van Home, K. C. und Yago, L. S. (Hrsg.): Sorbent Extraction Technology, 68, Analytichem, Harbor City, 1985

29. Blevius, D. D., Burke, M. F., Good, T. J., Harris, P. A., van Home, K. C. und Yago, L. S. (Hrsg.): Sorbent Extraction Technology, 72, Analytichem, Harbor City. 1985

30. Analytichem (Hrsg.): Proceedings of the 2nd International Symposium of Sample Preparation and Isolation using Bonded Silicas, Analytichem. Harbor City, 1985

31. Reid, S. J. und Good, T. J.: J. Agric. Food Chem. **30**, 775 (1982)

32. Chladek, E. und Marano, R. S.: J. Chromatog. Sci. **22**, 213 (1984)

33. Fröhlich, 0.: Diss. Univ. Würzburg, in Vorbereitung

34. E. Merck AG: Verfahren zur Gewinnung oder Anreicherung von Aromastoffen, EP 0 082 284 (1983)

Fast Quality Control by Headspace Analysis

R. Wittkowski

10 Fast Quality Control by Headspace Analysis

R. Wittkowski, Berlin

10.1 Introduction

It is now indisputable that gaschromatography, especially capillary gaschromatography, for separation of substances in multicomponent mixtures, has proved to be successful over the last two decades and moved to the forefront.

Since samples intended for gaschromatography have to be either in liquid or gaseous phase, analysis of food constituents often needs time-consuming extraction or comparable isolation procedures to separate non- or less volatile components.

Headspace gaschromatography (HSGC) has taken its place among the rapid methods, not because of the gaschromatography, which surely is a contributory factor but rather due to the reduction of sample pre-treatment. The sample preparation is reduced to simple steps such as filling and standard addition. All subsequent analysis steps, starting from injection to separation and quantitative evaluation, can be done automatically. Therefore, in on-line operations, HSGC belongs to the organisation type of rapid methods.

With HSGC, components with a sufficient volatility located in the headspace of a sample are analyzed. Headspace, according to Wyllie et al. [1], is the gaseous phase, surrounding a sample in a closed system and balanced in accordance with the thermodynamic rules.

Contrary to the classical gaschromatography with a peak area depending only on the concentration of substance, the vapour pressure of the substance as well as the activity coefficient within the prepared mixture play an important role in headspace analysis. Due to the temperature relation of the vapour pressure, maintaining a constant temperature during sampling is necessary [2].

Furthermore, it is important to keep the time of equilibration constant, i.e. the time until the equilibrium balance between liquid or solid and gaseous phases is reached. This time is dependent on the viscosity of a sample.

All these influencing factors complicate quantitative analyses and require more effort for calibration than the classical gaschromatography. However,

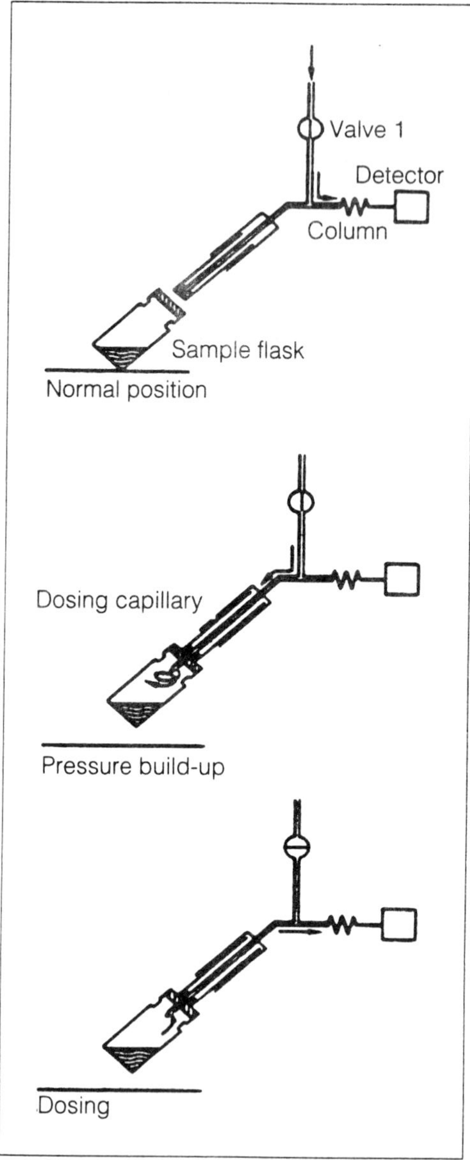

Fig. 1 Sample Taking with Automatic Injection Systems
(Source: Kolb, 1985, p. 12 [3])

the HSGC is used mainly as rapid method for series analyses. Optimization steps are necessary very rarely, especially since the matrix is usually known and well defined.

10.2 Methodology

The gaseous substances are transferred onto the separation column by gas-tight syringes. The solid or liquid analytical sample is filled into a glass bottle closed by a septum and tempered in a water bath. The temperature is to be kept constant until a thermodynamic equilibrium between gaseous and dissolved compounds is obtained (equilibration). The sample valve is connected to an injection needle that can be lowered hydraulically. By inserting the needle, pressure is built inside the sample vessel. Subsequent to closing the carrier gas valve, an aliquot of the gaseous phase is applied on the column by pressure balance (Fig. 1). Today, this equilibration headspace

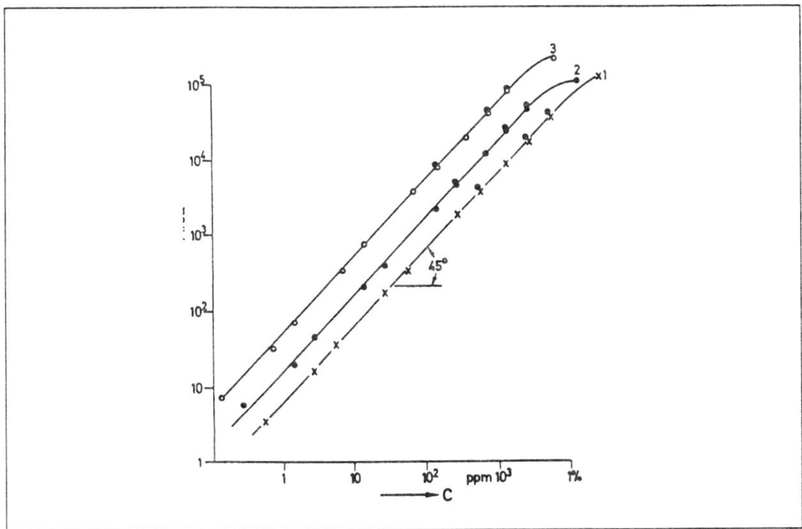

Fig. 2 Linearity of Headspace Analysis of Chlorinated Hydrocarbons in Aqueous Solution
(Source: Kolb, Auer, Pospisil, 1983, p. 444 [4])
1 = chloroform
2 = trichloroethylene
3 = tetrachloride hydrocarbon

gaschromatography (E-HSGC) is applicable without any problems for series analyses by means of autosamplers. Quantification of such samples is done by external standards of the same matrix or by spiking with internal standards.

High reproducibility of automatic injection systems, available on the market today, makes the addition of internal standards for correction of the injection error unnecessary. Furthermore, the E-HSGC offers a sufficient linearity for a large concentration range, as Figure 2 clearly shows.

10.3 Quantitative Analysis

For quantitative determination of volatile substances in liquid foods, defined compounds are added as standards to the sample. This excludes all matrix effects and requires calibrations only for detector response, as for the classical gaschromatography. Then only a repeat determination (with and without internal standard) is necessary for calculation of the required concentration using the difference of both determinations.

External calibrations of the same matrix are practicable, but limited to liquid samples, as for the method of standard addition. It is a rare case when a matrix, for example, in the aroma analysis of alcoholic beverages can be simulated in the form of an aqueous ethanol solution.

10.4 Quantitative Analysis by Multiple Headspace Extraction

On the other hand, an homogeneous distribution of added standards is not possible for solid samples or samples with inhomogeneous matrix, of which most of our foods are composed. This limits the above mentioned operation techniques. To execute quantitative determinations in such samples, the method of "Multiple Headspace Extraction (MHE)" has been developed.

It is based on a stepwise gas extraction with intermediate headspace analysis (Fig. 3). After each total extraction, the remaining concentration is measured [4-8]. The sample vessel remains under pressure after each injection. The uninjected portion of the gaseous phase above the sample is expelled into the atmosphere. This procedure can be continued practically up to the complete extraction of the volatile compounds. The concentration of a substance in the sample is the sum of all peak areas. The matrix effect is no longer important: this is the most decisive advantage of this operation method.

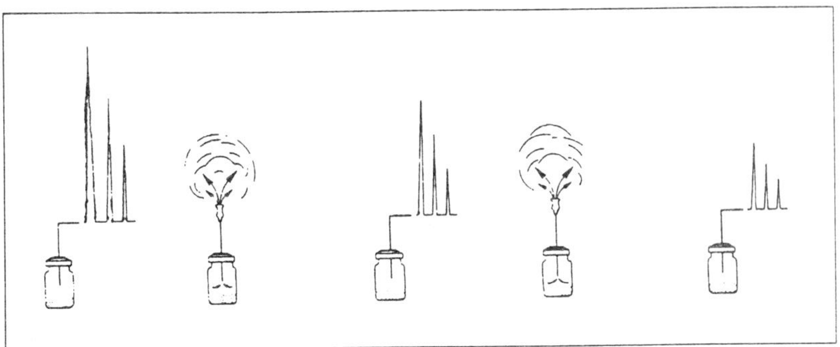

Fig. 3 Principle of Multiple Headspace Extraction (MHE)
(Source: Kolb, Auer, Pospisil, 1983, p. 448 [4])

In Figure 4, the example of the analysis of chlorinated hydrocarbons in water is used to show that the concentration decrease of the compound in the headspace is strictly exponential, if the time intervals between the analyses are kept constant. Therefore, a mathematical extrapolation is possible, and only two such determinations are sufficient. The calculation is made according to

$$\Sigma \; A_i \; = \; \frac{A_i^2}{A_1 - A_2}$$

where $\Sigma \; A_i$ is proportional to the total concentration. Internal or external calibrations are only necessary for the detector response.

10.5 Purge and Trap Method

The E-HSGC-methods, described above, are so-called static methods, applicable only if there is a sufficient amount of substance in the headspace. However, if the concentration of volatile components is too low, or if the compounds are less volatile, a dynamic headspace analysis (D-HSGC) is applied, the so-called "Purge and Trap Method".

The principle of this method is a continuous gas extraction of the volatile compounds (purge). The entrained components are trapped either by adsorption to solid adsorbents as Tenax, XAD, activated carbon, or by cold trapping. An enrichment effect up to factor 30-50 can be obtained. By quick heating of these traps, the compounds are transported into the gaschromatographic system [9-14].

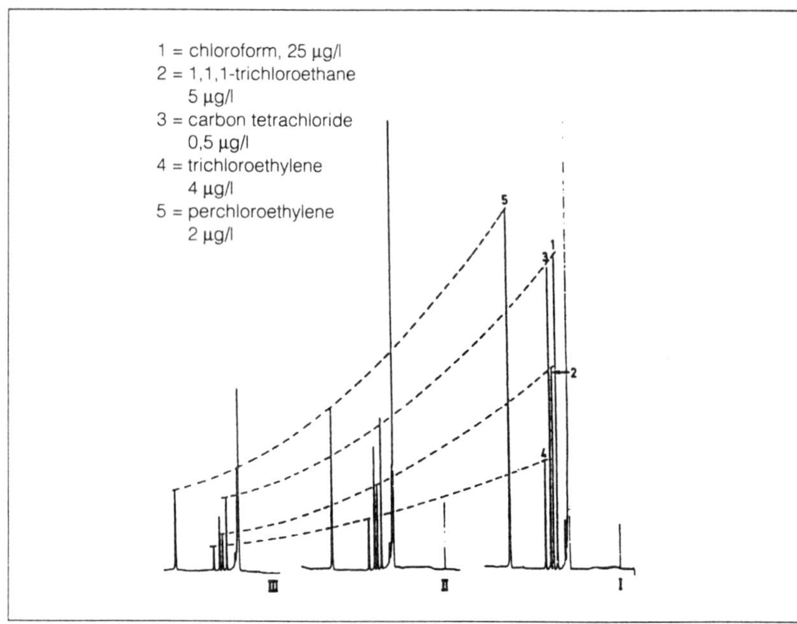

Fig. 4 Multiple Headspace Extraction (MHE) of Chlorinated Hydrocarbons from Aqeous Solution
(Source: Kolb, Auer, Pospisil, 1983, p. 450 [4])

For liquid samples, glass vessels are used, which are open on the top and closed at the bottom by a frit. A carrier gas is passed through the frit, extracting the volatile components.

Automatization of the D-HSGC method is more complicated and expensive than for E-HSGC.

10.6 Application

Originally, headspace chromatography was developed for alcohol determination in blood [15,16]. Since then, this analytical method is applied in several areas, wherever volatile compounds are to be determined:

A.   Aroma analysis
B.   Analysis of residues of technological auxiliaries

C. Water analysis
D. Production control

From the view of food production and food analysis a broad range of applications has opened. Especially aroma analysis is predestined for this methodology. Here, substances can be analyzed which, due to their volatility, arouse our interest because of their sensory susceptibility.

For quality control of raw material (e.g., spices) as well as for evaluation of a finished product, aroma characterization is important. With exact knowledge of the correlation of gas phase composition and sensory results the personnel intensive testing can somewhat be reduced in the future, especially because automatization can also be applied.
For that, identification of all individual components is not absolutely necessary. Comparison of chromatographic fingerprints will be sufficient in most cases for recognition of quality deviations. This includes the advantage of abandoning the use of internal and external standards.

Figure 5 shows an example of the correlation between sensory and analytical results, the chromatograms of two peppermint samples that were tested sensorically parallel by an expert committee. Sample A was evaluated as bad, sample B as very good. Analogous to this result, the quantitative compositions of the gaseous phases are distinguishable, but not the qualitative one. Notable in this example are the different ratios for alpha- and beta-pinene as well as for menthone and menthol.

Volatile compounds can contribute to a good as well as to an off-flavor. Such unwanted aroma deviations can be found in almost all foods. Examples may range from normal aroma off-notes as phenolic off-notes in beer, varnish- and cork flavor in wine, up to odor detectable deterioration in meat and cold cuts, fish, fruits and vegetables. The responsible compounds are components of the gaseous phase, and therefore analytically detectable by HSGC. For objective evaluation of rancidity of fats, is has been proposed to determine either the total aldehyde content or hexanal as the characteristic substance by HSGC [18]. Hollingworth et al. [19] found that sensory deterioration detection of smoked salmon also resulted in a higher ethanol concentration in the gaseous phase, and therefore is applicable as deterioration indicator.

However, it should be taken into consideration that some major contributors to an aroma are not always sufficiently concentrated in the headspace due to their very low odor threshold value. Some sulphurous substances, for example, belong to that group. In such cases enrichment methods with subsequent classical gaschromatography or sensory tests are indispensable.

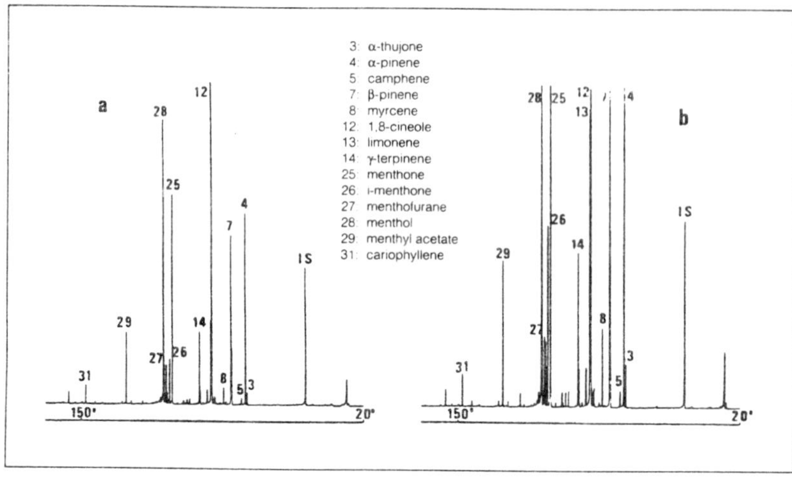

Fig. 5 Headspace Chromatograms of Two Peppermint Extracts
(Source: Chialva, Doglia, Gabri, Ulian, 1983, p. 436 [17])

The main portion of those applications mentioned in point B is the quantitative determination of solvents in foods, i.e. technological auxiliaries in food production or processing. Typical examples are residues of 1,2-dichlorethylene in coffee powder [20] or perchlorethylene in chicken eggs from fat-extracted chicken feed [21].

Figure 6 shows the analytical example of a residue content of 1,2-dichlorethylene in soluble coffee. By three-stage multiple headspace extraction, 73 ppm of chlorinated hydrocarbons were detected. For samples containing a certain amount of residual water, as in this example, addition of a sufficient quantity of water is recommended, because in each analysis part of the water is lost, resulting in a change in the adsorption behaviour of the sample in regard to the solvent to be determined. The added water as a displacer has the task of keeping the sample matrix constant during the MHE.

In addition to the analysis of residues of technological auxiliaries, the detection of monomers migrating from plastics into foods is important. Figure 7 shows the peak of 0.1 ppm monomer vinylchloride from a sample of edible oil [2]. Other monomers as styrene, benzene, toluene, etc. are detectable in the gaseous phase of polystyrene [22] and can migrate especially into foods with a high fat content.

For water analysis (C), quantification of chlorinated hydrocarbons is important. MHE is preferred in this case, and for extremely low concentrations the purge and trap method can be applied. Examples are shown in Figures 2 and 4.

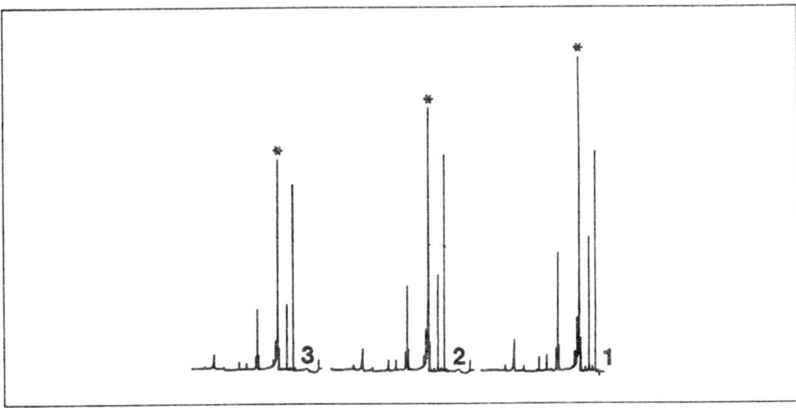

Fig. 6 Determination of 73 ppm trans-1,2-dichloroethylene in Coffee Powder by MHE
(Source: Kolb, 1985, p. 14, [20])

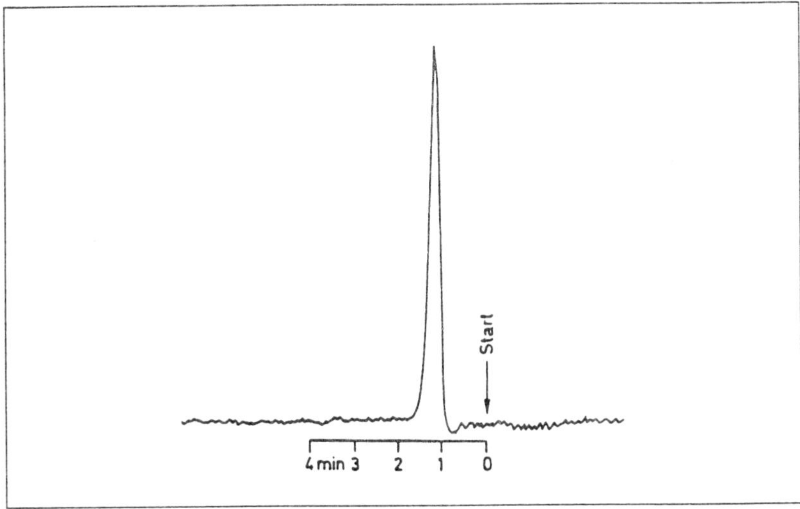

Fig. 7 Determination of 0.1 ppm VC in Edible Oil (Test Mixture)
(Source: Hachenberg, 1975, p. 4 [2])

In the application range mentioned above, under D, those technological production processes are worth mentioning, where volatile intermediates or final products show the respective phase (state) during the course of processing. Comberbach et al. [23] for example developed a method allowing gaseous samples to be taken at any given time during a fermentation process. The samples are fed on-line to a gaschromatographic separation. By determination of the acetone/butanol ratio in the gaseous phase, the course of fermentation can be followed. Similar procedures are possible, in principle, for other processing methods, e.g. for cheese ripening.

10.7 Conclusion

Depending on matrix and substance concentration, quantitative determinations of volatile compounds by different techniques of HSGC are possible, using either the static methods of equilibration headspace chromatography (E-HSGC), the technique of multiple headspace extraction (MHE), or the dynamic process (D-HSGC) of the purge and trap method. In comparison to the classical gaschromatography the most important advantage is the avoidance of isolation and clean-up procedures. Especially for solid substances and samples with inhomogenous matrix, quantitative results are obtained relatively easy by multiple headspace extraction. This method belongs to the group of the rapid methods because sampling can be done automatically in on-line operation with gaschromatographic separation. Therefore, HSGC is not only interesting for quality control but for control of technological production processes, too.

This methodology has excellent prospects to move into the analytical range of aroma quality control, even if only by fingerprint chromatogram, as soon as aroma quality standards are determined and defined. In this case, a technique would then be available for objectively evaluating aromas and off-flavours independently of sensory tests.

REFERENCES

1. WYLLIE, S. G., ALVES, 5., FILSOOF, M., JENNINGS, W. G. (1978) in: Analysis of Food and Beverages (Charalambous G., Ed.), New York, Academic Press Inc., P. 1

2. HACHENBERG, H. (1975): Applied Chromatography No. 25, Überlingen, Bodenseewerk Perkin-Elmer & Co. GmbH

3. KOLB, B. (1985): Applied Chromatography No. 44, Überlingen, Bodenseewerk Perkin-Elmer & Co. GmbH

4. KOLB, B., AUER, M., POSPISIL, P. (1983) in: Proceedings of the 5th International Symposium on Capillary Gaschromatography, Riva del Garda, Italy (J. Rijks, Ed.), Elsevier Scientific Publication Company, Amsterdam, 441

5. KOLB, B. (1982): Chromatographia **15**, 587

6. ETTRE, L. S., KOLB, B., HURT, S. G. (1983): Am. Lab. **10**, 76

7. ETTRE, L. S., JONES, E., TODD, B. S. (1984): Chromatogr. Newslett. **12**, (1), 1

8. KOLB, B. (1985) in: Essential oils and aromatic plants, (A. Baerheim Svendsen and J. J. C. Scheffer, Eds.), Martinus Nijhoff/Dr. W. Junk Publishers, Dordrecht, Netherlands

9. VENEMA, A. (1986) in: Proceedings of the 7th International Symposium on Capillary Gaschromatography, Nagara, Japan (D. Ishii, K. Jinno, P. Sandra, Eds.), The University of Nagoya Press, 92

10. NOY, T., VAN ES, A., CRAMERS, C., DOOPER, R., RIJKS, J. (1986) in: Proceedings of the 7th International Symposium on Capillary Gaschromatography, Nagara, Japan (D. Ishii, K. Jinno, P. Sandra, Eds.), The University of Nagoya Press, 64

11. ONDA, N., KOLB, B., AUER, M., SHIRAI, F. (1986) in: Proceedings of the 7th International Symposium on Capillary Gaschromatography, Nagara, Japan (D. Ishii, K. Jinno, P. Sandra, Eds.), The University of Nagoya Press, 110

12. ISHII, D., ITO, Y., WATANABE, S. (1986) in: Proceedings of the 7th International Symposium on Capillary Gaschromatography, Nagara, Japan (D. Ishii, K. Jinno, P. Sandra Eas.), The University of Nagoya Press, 129

13. WYLIE, P. L. (1986): Chromatographia **21**, 251

14. KOLB, B.. LIEBHARDT, B. (1986): Chromatographia **21**, 305

15. MACHATA, G.: Microchim. Acta (1968) **262**, 2

16. MACHATA, G.: Blutalkohol (1967) **4**, 252

17. CHIALVA, F., DOGLIA, G., GABRI, G., ULIAN, F. (1983) in: Proceedings of the 5th International Symposium on Capillary Gaschromatography, Riva del Garda, Italy (J. Rijks, Ed.), Elsevier Scientific Publication Company, Amsterdam, 430

18. DIRINCK, P. J., KUNTOM, A. H., SCHAMP, N. (1984) in: Progress in Flavour Research (J. Adda, Ed.), Proceedings of the 4th Weurman Flavour Research Symposium, Dourdan, France, Elsevier Science Publishers B. V., Amsterdam (1985)

19. HOLLINGWORTH jr., T. A., THROM, H. R. (1983): J. Food Sci. **48** A, 290

20. KOLB, B. (1985) in: Essential Oils and Aromatic Plants (A. Baerheim Svendsen, J. J. C. Scheffer, Eds.), Martinus Nijhoff/Dr. W. Junk Publishers, Dordrecht, The Netherlands

21. KOLB, B., AUER, M. (1981): Lebensmittelchemie und gerichtl. Chem. **35**, 92

22. GALIARDI, P., VERGA, G. R. (1983) in: Proceedings of the 5th International Symposium on Capillary Gaschromatography, Riva del Garda, Italy (J. Rijks, Ed.), Elsevier Scientific Publication Company, Amsterdam 418

23. COMBERBACH, D. M., SCHARER, J. M., MOO-YOUNG, M. (1984): Chromatogr. Newslett. **12**, (1), 4

Application of Infra-Red Spectroscopic Methods

L. Rudzik

11 Application of Infra-Red Spectroscopic Methods

L. Rudzik, Hannover

11.1 Introduction

In 1800, infra-red radiation was discovered by Sir William Herschel. But it almost took one century before the first systematic IR-spectra were taken by W.W. Coblentz in 1905. Another half century passed for development of spectrophotometers that can be handled under common laboratory conditions as a matter of routine. Apart from the energy dispersive IR units, employed up to now, equipment was introduced in the last years (since 1963), in which the infra-red spectrum is produced by interferometry and subsequent Fourier transformation (FT-IR instruments). Both types are sample adjusted and advantageous for measurements in the food industry.

In the last ten years, the IR spectroscopy has made enormous progress due to micro-electronics and electronic data processing for spectra manipulation, evaluation and identification. It can now be employed for qualitative as well as quantitative analysis in the food industry, too. Since infra-red analysis can be performed without or with little sample preparation and the IR spectrophotometers operate very quickly, the IR analysis is especially suitable for rapid methods without drawbacks regarding analysis accuracy (as with many other rapid methods).

The range of the infra-red (IR) radiation is located in the total spectrum of the electro-magnetic radiation between the micro wave radiation and the visible light. Under certain conditions the IR radiation can excite molecules. This phenomenon is based on the fact that the distances of the atoms in the molecules are not fixed but assume equilibrium positions allowing vibrations with stretching or deforming of the individual bonds.

Prerequisites for an IR excitation or IR adsorption are the correspondence between the frequency of the radiation supplied and the natural vibration of the molecule, and that the molecular vibrations are related to a change of the electrical dipole moment. If a sample of a certain substance is radiated with a continuous spectrum of IR light, defined absorption bands (IR spectrum), characteristical for the structure elements of the molecule, are obtained at resonance.

For infra-red spectroscopy three of wave number ranges or wavelength ranges are distinguished, depending on the kind of excitation vibrations:

Far IR-range	(FIR)	$50 - 400\ \mathrm{cm^{-1}} \triangleq$	$200-25\ \mu m$
Medium IR-range	(MIR)	$400- 4000\ \mathrm{cm^{-1}} \triangleq$	$25-2.5\ \mu m$
Near IR-range	(NIR)	$4000-12500\ \mathrm{cm^{-1}} \triangleq$	$2.5-0.8\ \mu m$

While the far infra-red range is not interesting for measurements in food, the MIR as well as the NIR methods, adapted to certain measuring problems and media, have proven very profitable.

For MIR spectroscopy, the measurement usually is made in transmission mode, for NIR spectroscopy reflection techniques are normally employed (NIRS, near infra-red reflection spectroscopy). Both techniques have their specific advantages and disadvantages.

The performance of the infra-red methods is illustrated using the following selected examples.

11.2 Application Examples for MIR Spectroscopy

The testing and development of new technological processes for production of food is valuable and sometimes necessary to gain insight as to the course of the technological process itself. Often the question occurs, which constituent is enriched, remains constant or is reduced. Take the ultrafiltration for example. In an UF plant, slim milk was concentrated. Every 20 min a sample was taken and the range of protein absorption was measured in a dispersive high performance IR unit with computer support. Additionally, for

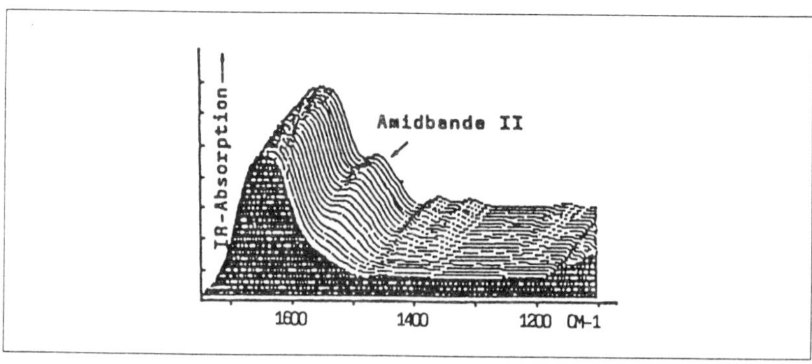

Fig. 1 3-D Illustration of IR Spectra of an Ultrafiltration

each sample the protein content was determined according to Kjeldahl as the reference method as well as with the Büchi rapid technique. The evaluation of the IR spectra was carried out according to the so-called Quest-technique, a predecessor of the Quant technique, commonly used today (both evaluation algorithms are from the Perkin-Elmer Company, Überlingen, FRG).

The 3D-illustration in Figure 1 clearly shows the protein enrichment, especially of the amide band II.

Table 1 compares the measuring values according to Kjeldahl, IR-Quest and a rapid method based on the principle of the Kjeldahl method. The usefulness of the IR method is very clear, especially under consideration of the fact that the IR measuring values are obtained in less than 5 minutes.

Tab. 1 Comparison of Measuring Values of Different Protein Analyses

	Time	Kjeldahl	Quest	Rapid Method*
Start	7.03 am	4.1	4.2	3.3 (?)
	7.20 am	4.3	4.4	4.3
	7.40 am	4.5	4.5	4.5
	8.00 am	4.1	4.2	4.0
	8.20 am	6.6	6.7	6.8
	8.40 am	7.9	7.9	8.1
	9.00 am	9.0	9.0	9.1
	9.20 am	9.9	9.9	10.2
	9.40 am	10.9	11.0	11.6
	10.00 am	11.5	11.7	12.1
	10.20 am	12.2	12.3	12.3
	10.40 am	12.6	12.6	13.1
	11.00 am	13.3	13.1	13.8
	11.20 am	13.8	13.8	14.4
End	11.40 am	14.5	14.5	15.4

* according to the Kjeldahl principle

For determination of constituents in e.g. milk and liquid substrates very effective analysis automates are available. They operate with filters in the MIR range. These routine procedures fail, however, if the measured substrate is not characteristical, if e.g. milk contains more urea than normal. For that case, another method is offered.

For some time now, another method has been employed successfully. Programs for the so-called curve fitting are used. Instead of measuring at certain frequencies, the data from the whole spectra or selected spectra ranges are utilized. A representative amount of standard spectra of known composition is registered and stored in the computer. Of course, at least as many standards are necessary as there are components to be determined. It is better to use more standards, if possible. After measuring the spectra of the sample, the computer using the standards calculates the curve most similar to the measured spectrum. For this calculation different proportions of the standard spectra are added. The highest degree of similarity between the measured spectrum and the synthetical one is stated in form of the least square fit. The sample composition is then calculated from the proportions of the standard spectra and their known concentrations.

11.2.1 Advantages and Disadvantages of the Method

This method shows significant advantages, compared to the commonly used base line methods:

1. All informations contained in the spectrum are utilized.
2. The method is universally employable, that means, it can be used for almost every substrate.
3. Intermolecular interactions play only a secondary role.
4. Even uncharacteristic substrates can be analyzed, assuming, the computer finds a suitable reference spectrum.
5. Measurements can be taken in a wide concentration range.
6. Because the RMS error (root mean square) is supplied as a numerical value, faulty measurements can be recognized immediately (that means, the computer did not find a suitable reference spectrum).
7. Deviations from the Lambert-Beer's law can usually be compensated.

A significant disadvantage compared to the filter units for routine operation, however, is the prolonged time needed for these determinations (scanning time and computer time).

11.2.2 Practical Examples

The IR spectroscopy is especially useful for identification of substrates. There are no two substances with an identical infra-red spectrum. Even if they are similar, they will differ in the so-called finger print area (approx. starting from 1,500 cm^{-1} and decreasing to lower wave numbers). The spectra of α-HCH and γ-HCH (lindane) will be compared here as an example. Both substances are chemically extremely similar. They differ only in the position of the hydrogen atom in the cyclohexane ring. This small distinction in the conformation is, however, clearly visible in the infra-red spectrum.

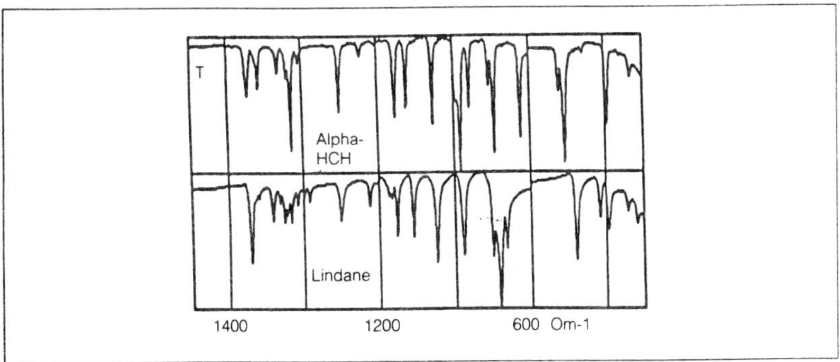

Fig. 2 MIR Spectra of α- and γ- HCH

If IR spectroscopy is used with computer support and a "search-program", the computer can identify an unknown substance in a short time using thousands spectra (if necessary) from a spectra library. For some companies in the food industry it is recommended to compile a library of their own in which all processed substances and auxiliary substances, important for this company, are stored.

Protein maturation can be taken as example for the characterization of a physical chemical process. First, it is necessary to describe the spectra recording technique [5]. Usually, if IR spectra in the medium infra-red range are to be taken from a solid or pasty substance, the analyst will mix the sample thoroughly with potassium bromide in the mortar and using a high pressure press (approx. 10–20 t) produce a transparent glassy sample pellet. For determination of protein this procedure may not be applied. Due to the high salt content of the pressed sample pellet matrix and the applied pressure, the protein structure will change resulting in interpretation errors due to artefacts when evaluating the spectra. It is much more advantageous to use the FMIR technique (frustrated multiple internal reflectance). For that the sample to be measured is distributed e.g. on a zinc selenide crystal (high refraction index) and placed in the radiation path of the IR unit. If the infra-red radiation strikes the boundary of the measuring medium with lower optical density (protein) with an angle of incidence larger than the critical one, the radiation is reflected 12-25 fold with only low intensity loss (depending on angle of incidence). For the wavelengths at which the optically thinner substance absorbs, the refraction index changes due to the abnormal dispersion and with it the attenuation of the radiation. This is as if the radiation penetrates a few Êm into the protein surface. The FMIR spectra are very similar to the conventional transmission spectra, despite the totally different mechanism. The absorption maxima shift for only a few wave numbers, depending on the angle of incidence. The main advantage is that the sample is not influenced or even denaturized prior or during the measurement. The protein is determined in its actual condition.

For infra-red characterization of the maturation course for protein a fresh sample sour curd (pure protein) is pressed into a Petri dish, incubated up to 4 days in the incubator at 30 ± 1 °C, and subsequently measured.

It is obvious that the absorption band intensities of the matured samples increase somewhat. This is easy to understand. In large protein chains the vibrations of atoms or atom groups are hindered sterically. During the maturation process the protein chains are split up into short segments, the sterically hindrance is partly eliminated and the atom groups are able to vibrate more freely. This effect is especially noticeable if e.g. end groups, capable of vibrations, are formed, e.g. by splitting the acid amide bond [as e.g. for the amide band II (1545 cm^{-1}) and the COOH vibration (1400 c^{-1})].

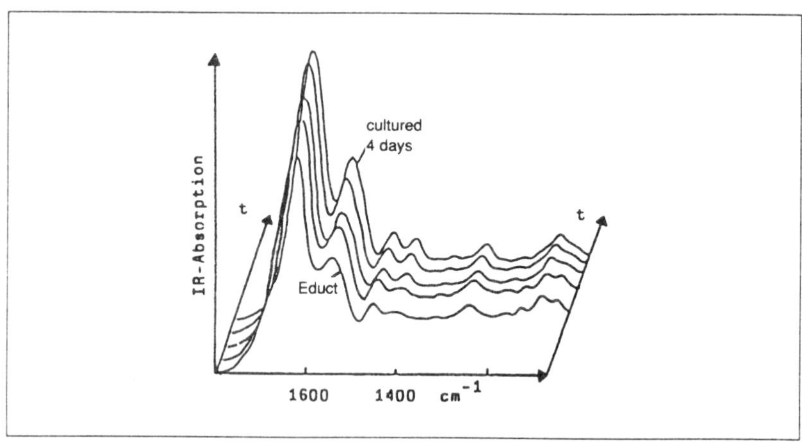

Fig. 3 3-D Illustration of the Maturation Course of a Sour Curd Sample

11.3 NIR Spectroscopy

Up to the 70's, little attention was paid to the NIR spectroscopy (near infrared reflectance spectroscopy) mainly because in the NIR range the overtones and combinations of the fundamental vibrations (MIR range) are measured and the spectra obtained are quite confusing and not very characteristic. Only after cheap and efficient computer systems became available could these spectra be evaluated. At present, the NIR spectroscopy is used especially in food analysis. Practically, there is no substrate that could not be analyzed for constituents with this method.

Especially advantageous for NIR spectroscopy is that no or only little sample preparations (grinding the sample) is necessary for many substrates. Furthermore, if the measuring unit is calibrated, the handling is so easy that assistants can perform the measurements.

Much more information is obtained by the NIR spectra than by MIR spectra, but a computer is needed for decoding this information.

Another advantage is that the data correlation for the NIRS method is very simple because the Lambert-Beer's law is completely obeyed, which means that the measurements taken are completely within the linear range.

Also, very efficient radiation sources and detectors are available. The optical systems are made of glass, making the units easy to maintain and robust. They can be employed "on-line" directly in the plant (at the location of the technological process).

Disadvantageous is, however, that surface effects (e.g. particle size) play a decisive role. Therefore, it is necessary, to standardize the surface. Grainy substances must be polished prior to use.

Result evaluation is only possible by computer, that means that good evaluation algorithms for the quantitative analysis are very important. In many cases, a high calibration effort is necessary for quantitative analysis. In extreme cases more than 100, in some cases up to 300 calibrations (with the corresponding wet chemical reference analyses) are necessary.

However, it can be stated today, that the quantitative analysis with NIRS methods supplies outstanding results for many fields. As to further successful industrial applications this is only the "tip of the iceberg".

In the past it was said that the NIR spectra cannot be or only partly be used for qualitative determinations due to their non-characteristic curves (overtones or combination vibrations). Today, this statement should be partly revised. Of course, it is true that less specific and less significant absorption bands are present in the NIR spectra. On the other hand, however, the information contained in these spectra is very useful. Through the use of special computer programs, qualitative determinations are possible with NIRS. Here the so-called discriminantes analysis is employed. Mark and Tunnel have described the theory of the discriminantes analysis for the Mahalanobis distances [6] extensively and vividly [7,8].

It has become possible to evaluate qualitatively even technological or sensoric properties such as the baking quality of wheat, the digestibility of hay or the "earthy" flavor in peas.

11.4 Summary

The NIR as well as the MIR spectroscopy can usefully be employed for quality assurance in the evaluation of food and raw materials as exceptionally rapid methods. But they are not, as sometimes said, competing rather than supplementing each other very well depending on the problem. However, since the application of infra-red spectroscopic methods in the food laboratory now on the increase, it must be noted, that much more development is necessary to enable these methods to reach their appropriate status in the evaluation of food.

REFERENCES

1. GÜNZLER/BÖCK: IR-Spektroskopie, Verlag Chemie 1975

2. HEDIGER, H.J.: Infrarotspektroskopie, Akadem. Verlagsgesellschaft Frankfurt/M (1971)

3. BELLAMY, L.J.: The Infrared Spectra of Complex Molecules, Capman and Hall (1975)

4. KENDALL, D. N.: Applied Infrared Spektroscopy, Reinhold Publishing Corporation (1966)

5. RUDZIK, L.: Deutsche Milchwirtschaft **35**, 1097 ff (1985)

6. MAHALANOBIS, P.C.: Proc. Natl. Inst. Sci. India **2**, 49-55 (1936)

7. MARK, H. L., TUNNEL, D.: Anal. Chem. **57**, 1449-1456 (1985)

8. MARK, H. L.: Fourth International Symposium on Near Infrared Reflectance Analysis (NIRA)–Technicon GmbH (1983)

9. PAUL, C.: Beiheft der 4. Dortmunder DASp-Diskussionstagung (9.11. Sept. 1986)

Application of NIR to Analysis of Dairy Products

G. Zaeschmar

12 Application of NIR to Analysis of Dairy Products

G. Zaeschmar, Leer

12.1 Introduction

Application of infra-red measurement techniques to dairy products is based mainly on the highly industrialized dairy industry's requirements for rapid analysis results for quality assurance and process control.

The existing reference methods, which supply, per definitionem, the "right" value (i.e. the officially acknowledged analysis result), often require too much personnel and time.

Until about 1960, the NIR spectroscopy was only applied by experts for very specific tasks. For example, in this IR range chemists examined low level transfers subjected to the intercombination ban. Later K. Norris and J.R. Hart began to investigate the usefulness of IR filter equipment for analyses of agricultural products (especially grain). Two important preconditions for successful NIR-analysis were called to mind, the reflectance technique (for examining solid state products) and the modern computer support. Later on NIR-reflectance spectroscopy was tested for measurements in milk and dairy products.

12.2 Discussion of the Method

As with all analytical methods, the NIR-spectroscopy has advantages as well as disadvantages.

The quantitative NIR-spectroscopy is not suitable under two preconditions:

- if the mass concentrations are low
- if only low sample amounts are available.

The fact that the absorption bands become weaker towards shorter wave lengths is disadvantageous. The intensities of the NIR-bands are about 10- to 100-times weaker than MIR-bands.

Additionally, the surface quality has a strong influence on the reflectance measurement, but surface standardizations are very difficult.

The same wave range, which is preferentially absorbed by diffuse reflection, is also very efficiently reflected. This decreases the contrast in the spectra and smooths them to a large extent.

Due to the many possible stimulation of vibrations in the NIR-range and the often occurring band overlay, a spectra interpretation often is impossible or at least very difficult.

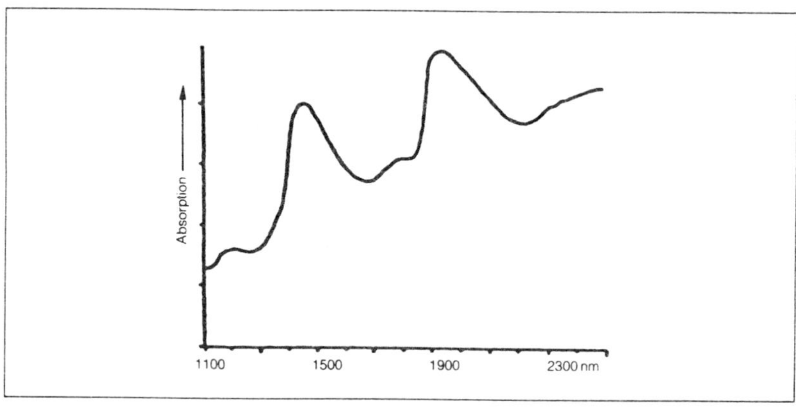

Fig. 1 NIR-Reflectance Spectrum of Milk (3.5% Fat)
(Infra Analyser, Technicon)

In the NIR-reflectance spectrum of milk for example (Fig. 1) only the two water absorptions at 1940 nm (= 2. overtone of OH-bending vibrations) and 1450 nm (= 1. overtone of OH-stretching vibrations) are visible. The milk constituents themselves can be seen faintly in the background. It is difficult to believe that the good quantitative results can be calculated from those not very clear spectra. But then, the advantages of the NIR-spectroscopy should be recalled:

The radiation source is about ten times stronger than the ones applicable for MIR. The detector is much more sensitive. The shorter wave lengths have higher dispersion coefficients, which lead to a better linearity when applying reflectance techniques. The Lambert-Beer's law (log $L_0/l = E \cdot c \cdot d$) is valid without limitation at reasonable concentrations. Furthermore, the measuring range is larger than for MIR. That means, an evaluation is possible on several locations in the spectrum. The amount of information obtained by NIR is also greater. This is the reason why it was possible to draw up such "exotic" applications such as measuring the pungency of pepper or the octane number of gasoline. The water bands do not interfere and can be measured directly (the first NIR units were designed for determinations of water content). Due to the low intensity of the NIR bands, cuvettes with larger thickness can be used to measure liquid media. This makes cleaning, maintenance and dimensioning more simple. Contamination becomes less possible. Finally, the signal to background noise ratio is better than in the MIR-range.

NIR-spectrophotometers used in the dairy industry must be calibrated using the results from the standard methods. Therefore, it is of primary importance to do many exact standard examinations.

On the other hand, the proper evaluation algorithm is of significant importance. Only if good computer programs are available correct values will be obtained. This is a wide field of activity for the different applications, because there is no program which could not be improved.

NIR and MIR are not, as sometimes believed, competing but complementing techniques, depending on the type of problem. The dairy industry can only gain from the availability of different techniques and different equipment. The development of MIR- and NIR-technique is far from being concluded.

12.3 Measurement Principles

Each constituent to be determined, such as protein. fat, water, etc. has a typical absorption and with that a reflectance spectrum for light in the near IR-range. Those reflectance spectra assure the optimal quantitative determination of the relevant constituent. IR-light of a defined wavelength is directed onto the test sample and the intensity of the diffused reflected light is measured. The different signals are evaluated, supported by a computer.

12.4 Operation

The operation and indication display is dust protected. With this display the constituents to be determined are selected and the results are indicated.

12.5 Function

The ground or homogenized test sample is put into open or closed sample cups, depending on the type of product. These cups are transported on a sample tray into the light path. The IR-light is reflected at the sample surface. A gold-coated spheric reflector, which includes two PbS detectors, is located directly above the sample surface. With this reflector the reflected light is lead with optimal yield to the detector. The spheric reflector also serves as internal standard in which the reflectance of the golden surface is measured for reference. The Technicon's Infra Analyzer 400 D (IA 400) operates in this way. It is available with maximal 19 filters or wave lengths, respectively, of which about six of them should be selected for computer support.

12.6 Calibration

The milk composition changes during the year due to the changing feeding conditions and the lactation-status of the milk cows. A representative selection of milk samples over one year is necessary. This milk is examined for fat and water content. The values are entered into the computer connected to the IA-400. These samples are then measured according to the regulations with the IA 400. Finally, the computers checks for correlation between

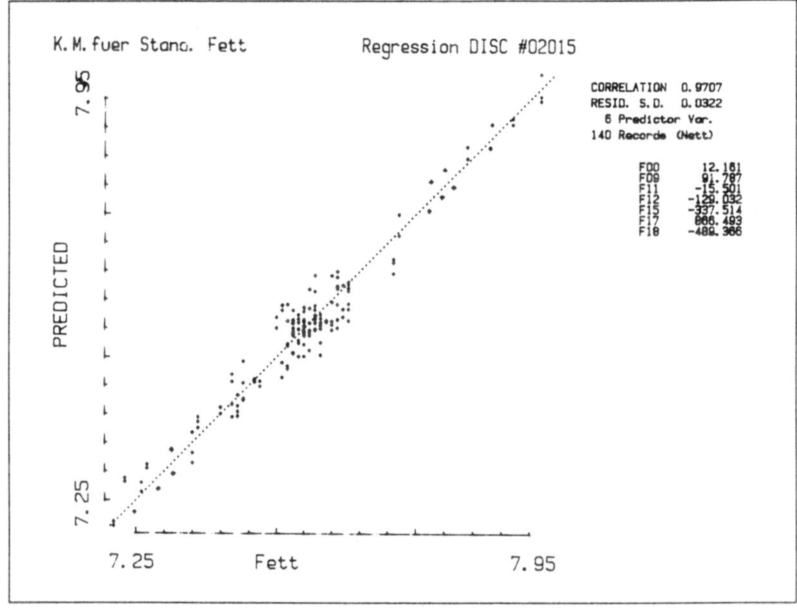

Fig. 2 Regression Line for Fat Determination in Condensed Milk

the entered results of the reference methods and the IA-400 data. Unsuitable sample data are eliminated. The software "best-wafe-program" allows the number of filters to be used to be reduced to six to eight.

These methods are based on statistical calculations using correlation coefficients, T-tests, etc. as decision criteria.

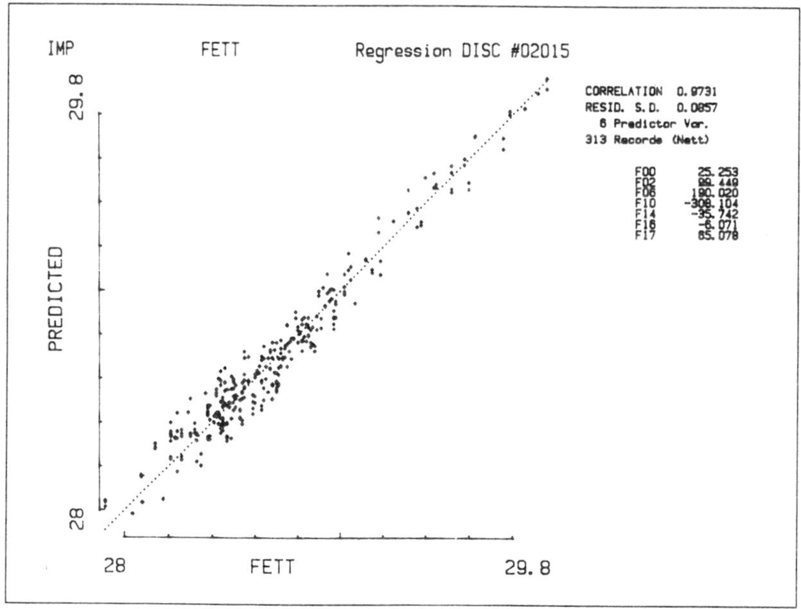

Fig. 3 Regression Line for Fat Determination in Whole Milk Powder

12.7 Measuring Liquids

For measuring liquid milk products another kind of "sample drawer" is necessary. This one contains a circulatory measuring cell in which the temperature can be controlled. Sample homogenization is necessary. This is done in an additional unit that also has the task, together with the IA-400, of rinsing the measuring cell.

In Figures 2 and 3 regression lines for fat determination in condensed milk (KM) and whole milk powder (IMP) are shown.

Starting from an "universal calibration" a considerable number of reference values is necessary to select a filter suitable for the special characteristics of the individual products.

Not only the local conditions are important (e.g., composition of raw milk) but also the technological conditions, i.e. it might be necessary to calibrate each of several spray towers in a dairy plant.

Assuming a well equipped autocontrol system, NIR-measurement of milk constituents can be an effective instrument for quality assurance. Reproducibility of results is better than for most collaborative trials with the reference method.

The repeatability is good. For some products the "accuracy" is often better than for the corresponding reference methods because compensation of certain disturbing factors is better with the NIR-method.

It is interesting to note that the problems of many reference methods first became apparent during the comparison examinations.

REFERENCES

1. RUDZIK, L, MIR- NIR Ein Methodenvergleich, Vortragsscript Vortrag Interlab Tagung 1984

2. Technicon Benutzeranleitung Infra Analyzer 400, Technicon GmbH, Bad Vilbel 1984

3. Vortragsscripte Seminar "Moderne Analysenmethoden", Milchwirtschaftliche Lehr- und Untersuchungsanstalt, Oldenburg, 19.6.1986

4. Vortragsscripte 2. Ahlemer Infrarot Seminar, Milchwirtschaftliche Lehr- und Untersuchungsanstalt, Hannover

Low Resolution NMR

P. J. Barker

13   Low Resolution NMR

P. J. Barker, Melbourne

13.1 Introduction

Since the very first Nuclear Magnetic Resonance experiments were performed in the late 1940s, the technique of NMR spectroscopy has become one of the most important analytical tools in research areas from solid state physics to biology and medicine. When the research chemist thinks of NMR, sophisticated structural analysis of a wide range of materials and substances comes to mind. When the analytical chemist thinks of NMR, he thinks of a technique of high sophistication, not ideally suited to performing routine analyses in a busy analytical laboratory with a high sample throughput. The cost of such instruments both in terms of the hardware and specialist personnel is also sometimes hard to justify. Development of new spectrometers has centered on higher resolution for studying increasingly complex phenomena.

However, there exists a class of NMR spectrometers of in NMR terms low resolution, which have an exceptionally wide range of uses in the analytical laboratory. Such spectrometers are readily at home in the hospital or the oil refinery, in the plastics industry or especially in the food industry. Development of these spectrometers has focussed upon ease of operation by non-specialist laboratory personnel and achievement of a high degree of accuracy and reproducibility of results, yet keeping costs low enough to provide an attractive and cost efficient alternative to other methods.

The idea of low resolution NMR in food industry applications is not new, for example the reviews on the topic, outlining the potential of the technique applied to measurement of moisture in foodstuffs, appeared as early as 1957 [1]. It has been only more recently, however, that the so-called pulsed NMR technique has been developed and applied as a rapid analytical technique, accurate, yet fast enough to be used in process control.

Unfortunately there are few textbooks suitable for the reader interested in low resolution NMR. Most textbooks are written only with high resolution NMR in mind. This is particularly disappointing for those interested in food industry applications because around 50% of papers published on the topic of NMR in the food industry over the past 10 years or so are concerned with low resolution studies. An excellent general review on NMR spectroscopy applied to the analysis of foods and beverages appeared in 1984 [2], but even so this is mainly directed to high resolution techniques rather than those suited to routine analyses.

As an example of the increasing recognition of low resolution NMR in the food industry, the American Oil Chemists Society held a short course titled "Applications of Low Resolution Pulsed NMR Techniques to Food Analysis" prior to the Phoenix National Meeting in May 1988. In addition, the American Chemical Society held a special symposium "Food and Agricultural Applications of NMR" in September 1988 in which low resolution studies also played a major role.

This article attempts to introduce the reader to the idea of NMR in the analytical laboratory without exhaustive theoretical discussion. A typical set of experiments is described for a usual type of analysis, but the most important section of the article is the summary of applications shown in Table 4 which is focussed upon time of measurement, sample preparation and accuracy of results.

13.2 NMR Methods – Continuous Wave (CW) or Pulsed NMR

There are two types of analytical NMR instruments commonly used in the food industry today. One type of instrument is based on the "continuous wave" or "CW" method of measurement, and the other type of instrument uses the "pulsed" NMR method. Both types of instrument have been proven with regard to accuracy and reproducibility in many laboratories over the past 15 or so years. The underlying principle of both types of measurement have been described in many books, reviews and articles [3] and there is no need to go further here. However, it is necessary to stress the differences between the two methods as only the pulsed NMR method is suitable for "rapid" determinations which is the topic of interest here.

The development of both types of analytical low-resolution instruments has been parallel to the development of high resolution instruments. Since the discovery of the NMR technique, the possibility for both types of measurement was recognized. In the early years however, the CW technique was more widely utilized because of difficulties in development of the hardware necessary for the pulsed experiment, and thus analytical CW instruments were available by the mid- to late 1950s. Starting in the mid 60s, when the first commercial pulsed instruments appeared on the market, the change-over to pulsed NMR among high resolution users is now almost complete and the advantages offered by the analytical pulsed instrument have ensured that there has been a steady changeover in the low resolution field in recent years.

13.2.1 What is NMR?

NMR spectrometers of the type under discussion function by measuring the quantity of hydrogen atoms in a sample of interest. In the food industry, the hydrogen atoms are most commonly associated with a fat (oil) phase, or with a moisture (water) phase in a processed or raw foodstuff. The nuclei of these hydrogen atoms behave like small bar magnets, and when the sample is placed in a magnetic field they try to align themselves parallel to the field thereby inducing a net magnetization in the sample. The orientation of nuclei may be changed by applying radiofrequency radiation to the sample, and detection of the energy absorbed by the sample leads to the NMR signal. The basic requirement of a permanent magnet, with a space between the pole caps to hold a sample is common to both measurement methods, but instrumental realization of the excitation and detection forms the difference between the two techniques, and leads to a large difference in the time scale of the experiment.

13.2.2 Continuous Wave (CW) NMR

The continuous wave (CW) technique is a field swept technique; as the magnetic field is swept, the nuclei are selectively excited by the constant radiofrequency radiation, and a frequent domain absorption spectrum is obtained. The area under the absorption curve is proportional to the number of hydrogen nuclei present. Unfortunately, the line widths of the absorption lines depend upon many factors, for example absorption lines from solid phase nuclei are very broad and difficult to isolate, lines from liquid samples are broadened by paramagnetic impurities, and more importantly when two liquid phases are present (water and oil) it is difficult to measure both. Normally, in this latter case the sample must be dried for several hours, or overnight, in an oven, adding considerably to the measuring time. Thus, although in general the results are excellent in terms of accuracy and reproducibility, for rapid measurements the CW technique is not optimal.

13.2.3 Pulsed NMR

In the pulsed NMR technique [4], the nuclei are excited by an intense pulse of radiofrequency (r.f.) radiation, lasting only a few microseconds. All the hydrogen nuclei in all phases are excited by this pulse. When the pulse is switched off, the nuclei return to their original state, emitting an NMR signal as they do so. The detected signal has a maximum intensity when all the nuclei have been rotated 90 degrees with respect to the direction of the static magnetic field. The duration of the r.f.pulse is adjusted to give this condition and is thus termed "90 degree pulse". The time evolution of the NMR signal after a single 90 degree pulse contains most of the important information for analytical purposes (see Figure 2) and it is characterized by the following properties:

The initial amplitude of the signal is proportional to the total number of hydrogen nuclei in the sample.

The signals due to nuclei in different chemical or physical environments decay at different rates, which means that for a sample with hydrogen nuclei in more than one phase, the observed signal is a superposition of more than one component.

Each component of the signal decays with a characteristic time constant, the so-called "relaxation time", thus signals due to hydrogen nuclei in solid phases decay far more rapidly than those in liquids.

In some cases further information may be obtained after the application of further radiofrequency pulses in a pre-determined sequence (a "pulse sequence"). In all examples, by measuring the NMR signal at appropriate time intervals after the r.f.pulse, all the information necessary to perform the desired analysis is acquired. This analysis is normally performed internally by the microprocessor of the instrument.

As an example of the time scale of the pulsed experiment, we can state that the relaxation time of the cocoa butter content of cocoa powder is around 100 msec. or 0.10 seconds, this means that most of the hydrogen nuclei have returned to their aligned state in less than 1 second after the initial radiofrequency pulse, i.e. the NMR signal has decayed to zero and the experiment is effectively over in this time. Of course, in practice we can improve the data by accumulating more than one scan, but even so it is seldom that more than 30 seconds are needed for an analysis.

13.2.4 A Typical Pulsed NMR Spectrometer

Figure 1 shows a low resolution pulsed NMR spectrometer of the type that may be used in the chocolate industry for example.

The two boxes house the electronics of the instrument, with the controls for operator input on the right, and the permanent magnet required for the ex-

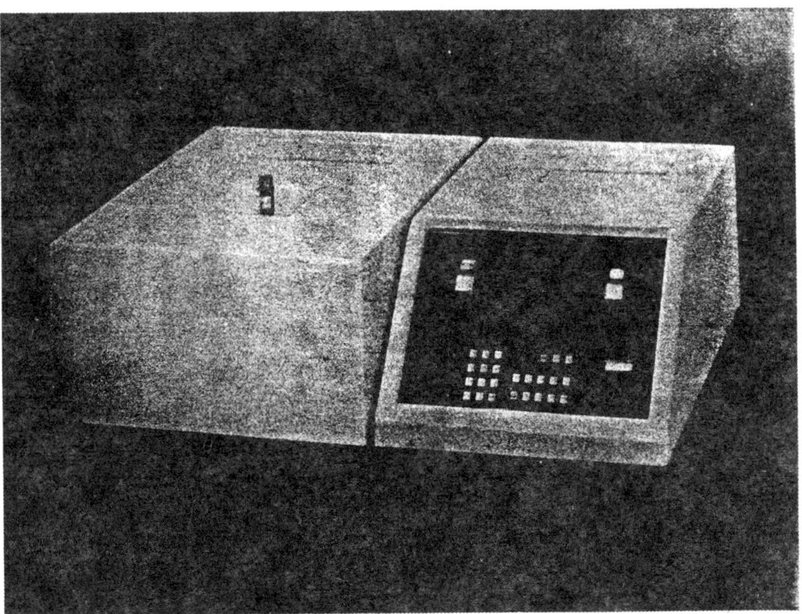

Fig. 1 A Typical Low Resolution Pulsed NMR Spectrometer – The Bruker 'minispec pc 120'

periment with the space for the sample tube on the left. The magnet box is rather heavy (60 to 80 kg) but may rest on a normal laboratory bench or table without difficulty. The measurement commences upon sample insertion. The front panel display shows the result after calculation and also gives visual assistance for the input of parameters. Results may also be printed out on the printer to the right of the keyboard. An rs232 interface is provided so that the data may be sent to a personal or mainframe computer for storage or further manipulation. For use in potentially hazardous areas, the magnet box may be situated up to 30 meters from the electric box.

13.3 Experimental Considerations

13.3.1 Sample Size

The sample size used in these instruments depends upon the application for which the instrument is to be used. This is because the measurement depends somewhat on the way in which the protons to be measured are distributed through the sample. Tubes may be from 7.5 mm to 40 mm in diameter and the measuring area is 30 mm deep. For some measurements, for instance the measurement of oil in sunflower seeds or cocoa beans, the largest size of probehead is optimal, but for measurements of margarine or chocolate powder a 10 mm or 13 mm probehead is sufficiently large.

13.3.2 Different Types of Measurement

For an instrument such as the one described above, the applications may be classified in one of three ways.

1. This type of instrument was originally developed for use in the edible fats industry and particularly for the measurement of the "Solid Fat Index" (SFI). Such a measurement is termed a "ratio" measurement and is a weight independent measurement. Essentially such an experiment measures the NMR signal at two different points after a single '90 degree' r.f.pulse (see Figure 2), one (S1) in the early part of the decay (proportional to the total amount of fat in both solid and liquid phases) and the other (S2) in the liquid part of the decay (proportional only to the total liquid

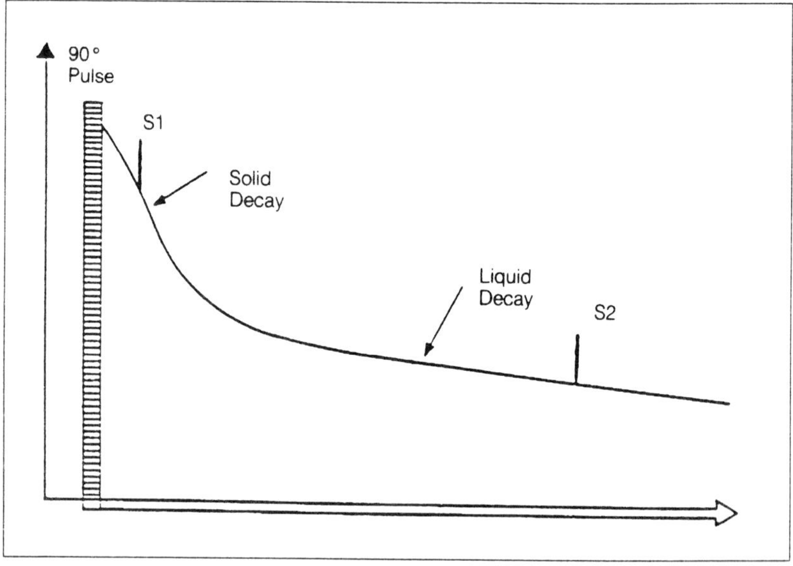

Fig. 2 Decay of the NMR Signal with Time Showing the Solid Part, the Liquid Part and Two Points where the Signal Amplitude is measured

fat content). The ratio of these signals is therefore directly proportional to the **real** solid/liquid ratio, a property of fundamental importance in determination of melting properties of fats.

2. It is also clear from Figure 2 that the signal amplitude at S2 may be normalized to the weight of the sample to give an effective "signal per gram" of sample. This observation provides the basis for a different type of measurement, where the absolute amount of oil in a sample of interest may be measured. This type of measurement is termed "absolute" measurement and in this situation, the "signal per weight" may be comparable between samples providing the sample tube is filled to a consistent depth (normally 30 mm). This principle forms a basis for the most rapidly expanding group of applications for pulsed NMR, essentially because of the speed of measurement and the ease of operation. Most of the results discussed here are "absolute" type measurements.

3. The third type of measurement is a little more sophisticated an entails using quite complex pulse sequences to actually measure the "relaxation time" of different types of sample. These relaxation times may be related to a variety of physical properties or may be used to calculate very accurately the initial signal amplitudes due to different components.

13.3.3 The "Spin Echo" Method

In the measurements above applications, where the effect is produced by single pulse, a '90 degree pulse', of radiofrequency radiation upon a group of hydrogen atoms of interest, were discussed. For certain types of absolute measurement it is optionally to use the so-called "spin-echo" method [4] to make an accurate analysis. Here, more than one radiofrequency pulse is applied to the sample to generate the NMR signal to be measured.

The sequence of events is outlined in Figure 3. At a time T (normally a few milliseconds) after the initial pulse another radiofrequency pulse is applied, a '180 degree pulse', and at time 2T the 'spin-echo' NMR signal appears.

In the normal way for 'absolute' type measurements described above, the signal amplitude (S1) may be normalized to the weight of the sample being measured. There· are two cases where the spin-echo technique is most commonly applied in the food industry. Firstly, in cases where paramagnetic impurities are present in samples, the shape of the echo signal changes but not the amplitude. This means that samples such as chocolate, which sometimes contain small quantities of iron, can be measured without difficulty.

The second case is a measurement of oil containing seeds. Here, when water is present in quantities up to 14 percent, the spin-echo is measured at 7 or 10 (or even 50) milliseconds, where it is certain that only the oil phase of the seed is measured, and there is no contribution from the water phase. The origin of the spin-echo, and the effect of further pulses on different spin systems are discussed in full in NMR textbooks, and further discussion is not warranted here.

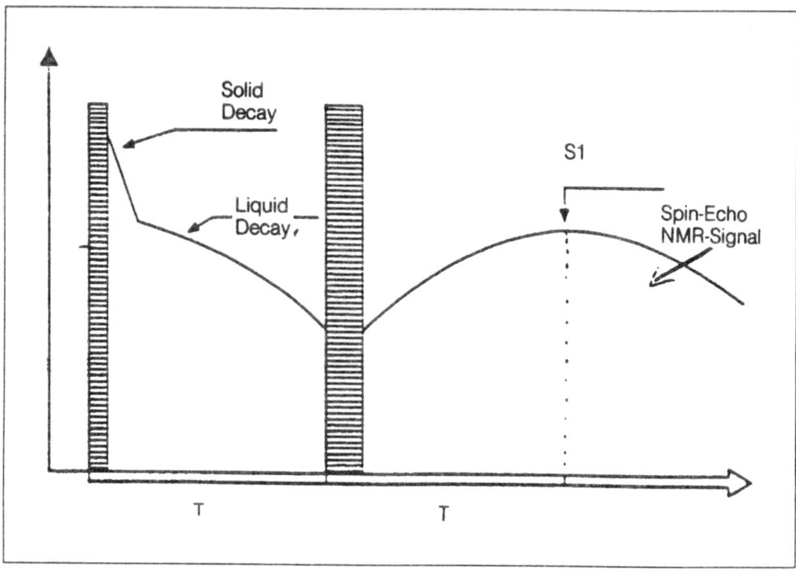

Fig. 3 Representation of the "Spin Echo" NMR Signal, showing the Measuring Point of Oil Content, the Amplitude of which is normalized to the Weight of the Sample

13.3.4 Calibration

In any new analytical application, it is obviously necessary to compare the data with that obtained from the existing standard method for the application. Only when the results for the new method have been proven over several hundred samples can work progress to use the new method with the same confidence as the standard one. Subsequently, new protocols for the analysis using new techniques may be drafted. To the instrumentalist this is often rather hard to understand, because the NMR measurement, for example, is a **direct** measurement proportional to the number of hydrogen atoms in a sample whereas of course many wet chemical analyses are not only indirect, but also have many potential sources of error.

In fact, to compare the NMR method with conventional methods of analysis is extremely useful, not only in terms of the reproducibility and accuracy of the instrument, but also to provide an effective demonstration of the savings in time and increase of sample throughput in the analytical and process control laboratory.

Calibration of the pulsed NMR instrument requires at least 3 samples where the quantity of interest is already known. The quantities should ideally cover the whole range to be encountered in the measurement. Standard samples for the margarine industry to calibrate the instrument for SFI measurements cover the whole range from 0% to 100% solids. In addition, the instruments which are microprocessor controlled are able to store many calibration curves for different products which may be recalled at a single keystroke. Thus, the instrument may be set up to measure different products, even with different measurement methods (see 13.3.2), in a matter of seconds, without further time consuming recalibrations or parameter optimization routines.

13.3.5 Sample Preparation

Preparation of samples for analysis by pulsed NMR varies from application to application. For 'absolute' applications obviously the sample must be weighed while for ratio or relaxation time applications the weight of the sample is not critical.

The major consideration in sample preparation is whether or not the sample must be tempered. Tempering of samples may be necessary for either or both of two reasons: firstly, the NMR signal amplitude is temperature dependent, thus a constant sample temperature will make the results consistent and reproducible; secondly, if an absolute measurement of total oil content is required, then steps must be taken to ensure that all the oil is in a liquid phase. This is illustrated for the measurement of cocoa butter in cocoa powder (Table 1). The difference in the accuracy of the measurement (directly related to the correlation of the calibration data) at 40 and 60 degrees C is due entirely to the fact that at 40 degrees C not all the cocoa butter is in the liquid phase. Therefore, cocoa powder samples are tempered at 60 degrees C for a short while before measurement. Melting curves for edible fats are measured according to a standard procedure involving the tempering of the samples at each of a range of temperatures prior to measurement. On the other hand, in measurements such as the determination of oil in seeds, the temperature is not so critical.

Sample preparation therefore is not complex. Whole grain products such as oilseeds, wheat, etc., can be measured without grinding in most cases, so possible sources of error can be kept to a minimum, and the preparation may proceed without requirement for skilled personnel.

13.4   Some Typical Results for the Food Industry

13.4.1 SFI of Edible Oils and Fats

Sample Preparation: tempering
Measuring time: 50 seconds (IUPAC standard method)

Pulsed low resolution NMR spectrometers were originally developed for the margarine industry as a routine method for SFI determination. While the method is well documented and forms the IUPAC standard method for measurement [5], no review such as this could be complete without mention of this analysis. Since the early 70s, the world demand for margarines has increased dramatically. With this increase the major areas of production have also shifted and will continue to do so. The application of pulsed NMR therefore continues to be an expanding one.

The traditional method of SFI measurement is by dilatometry. This is an indirect method based upon the expansion of melted fat. While the method is quite flexible and suitable for laboratory scale determination, it is time consuming and involves a large number of actions by experienced personnel. The method also involves systematic errors arising from the physical properties of the fats themselves. These factors combine to give a measurement with a standard deviation of as high as 2% solids content absolute. The pulse NMR measurement is carried out in the manner described in section 13.3.2 using a single r.f.pulse, it is a direct method and while tempering procedures are the same as for the dilatometry method the actual measurement only takes a few seconds resulting in a much higher sample throughput. In addition, the standard deviation of the measurement is reduced to 0.3% or better. An objective and realistic comparison between pulsed NMR and classical methods may be found in recent articles by Rossell and Waddington [6]. Successive measurements at different temperatures allow the melting properties of fats to be evaluated, providing further valuable information. Figure 4 shows the melting curves obtained from refined and hydrogenated coconut oils.

The obvious advantages associated with the pulse NMR technique have led to its worldwide adoption in the edible fat industry as the method of choice for SFI measurement.

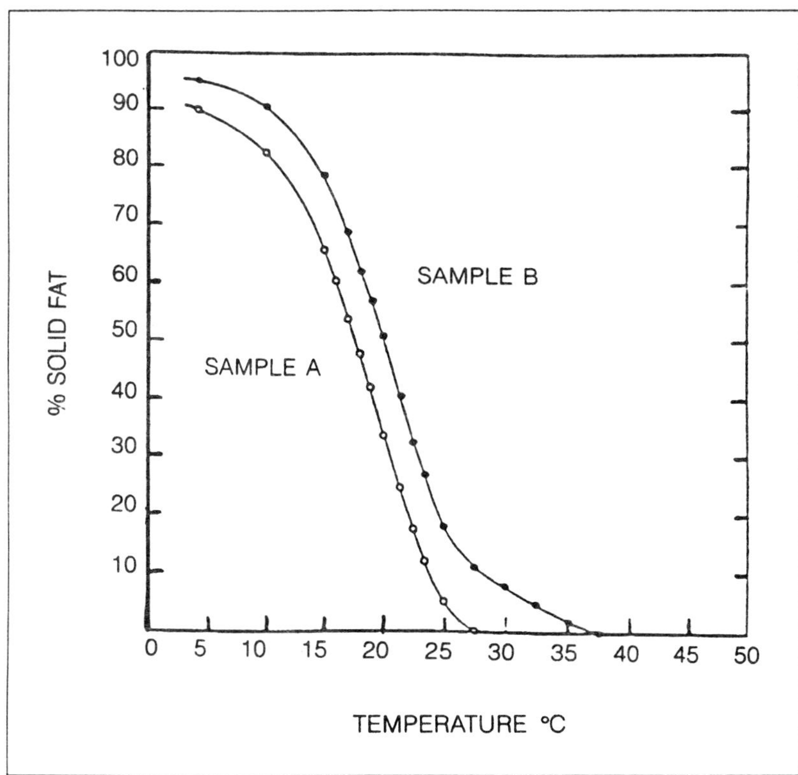

Fig. 4 Melting Curves for Refined (A) and Hydrogenated (B) Coconut Oil plotted after successive SFI Measurements following the IUPAC Experimental Protocol

13.4.2 Measurement of Cocoa Butter in Chocolate Powder

Sample preparation: weighing, tempering
Measurement time: 20-30 seconds per sample

In this measurement, a simple, single spin-echo pulse sequence is used and the amplitude of the spin-echo is normalized to the weight of the sample.

Sample preparation is minimal and involves merely tempering of the sample in an electric heating mantle or aluminium block at 60 degrees C for 10-15 minutes. Once the instrument has been calibrated and the instrumental parameters have been defined, the time of measurement for subsequent samples is of the order of 20 seconds per sample.

Tab. 1 The Effect of Temperature upon Accuracy of Measurement of Cocoa Butter on Chocolate Powders

sample	% fat (soxhlet)	40 Deg C		60 Deg C	
		% NMR	Error	% NMR	Error
1	9,3	10,8	+1,5	9,4	+ 0,1
2	10,8	11,9	+1,1	11,3	+ 0,5
3	15,3	16,4	+1,1	15,3	0,0
4	20,2	20,5	+0,3	20,1	− 0,1
5	21,1	21,4	+0,3	21,1	0,0

Table 1 shows the effect of temperature upon the correlation of the results; the differences are due to the fact that at 40 degrees C not all the cocoa butter is in a liquid state.

13.4.3 Moisture Determination in Maize Gluten

Sample preparation:　　　　　　　　weighing
Measurement time:　　　　　　　　20 seconds

This is a typical measurement for moisture in the food industry. Gluten contains only a small amount of oil and the moisture may be measured by the simple one pulse method with a measuring point 70 milliseconds after the r.f.pulse. Correlation for both coarse and fine gluten is good, and the results are tabulated in Table 2.

13.4.4 General Applications

In a short article such as this lack of space prevents complete description of all the many and varied applications of pulsed NMR in the food industry. We can summarize a cross section of applications, however, in tabular form. The Table 3 not only describes many different measurements but also emphasizes the measurement time which is of fundamental importance to this chapter. Where possible, preparation procedures are given as are the literature references to each measurement.

Tab. 2 Measurement of Moisture in Coarse and Fine Maize Gluten

Sample	Fine ground gluten			Coarse ground gluten		
	% drying	% NMR	Error	% drying	%NMR	Error
1	10,2	9,9	– 0,3	9,9	10,0	+ 0,1
2	9,3	9,2	– 0,1	9,1	9,0	– 0,1
3	10,0	9,8	– 0,2	10,0	10,1	+ 0,1
4	9,2	8,7	– 0,5			
5	9,3	9,3	0,0	9,1	9,3	+ 0,2
6	10,2	10,0	– 0,2	10,4	10,4	0,0
7	9,9	9,7	– 0,2	9,8	9,9	+ 0,1
8	9,3	9,4	+ 0,1	9,6	9,5	– 0,1
9	5,4	5,6	+ 0,2	5,6	5,6	0,0
10	8,9	8,9	0,0	9,3	9,4	+ 0,1

13.4.5 Future Developments

New applications for NMR analysis are being discovered constantly and future developments obviously focus around these discoveries. However, already it is becoming clear that different types of measurement may also be made. In the margarine industry for example there is much interest in the direct, rapid measurement of water content, and also in the water droplet size. Both of these measurements are now possible using low resolution pulsed NMR; the former using a refinement of existing methodology, and the latter using a method common in high resolution NMR but not utilized in low resolution methods until now. This latter method uses the "Pulsed Field Gradient Technique" which requires further major hardware accessories adding around a quarter to the overall system cost, which is comparatively little in comparison with the time and money saved in the long term.

Low resolution pulsed NMR also shows good potential in its adaptability to real **"on-line"** measurements. Initially this will probably be in the measurement of moisture in powders or grains, but may well develop to applicability to all analyses.

There is also considerable interest in the possibility of "absolute" type measurements which may be made independently of the weight of the sample. This would eradicate another possible source of error (weight) and greatly simplify the sample preparation procedures thus still further reducing the overall measurement time and increasing sample throughput. Related to this is the development of applications where more than one measurement is made in a single experiment, such as simultaneous measurement of oil and water in oilseeds.

The possibility of using the pulsed NMR instrument interfaced with a personal computer is also currently attracting much attention, not only in terms of data manipulation and storage, but also to effect complete control of the instrument from the PC.

Recently, the first articles have appeared on the use of low resolution NMR in the frozen foods industry. NMR appears to have a big potential in this major branch of the food industry and early results have been summarized by Weisser and Harz [32].

One of the more significant developments recently has been a low resolution pulsed NMR method for the determination of alcohol in beverages and fermentation musts [33]. Although the theory of the method is based upon principles developed in high resolution NMR, a low resolution spectrometer of the type, depicted in Figure 1, was utilized throughout. The method presents an excellent example of the potential rewards of bridging the gap between the fundamental requirements of the food chemist and the specialist NMR spectroscopist.

13.4.6 Research Applications

In addition to the many and varied applications of pulsed NMR listed in Table 3, use of these instruments is increasing in the research laboratories of major food companies. This is not only to develop new analytical applications 'in House' but also to do fundamental research on the nature of foodstuffs. Probably the most widely investigated area to date is the study of water binding in foodstuffs [31,32]. While the discussion of these applications is beyond the scope of this review, this illustrates further the versatility of low resolution NMR in general, and the pulsed technique in particular, within the food industry.

13.5 Summary

In conclusion, it can be said that low resolution NMR in general is a valuable and accurate analytical tool. In particular, the pulsed NMR method is ideal for rapid measurements, and as the drive towards automation gathers momentum, more and more laboratories will adopt the NMR instrument as standard. The current cost of such an instrument is comparable to a single years' salary for one analytical chemist, yet in terms of overall costs, the savings are considerable in terms of time, money and laboratory facilities when compared with the increase in efficiency. In addition the fundamental aspects of Laboratory Safety, such as the absence of inflammable and noxious solvents which are further advantages of instrumental methods cannot be stressed strongly enough.

Tab. 3 A Summary of Some Applications of Low Resolution Pulsed NMR
Values are quoted only when given in the original reference. This list does not attempt to be complete, just to illustrate a wide variety of products and applications

Product	Measurement	Time of Measurement	Preparation	References
cornstarch	moisture	30 sec	weigh, temper	8,9
Processed starch products: flour, semolina, rice flour, cornflour ground barley, potato starch, ground beans, lentils, gluten, enriched pasta, breadsticks, bread, etc.	moisture	30 sec	not given	9
rice	moisture	4 mins	weigh	10
	+ oil	(both)	weigh	10
brown sugar	moisture	1 min	weigh	11
glucose	moisture	20 sec	weigh	12
sugar beet (pulp)	moisture	1 min	weigh	13
Confectionery products chocolate	moisture	20 sec	weigh, temper	14, 15
	oil	(each)	weigh, temper	
bisquits and cookies	moisture	20 sec	weigh, temper	16
	oil	(each)	weigh, temper	
marzipan	moisture	2 mins	weigh, temper	20
	oil	(both)		

Tab. 3

Product	Measurement	Time of Measurement	Preparation	References
Dairy Products milk powders, skim milk, dried milk, milk shakes, milk drinks, baby foods, etc.	moisture milk fat	20 sec (each)	weigh weigh	17
cottage cheese cheese (dried)	moisture fat	10 sec 20 secs	weigh weigh, temper	18 19
Oilseeds rape, soya sunflower, linseed, poppy mustard, peanuts, almonds	moisture oil	20 sec (each)	weigh weigh	20,21 22, 23 24, 25 26
Other measurements edible oils and fats (margerine etc) olive husk cereal grains	SFI melting prop. iodine number oil moisture	20 sec 20 sec 4 mins 20 sec	temper temper temper weigh	5,7 30 28 27

REFERENCES

1a. CONWAY, T. F., COHEE, R. F. und SMITH, R J.: **Food Eng.,** 1957, **29**, 80

1b. CONWAY, T. F., COHEE, R. F. und SMITH, R. J.: **Mfg. Confectioner,** 1957, **37**, 27

2. HORMAN, I.: in "Analysis of Foods and Beverages", Academic Press Inc., 1984, S. 205–264

3. For a survey on both methods refer to:

   MARTIN, M. L., DELPEUCH, J.-J. und MARTIN, G. J.: "Practical NMR Spectroscopy, Heyden Verlag, 1980

4. FARRER, T. C. und BECKER, E. D.: "Pulse and Fourier Transform NMR—An Introduction to Theory and Methods" Academic Press, 1971

   FUKUSHIMA, E. und ROEDER, S. B. W.: "Experimental Pulse NMR—A Nuts and Bolts Approach", Addison-Wesley, 1981

5. "I.U.P.A.C. Standard Methods for the Analysis of Oils, Fats and Derivatives", 6th edition, Pergamon Press, Oxford, UK, 1st Supplement part 6, 1982

6. ROSSELL, J. B.: in "Analysis of Oils and Fats", R J. Hamilton and J B. Rossell, Eds., Elsevier, 1985, S. 77

   sowie: WADDINGTON, D.: ibid. S. 384

7. LAMBELET, P., DESARZENS, C. und RAEMY, A.: **Lebensmit. Wiss. u. Technol.,** 1986, **19**, 77

8. BROSIO, E., CONTI, F., LINTAS, C. und SYKORA, S.: **J. Food Technol.,** 1978, **13**, 107

9. GAURA, R. M.: **Minispec NMR Applications Note**, IBM Instruments Inc., Danbury, Connecticut, USA

10. SHIH, J M.: **ibid**.

11. DEJOVIN, J. A., SHIH, J. M. und WILSON, J.: **ibid**.

12. JONES, S. A.: **Minispec Applications Note 15**, Bruker Analytische Meßtechnik, Am Silberstreifen, D-7512 Rheinstetten 4

13. HARZ, H. P, und WEISSER, H.: **Minispec Applications Note 12**, Bruker Analytische Meßtechnik, Am Silberstreifen, D-7512 Rheinstetten 4

14. VAN PUTTE, K., VERMAAS, L., VAN ENDEN, J. und DEN HOLLANDER, C.: **J. Amer. Oil Chem. Soc.,** 1975, **52**, 179

15. DESARZENS, C. und MICHEL, F.: **Rev. Francais des Corps Gras,** 1982, **29**, 419

16. BARKER, P J.: **Minispec Applications Note 31**, Bruker Analytische Meßtechnik, Am Silberstreifen, D-7512 Rheinstetten 4

17. BARKER, P. J: **Minispec ApplicntionsNote 32,** Bruker Analytische MeBtechnik, Am Silberstreifen, D-7512 Rheinstetten 4

18. HESTER, R. E. und QUINE, D. E. C.: **J. Food Technol.,** 1976, **11**, 331

19. JONES, S. A.: unveroff.

20. BARKER, P. J und JONES, S. A.: **Minispec Applications Note 33**, Bruker Analytische Meßtechnik, Am Silberstreifen, D-7512 Rheinstetten 4

21. TIWARI, P. N., GAMBHIR, P. N. und RAJAN, T. S.: **J. Amer. Oil Chem. Soc.,** 1974, **51**, 104

22. MADSEN, E.: **J. Amer. Oil Chem. Soc.,** 1976, **53**, 467

23. TIWARI, P. N. und BURK, W: **J. Amer. Oil. Chem. Soc.,** 1980, **57**, 119

24. JONES, S. A.: **Minispec Applications Note 17**, Bruker Anal. Meß.

25. SRINIVASAN, V. T: **J. Amer. Oil Chem. Soc.,** 1979, **56**, 1000.

26. DAUN, P. und BIELY, J.: **J. Amer. Oil Chem. Soc.,** 1980, **57**, 380

27. BARKER, P. J. und BRUNSCHWEILER, D.: unveröff.

28. BROSIO, E., CONTI, F., DI NOLA, A., SCORNO, O. und BALASTRIERI, F.: **J. Food Technol.**, 1981, **16**, 629

29. BROSIO, E., CONTI, F., DI NOLA, A. und SYKORA, S.: **J. Food Technol.**, 1981, **16**, 67

30. DI NOLA, A. und BROSIO, E.: **J. Food Technol.**, 1983, **18**, 125

31. BROSIO, E., ALTOBELLI, G., YU, S. Y. und DI NOLA, A.: **J. Food Technol.**, 1983, **18**, 219

32. WEISSER, H. und HARZ, H. P.: in **"Engineering and Food"**, Ed. B. M. McKenna, Elsevier, 1984, Bd. 1, 445

33. GUILLOU, M. und TELLIER, C.: Anal. Chem., 1988, 60, 2182

Rapid Determination Methods
for Drugs and
Fattening Substances in Animals

H. Büning-Pfaue

14 Rapid Determination Methods for Drugs and Fattening Substances in Animals

H. Büning-Pfaue, Bonn

14.1 Introduction

Many methods for drug residue analysis are characterized by a complicated sequence of solvent extractions, liquid/liquid separation, and possible fractionations. It is known that an extensive instrumentation is necessary to simultaneously verify the presence of all the presumed active ingredients.

In this situation the question for faster, but even more sensitive general analyses arises. Screening methods may be used for separating residue-contained or suspicious samples from the mostly larger amount of "clean" samples in short time. However, such methods are rarely successful because the low specific concentration of drugs in a complex sample matrix do not allow simple physical-chemical investigations as spot tests.

Rapid detection methods for drugs are therefore defined as procedure steps contributing towards acceleration of the residue analysis. Among these rapid detection methods are:

a) Simplifications of the sample preparation, which includes shortened extractions of active ingredients and clean-up of the raw extract of sample matrix fractions,
b) Group determinations by thin layer chromatography (TLC)
c) Immunological tests, used for example for detection of steroid compounds used as fattening substances, and
d) Rapid tests for confirmation of the identity of the detected residues, for which a variety of instrumental methods are available.

This article is concerned with the above mentioned rapid methods and the practical examples described as rapid methods, which are in some cases improvements of already known investigation sequences of the drug residue analysis.

14.2 Twelve Plate Inhibition Test

One exception among screening methods is a drug residue analysis that has been introduced recently for the group of antimicrobial drugs [1].

In this method the previously known inhibition test is extended to a twelve plate system using different media, media additives and test strains, resulting in a specific reaction profile for each individual active ingredient which enables screening and substance classification simultaneously.

Penicillin simultaneously inhibits the micrococci and B. subtilis, employed in this test but not B. cereus. The effect of inhibiting can best be reduced by addition of penicillinase to the medium.

The sulfonamides are best detected by the synergistic action with trimethoprim. This inhibition is cancelled by p-amino benzoic acid. The aminoglycosides streptomycin and neomycin restrict the growth of B. subtilis more than the growth of B. cereus. In both cases, heparin is used as additive for specifically reducing the action of the aminoglycosides. The aminoglycosides are then differentiated by the fact that the B.luteus strain employed in the last test is resistant to streptomycin.

This test has been proven in practice over the last three years and partly fulfills requirements regarding detection limits.

14.3 Sample Preparation /Drug Extraction

A rapid and commonly used method for drug extraction from samples of animal origin is now described:

Sufficient sample is homogenized in aqueous solution at a defined pH-value, often using a buffer solution. An organic solvent is then added and an intensive dispersion of the phases (e.g., by Ultra Turrax) follows. The organic phase, containing the extracted active ingredients, is separated by centrifugation to yield raw extract.

Using solvents immiscible with water has an associated time advantage because additional manipulation for separation of water from the extraction phase via addition of NaCl or dichlormethan is not necessary. These steps are necessary if solvents such as acetonitrile are employed.

Four very different examples are described to illustrate the method:

I. Extraction of Trenbolon from Meat [2]

The meat sample is homogenized with a pH 7.2 buffer. Trenbolon, which can be determined by radioimmunology, is extracted with butyl methyl ether. Subsequently only defattening and clean-up of the raw extract on a RP 18 phase is necessary.

II. Extraction of Levamisole from Milk [3]

The milk sample is acidified with HCl and anthelminticum can then be extracted with hot methanol. Clean-up by Extrelut follows.

III. Extraction of Propionylpromazine from Meat [4]

The author has thoroughly investigated the dependence of extraction efficiency upon the pH value of the buffer medium. Excellent extraction of the tranquilizer is achieved at pH 10.5 with a solvent mixture of diethyl ether/2-propanol.

IV. Extraction of Chloramphenicol, Sulfadimidine and Furazolidone from Fish [5]

Samples of fish muscle and fish organ tissues are homogenized in a buffer of pH 5.2 and extracted twice with ethyl acetate.

The above examples clearly show that rapid drug extraction at a defined pH value has proven effective for very different classes of drugs.

14.4 Sorbent Extraction / Solid-Liquid Extraction

With each extraction, significant amounts of sample matrix passes into the solution containing the drug residues. These matrix contaminants often interfere with chromatographic separations, rendering liquid/liquid extraction a time and solvent consuming method of clean-up.

On the other hand, sorbent extraction, already employed in many laboratories, provides a rapid technique for sample preparation. This solid/liquid extraction allows clean-up of the raw extract, and enrichment and fractionation of the residues simultaneously. The principles of this fast clean-up method is outlined in Figure 1.

Solid/liquid extraction has three distinct steps. Firstly, the sample raw extract is loaded onto a solid separation medium (e.g., silica gel with a chemically modified surface) in form of a mini column. Drugs are adsorbed at this surface through polar interactions between functional groups and adsorbent polar groups. Substances with molecular weights above 15,000 (e.g. matrix proteins) do not pass into the pore system of the solid phase.

Secondly, the matrix components are flushed out with an appropriate solvent, leaving only the active ingredients adsorbed upon the solid phase.

Finally, the residues are eluted from the column using a suitable eluent.

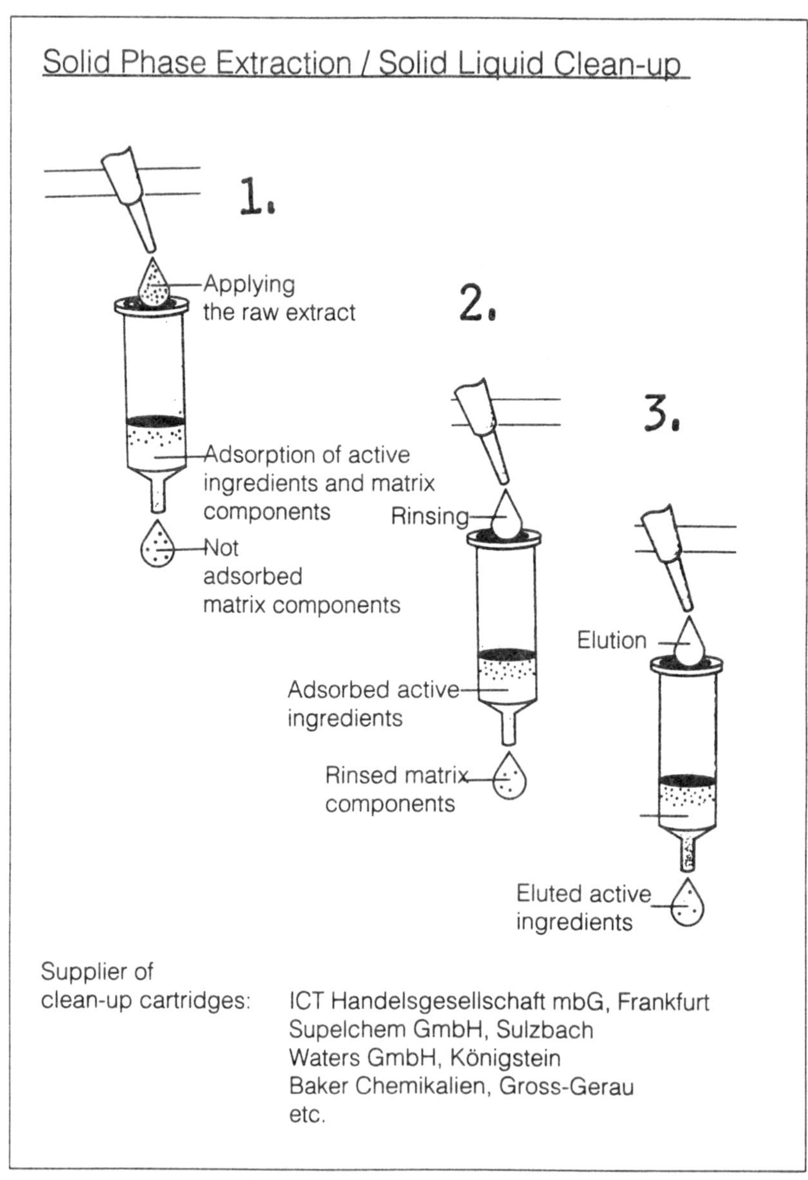

Fig. 1 Scheme of Solid/Liquid Extraction with Subsequent Steps for Clean-Up of the Sample Raw Extract

Manufacturers of commercially available mini-column cartridges are listed at the foot of the diagram. The principle of the separation method may be illustrated by three examples:

I. Clean-up of a Sulfonamide/Meat-Raw Extract [6]

The acidified raw extract is placed onto a pretreated cation exchange mini-column. The interfering matrix proportions is rinsed out with a methanol-water mixture. Ammonia vapour is then passed through the column causing a pH change which then allows elution of the sulfonamides by methanol.

II. Clean-up of a Chloramphenicole/Milk Raw Extract [7]

After centrifugation, the raw extract is applied to a RP 18 cartridge. Flushing with water is sufficient to remove the milk matrix. The active ingredients are then eluted with a mixture of acetonitrile and water.

III. Clean-Up of a Furazolidone Fish Raw Extract [8]

The solid phase for this clean-up is silica gel adjusted to a defined humidity. Matrix material is removed by rinsing with dichloromethane and the furazolidone is eluted with methanol

A polar silica mini column in combination with an RP 18 cartridge has been proposed by Haagsma [6] for fish samples prepared with sulfamidine. The findings led to some cautionary remarks upon the employment of solid separation media. They noted that, unlike the muscle samples, the organ samples showed 30 - 40% lower recovery rates due to retainment of active substances on the adsorbent. The decreased performance of the solid separation phase was caused by a change in the matrix layer on the adsorbent surface.

Therefore, when using this approach, cartridges should be pretested when new batches are received or when manufacturers have been changed. For samples of different composition, controls are necessary for the clean-up of the same active ingredients in order to ensure reproducible recovery rates.

14.5 Thin Layer Chromatography (TLC)

Thin layer chromatography allows rapid functional group analysis, especially when high performance separation layers are used. Two examples are described here:

I. Thin Layer Chromatography of Sulfonamides [9]

In this case, cleaned-up ultrafiltrates of the raw extracts of muscle or kidney tissue can be used directly for the TLC analysis. The analysis is carried out in two dimensions, using different solvents and even in the trace range of about 0.1 ppm 23 sulfonamides can be detected.

II. Thin Layer Chromatography of (Methyl)-Thiouracil [10]

An ethyl acetate extract from muscle or thyroid tissue is defattened using petroleum ether, followed by further clean-up on a polar Sep-Pak cartridge. The resulting extract is subjected to high performance thin layer chromatography and after the separation the compounds are derivatized and detected under UV light.

While the test for thiouracil is relatively simple, it can still be useful even if treatment with the fattening substances had been stopped by the animal breeder long time prior to slaughter in ignorance of the immediate diminishing effect. Traces of these drugs may be detected even if the animal has excreted the additional water (which would have increased the slaughter weight significantly) prior to slaughter. Both examples clearly demonstrate that TLC remains of considerable importance for rapid general analyses.

14.6 Immunological Tests

The sensitivity obtained by TLC, however, is not good enough in, for example detection of anabolically and estrogenically effective fattening substances. This is applicable in the case of muscle tissue in which trenbolon - an actual steroid fattening substance - is present in the ppt range. The mycotoxin derivate zeranol is another substance, the presence of which should be assumed as an anabolic fattening substances.

While it is obviously difficult to find a pellet of fattening substance about the size of a grain of rice, which may be implanted near the breast or near the ear of calves, radio immuno assay (RIA) is an ideal rapid method of detection. It is possible to analyse more than 20 sample per day per analyst. The principles and methods of RIA have been described previously and only general publications are referred to [11].

In future, enzyme immuno assay (EIA) will gain more importance as a rapid method compared with RIA. In EIA an enzyme coupling to a reaction partner is effected instead of the tritium labelling used in RIA.

The enzyme reaction, an indication for an antigen-antibody reaction having occurred, is much faster to carry out and easier to evaluate. This is possible because the reaction partners are impregnated upon the surface of multititer plates used for the EIA. Many measurements may be evaluated simultaneously in a multi-channel photometer within a minute, thus significantly reducing the material and equipment effort required compared with RIA. In addition, EIA may be used for detection of fattening substances for which the tritium labelled compounds (necessary for RIA) are not yet commercially available.

14.7 Selected Ion Monitoring (SIM-Technique)

A rapid method for result confirmation in drug residue analysis is the selected ion monitoring method, carried out with a gas chromatograph/mass spectrometer (GC/MS) system. In this technique, only a few pre-selected fragments within a mass spectrum are measured [see also ref. 12]. After measuring times as short as 50 milliseconds, so-called fragmentograms with the characteristic signals are obtained, even in the presence of background interferences, a situation characteristic of drug residue analysis, where ideal chromatographic separation of matrix components from the extract solution are hardly ever possible. Therefore, SIM is a preferred method for rapid result confirmation for residue analysis.

Figure 2 shows a mass fragmentogram with the mass ion m/z 256 from fragmentation of the metabolite N-4-acetyl sulfamidine, detected in addition to

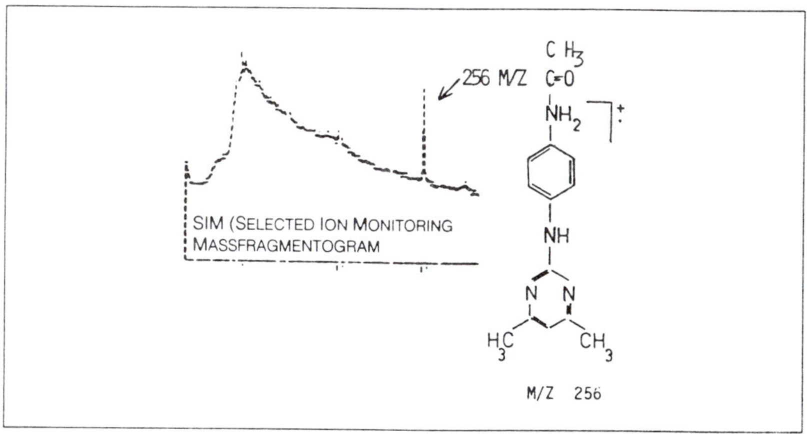

Fig. 2 Mass Fragmentogram with the Mass Ion 256 m/z from the Fragmentation of the Metabolite N-4-acetyl sulfadimidin

the unchanged drug as residue in carp. Several characteristic fragments are detected from their comparative signal intensity towards and verify the residue result.

14.8 UV Spectra Comparison

Another less demanding instrumental identity test with known drugs is the UV spectra comparison, carried out with corresponding standard solutions.

This test offers reliable results in the trace range of the residues but there are some reservations over the use of diode array detectors, which do not offer the same sensitivity of the simple UV detectors commercially available today, which can also be used for flow and stopped-flow applications.

14.9. Combination of Several Detection Systems

With further instrumental effort it is possible to obtain more rapid result confirmation for drug residues. Thus two or even three different detection systems are utilized subsequent to chromatographic separation, with each type of detection indicating the presence of the same substance being used as qualitative confirmation.

Figure 3 shows two chromatograms for sulfadimidine, taken with an UV detector and an electro chemical detector. Double detection could be achieved because the drug shows UV absorption and can be oxidized at a predetermined voltage (the oxidation potential).

14.10 Electro Chemical Detection

This measuring principle is based on the fact that each substance has a characteristic voltagram (current/voltage-curve). Thus the appearance of a signal at a particular polarization voltage may be a rapid proof of identity when an internal standard is employed.

A good example is in the detection of the coccidiostaticum nicarbazine residue in eggs. Here, UV detection fails but electro chemical detection succeeds and is more selective, especially at negative polarization voltages.

Both chromatograms in Figure 4, with nicarbazine and nifursol (as internal standard) at two different polarization voltages, illustrate the principle, where both substances show a characteristic ratio of peak areas at polarization voltage of -580 and -780 mV. The following simple working method is thus used in detection of nicarbazine: Firstly, a known amount of nifursol is added to the sample suspected of containing nicarbazine; two chromatograms are obtained at different polarization voltages; the difference in the respective peak area ratios is characteristic for the presence of nicarbazine; if the experimental correspond to those calculated the presence of nicarbazine is confirmed.

This method is very rapid and convenient compared to the alternative complex HPLC separation, fractionation and GC identification previously thought unavoidable for detection of nicarbazine [13].

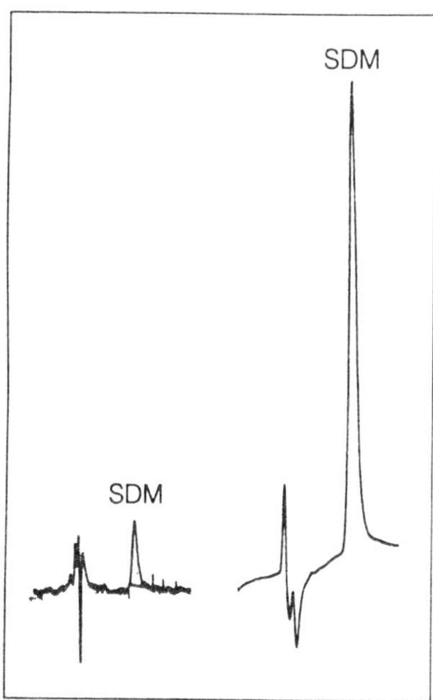

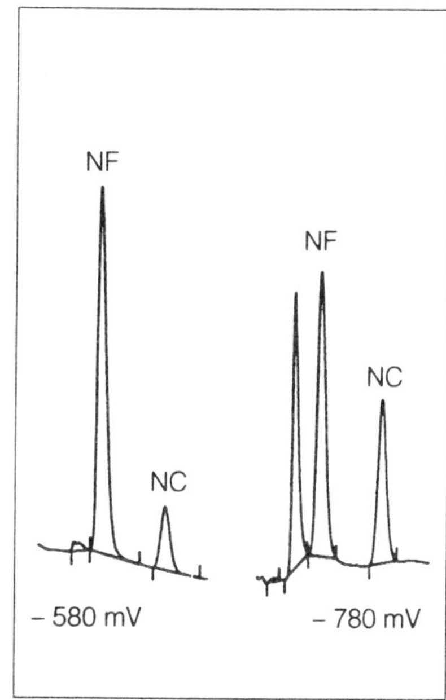

Fig. 3 Detection of 0.1 ppm Sulfadimidin by UV Detection (left) and Electro Chemical Detection (right)

Fig. 4 Electro Chemical Detection of Nicarbazine (NC) and Nifursol (NF) from the Same Sample Solution at - 580 mV and -780 mV Polarization Voltage

REFERENCES

1. LUND, C.: Proceedings 2nd World Congress Foodborne Infections and Intoxications, Vol. 11, 819, Berlin, 1986

2. RAPP, M. und MEYER, H. D.: Archiv f. Lebensmittelhyg. **36**, 31, 1985

3. ÖSTERDAHL, B. G. u. a.: J. Chromatogr. Biochemical Appl. **337**, 151, 198s

4. ARNETH, W.: Fleischwirtsch. **66**, 922, 1986

5. BÜNING-PFAUE, H. und SCHMIDT, TH.: Archiv f. Lebensmittelhyg. **36**, 87–93, 1985

6. HAAGSMA, N. und WATER, C.: J. Chromatogr. **333**, 256, 1985

7. BERGNER-LANG, B. und KÄCHELE, M.: Dtsch. Lebensm. Rundsch. **81**, 278, 1985

8. SCHMIDT, TH. und BÜNING-PFAUE, H.: Dtsch. Lebensm. Rundsch. **81**, 239, 1985

9. SCHLATTERER, B.: Z. Lebensm. Unters. Forsch. **176**, 20, 1983

10. POCHARD, M. F. u. a.: Analysis. **11**, 499, 1983

11. KYREIN, H. J.: Z. Lebensm. Unters. Forsch. **177**, 415, 1983

12. BORIES, G. S. F. u. a.: J. Assoc. Off. Anal. Chem. **66**,1521, 1983

13. MALISCH, R.: Z. Lebensm. Unters. Forsch. **183**, 253, 1986

Enzymatic Rapid Methods

G. Henniger

15 Enzymatic Rapid Methods

G. Henniger, Tutzing[1]

When talking about enzymatic rapid methods, the terms **"enzymatic"**, **"rapid"**, and **"methods"** have to be defined:

15.1 Enzymatic

"Enzymatic analysis" is, first of all, the **determination of substances by means of enzymes.** In this case, enzymes are used as reagents for specific catalysis of chemical reactions. Enzymatic analysis is, therefore, a special form of chemical analysis. Because enzymes react mainly with naturally occurring substances (these are metabolic intermediates, metabolites), enzymatic analysis is limited to these substances - the substrates of the enzymes.

Enzymatic determinations of substrates are carried out in research (universities, technical schools, but also in industry for the development of new products), for accurate and precise determination of food ingredients, and for control of changes during technological processing and storage. Chemical methods are not always suitable because they are not always specific enough. Modern physicochemical separation techniques with subsequent detection often require too much effort, especially regarding cost for equipment and personnel, or when the specificity of the method and with that the "quality" of results must be verified by additional experiments.

Enzymatic substrate determinations are further used for measuring metabolic end products of microorganisms in order to make an assessment regarding the "hygienic situation" of a sample. Typical examples for this kind of application are the determinations of pyruvate, lactate, and ammonia in milk. Substrate determinations usually are very quick, they require only a few minutes whereas the "classical" microbiological "germ count" methods need one to two days (actually no bacteria, but colonies are counted). Additionally, substrates are distributed more homogeneously in a sample than microorganisms and a homogeneous sample for analysis is produced more easily for substrate determinations.

Another advantage of enzymatic substrate determination is the possibility to judge the "history of a sample", even if the sample had been heated during

1   Boehringer Mannheim GmbH, Biochemica Werk Tutzing, D-8132 Tutzing/Obb.

technological processing, for example: during heat treatment, the substrates remain unchanged while microorganisms are killed, making a germ count senseless.

Enzymatic analysis includes **measurement of enzymatic activities**. Such measurements are made for checking and evaluation of heat treatment during the technological processing, e.g., pasteurization of milk and blanching of vegetables. These methods are based on the temperature sensitivity of enzymes: heat treatment causes destruction of the enzyme structure with a loss of their ability to catalyze, i.e. to speed up, chemical reactions. Therefore, high temperature treatments which are forbidden or limited can be detected, e.g. for honey. Honey can be declared as "natural" only when the enzyme diastase is sufficiently active.

Measurement of enzyme activities offers further evaluation possibilities: If enzyme activities can be found in heated samples, one can conclude that the sample has been contaminated by microorganisms after heating, because metabolism of microorganisms "works" by means of enzymatic reactions, as already known. Precondition is that no reactivation of heatdenaturated sample enzymes had taken place. Measurement of such enzyme activities requires only a short time as do substrate determinations. An assessment based on colony counts, however, is possible only after one or two days.

The enzymatic substrate determination is limited, either no enzymes are known in nature which can transform a certain substrate, or it is not possible to follow an enzymatic reaction due to insufficient measuring techniques, e.g., no suitable indicator reaction is available. These limits can be overcome when enzymatic methodology is combined with immunological methods. The antibody-antigen-reaction is known to be highly specific.

Furthermore, there are antibodies for all "chemically defined" compounds, even for those which do not occur naturally and also of compounds which might be synthesized in the future. When using enzymes as marker substances (instead of radioactively labelled compounds) the range of application for enzymatic anaylsis is broadened widely, it is practically unlimited. Enzyme-immuno-assays (EIA) are used for routine determinations e.g. of hormones in clinical chemistry. In food analysis they are not routinely used.

With these techniques compounds such as low-molecular toxins in food or drugs (drug residues), e.g., in meat, or compounds of high-molecular weight such as proteins could be determined.

15.2 Rapid

The term "rapid" is a statement made on the speed of a process. Rapidity, i.e. the time between sampling and obtaining the result, must always be seen relatively. In some cases, one minute might already be too long, in other cases, however, one day might be rapid.

Rapidity, that is, the time requirement for carrying out an analysis, should not be seen individually and isolated. It is closely connected with:

- **Specifity:** Only specific methods can give accurate results, unspecific methods must lead to incorrect results when analyzing such "complex samples" as food. Incorrect data can be the cause for disastrous misinterpretations and misleading conclusions. Non-specific or less specific methods may be used for industrial process control when the product to be examined is known regarding its composition and when the analysis is aimed at determination of uniform composition of the different batches. However, if kind and content of sample constituents are not known, e.g., when examining competitor's products, and especially in the official food analysis, then accurate analytical results are important and necessary. Only specific methods can be used for this purpose.

- **Precision:** Yes/No-decisions on whether a sample is acidic or alkaline can be made within seconds. For an assessment more precise as to whether a solution has a pH-value of 4 or 5, for example, less than one minute is necessary for comparing the test strip with color scales. Determination of the pH-value with a precision of Ò0.01 pH-units requires a pH-meter: adjustment, measurement and control take several minutes. This example of pH-measurement shows that the effort for a measurement is directly proportional to the required precision: the more precise the result, the greater the experimental and time effort.

- **Instrumentation:** The decision to by an analytic is based not only on cost of purchase, maintenance and repair but also the cost of personnel, training, and finally also on the time required for sample preparation and performance of the measurement itself. Test strips and electrodes can directly be used in turbid or colored samples which require no special sample preparation. With infrared reflection measurements even solid samples can be examined; however, the necessary calibration standards require very extensive analysis. Photometric measurements require only simple sample preparation when specific analysis methods are used. Modern physicochemical separation techniques with subsequent detection of the separated substances require not only "highly sophisticated equipment", but also very careful and often time and material consuming sample preparation.

● **Sensitivity**: Generally, measurements of analytes in the ranges g/100g, g/kg, or g/l do not cause too many difficulties. Most of the valuable constituents in food are found in these ranges. Analyses become more difficult and time consuming if carried out in the ppm-range. Many methods fail here, or the corresponding analytes must be isolated or concentrated prior to measurement.

● **Economy**: When viewing the costs, one must see on one side the cost for employees, equipment and materials, and on the other side, the cost of the product which is to be evaluated using the analytical results. Destruction or changes of the product due to false or incorrect analytical results can be very expensive, much more expensive than, e.g., more sophisticated analyses. For tax purpose or customs declarations such high requirements regarding precision and accuracy of analysis results are made (analytical errors can be seen in direct proportion to a certain amount of money) that they can hardly be fulfilled by analysis.

● **Coping with high sample throughput:** When considering the time requirement for analysis, one must discriminate between the real working time and the incubation waiting period. Furthermore, it must be considered, that the analyses may have to be carried out one after the other, because the analysis period depends on the working period or because a piece of necessary equipment is a limiting factor. This occurs usually in physicochemical separation techniques for which the subsequent determination can only be started after the previous "run" has been completed. Also, it should be considered whether all samples to be determined are available for analysis at the same time, allowing use of automatic analysis procedures or whether the samples are taken over a longer period, e.g., a working day, and further, at what time results are needed. If analysis results are needed for immediate decision in production, the analysis must be carried out directly after sampling; sample collection and analysis of all samples in series or by means of an automat is not possible in this case.

15.3 Methods

A **method**, e.g., in clinical chemistry, is defined (according to DIN 58 937, part 4, draft 1985) as " *the analysis procedure based on defined, namely biological, chemical and physical principles as well as mathematical theories for determination of ingredients and quantities in test samples"*.

Distinctions are made between reference methods, commonly used methods and rapid methods:

15.3.1   Reference Methods

These are methods in which accuracy and precision are very important. In general, these methods are very time consuming and require "expensive" equipment. Economy is another important criteria because employing reference methods should not be limited to special laboratories only. However, it is not necessary that all laboratories, even the small ones, should be able to work with reference methods. Usually specially educated personnel is necessary for performing reference methods.

Accuracy of analysis results is very important in cases of indecision and especially in cases of conflict. Reference methods are issued e.g. by institutes for standardization (in Germany DIN, Deutsches Institut für Normung) or in method collections published either by food regulation and inspection authorities or by national or international analysis committees. These methods are often referred to in the title as reference methods.

15.3.2   Commonly Used Methods

These are methods with less stringent requirements regarding precision and accuracy, or methods "designed" for a certain sample material, and therefore, not generally applied. With commonly used methods a larger number of samples can be examined in relatively short time. No unusually high degree of education is required of the staff. The equipment, required for these methods, can be found in each "normal" laboratory.

15.3.3   Rapid Methods

In the past, a rapid method was defined as a method requiring less time and effort than the method which produced the "correct" results. Today, rapid methods are screening or searching methods used to locate those samples that should be examined more intensively by means of a "better" method, e.g., by a quantitative determination with high accuracy and high precision. In general, test strips are regarded as rapid methods. This is of course only a limited view.

Among the rapid methods are the methods using color comparisons, e.g., the use of comparators after a chemical reaction has taken place. Where are the boundaries between a rapid method and a commonly used method? Should the term "rapid method" include the quantitative measurement or should rapid methods give only semi-quantitative results? The author defines rapid methods as methods that can be carried out by semi-skilled/ trained personnel, including calculation of the result. "Technical analysis methods" such as the different chromatographic or spectroscopic techniques can be carried out quickly, but should not be called rapid methods in official method collections or standardizations.

15.4 The "Quality" of Enzymatic Methodology

Enzymatic methods differ significantly from chemical and physicochemical methods: they are substrate- and not group specific: enzymes react only with specific substances. Today, the statement made by Nobel prize winner Otto Warburg, that enzymes are the "most specific" reagents known and therefore, the results obtained are the "most accurate" ones, is still valid. Enzymes react only with their substrates, other ingredients of the sample and additives such as antioxydants or preservatives do not react.

Knowledge of enzyme specificity for a defined substrate is necessary because, in addition to highly specific enzymes, there are also less specific, even non-specific enzymes. Less specific enzymes can be used for analysis if a reaction chain contains a specific enzyme. Non-specific enzymes are only employed for auxiliary reactions or for sample preparation. Often several possibilities for the determination of a substrate are available. In those cases it is the task of the analyst to find the best, that is, the most specific determination.

Interactions between sample ingredients do not occur during enzymatic analysis due to the very gentle reaction conditions so that no matrix effects occur in enzymatic analysis. The sample matrix is only required for sample preparation, which is carried out to bring the analyte into a solution suitable for analysis.

Furthermore, suitable laboratory equipment (photometer, cuvettes, pipettes) and careful analysis allow very precise results. The variation coefficients are usually between 0.5 and 2%.

The enzymatic method meets all requirements for a reference method:

Enzymatic methods are mostly reference methods.
Many enzymatic methods are standards, or standardized and recommended methods (e.g., DIN Germany, NBN Belgium, NEN Netherlands, NF

France, and ISO International Organization for Standardization; 35 of the German Food Law, LMGB; chapter 61 of the Swiss Food Manual; Italian and Spanish Laws).

Enzymatic methods are used for the analysis of practically all sample materials from the food and non-food sector, but also in clinical chemistry, biochemistry, and biotechnology. Enzymatic analysis is independent of the sample material, but dependent on commercially available pure enzymes and on instructions proved in routine analysis.

Today, enzymatic methods are used in **research and development, industry** (raw material-, produc-tion-, and end-control as well as examination of competitor's products, and for development of new products), **official food control** (control of food and other consumer goods according to legal regulations), **military** (comparison of quality and cost, storage and stability investigations) and **customs** (fixing of tariffs). Enzymatic methods are used in all field of analysis:

Enzymatic methods are commonly used methods.

Enzymatic methods are very flexible in regard to the sample material, which determines only the technique of sample preparation. The measurements are done photometrically, one of the most simple techniques, as far as practicability in routine analysis is concerned. Also, the cost for personnel, equipment and other materials are favourable, satisfying all economical requirements.

Enzymatic methods are fast to perform, usually much faster than common chemical reference methods. However, they require more time than the "simple" test strip methods. The simple techniques for sample preparation are quick, the actual measurement of enzymatic reactions is also fast.

Depending on the method, one determination takes between less than ten minutes to one hour. Each further determination takes about two to five minutes more, since series measurements are much more efficient. Distinctions between actual working time and the total analysis time must also be made for enzymatic determinations. Usually, the incubation period, that is the waiting period, is longer than the actual working period of approximately 5-10 minutes. It decreases with each determination when working in series. The incubation period can be used for further determinations.

15.5 Conclusion: Enzymatic Methods are Rapid Methods

The enzymatic substrate determination is one of the few methods or even the only one to fulfill all conditions of **reference methods, commonly used methods and rapid methods at the same time:**

● Enzymes are reagents with the highest specificity. Reactions catalyzed by them show results with the "best" accuracy.

● Only a few, easily controllable, simple steps for sample preparation offering low error possibilities, as long as the characteristics of the analytes are known by the analyst.

● The performance of the assay requires pipetting, reading the photometer, and calculating the result. Suitable "utensils" which are also low-priced, as well as proper know-how, minimize the error possibilities.

● The precision of enzymatic determinations is good; random errors, occurring during the determinations, result in data with a variation coefficient of approximately 0.5 to 2%.

● Enzymatic measurements can be very sensitive; even measurements in the ppm-range are possible without excessive problems.

● Interferences occurring in food analysis are rare, they can be recognized quickly and simply, and usually eliminated.

● The enzymatic analysis is very economical. Cost for personnel, depending on education and experience, of equipment (purchase, mainten-ance and repairs), and of materials is comparably low.

● Furthermore, the enzymatic analysis today fulfills all requirements regarding laboratory safety and environmental protection. There are no disposal problems because biochemical reagents are not dangerous.

● Enzymatic determinations can be carried out in short time. Distinction is to be made between the actual working time and the waiting period. Furthermore, the time factor should not be evaluated isolated from other features.

The analytical principle of the enzymatic analysis is free of disadvantages. The system, which the living cell has utilized for million of years, has been adapted for use in the cuvette and employed for analysis.

15.6 The Enzymatic Methods

15.6.1 Carbohydrates

Glucose belongs to the compounds which is measured often and usually together with other carbohydrates (e.g. fructose, sucrose, maltose, starch). The determination is done with the specific "hexokinase method" [1 and 2]:

$$
[1] \quad \text{D-glucose + ATP} \xrightarrow{\text{hexokinase}} \text{ADP + glucose-6-phosphate}
$$

$$
[2] \quad \text{Glucose-6-phosphate + NADP +} \xrightarrow{\text{G6P dehydrogenase}} \text{D-gluconate-6-P} \\ \text{+ NADPH + H}^+
$$

Other enzymatic methods are not recommended, e.g., because of insufficient specificity (glucose dehydrogenase) or interference by ascorbic acid and other reducing compounds in the sample (glucose oxidase/peroxidase method used, e.g., for glucose test strips). A subsequent measurement of fructose is not possible.

The measurement of **fructose** is carried out after the glucose measurement according to [3,4 and 2]:

$$
[3] \quad \text{D-fructose + ATP} \xrightarrow{\text{hexokinase}} \text{ADP + fructose-6-phosphate}
$$

$$
[4] \quad \text{Fructose-6-phosphat} \xrightarrow{\text{phosphoglucose isomerase}} \text{glucose-6-phosphate}
$$

The enzymatic method is advantageous not only if "reducing sugars" are of interest, but also for example when the ratio of glucose to fructose (in order to recognize adulterations) is of interest.

Measurement of **galactose** is carried out according to [5]:

$$
[5] \quad \text{D-galactose + NAD +} \xrightarrow{\text{galactose-DH}^*)} \text{D-galactonic acid + NADH + H}^+
$$

The disaccharide **sucrose** is measured after enzymatic hydrolysis (inversion) [6] via glucose [1 and 2]:

*) DH = Dehydrogenase

$$\text{[6]} \quad \text{Sucrose} + H_2O \xrightarrow{\quad \beta\text{-fructosidase} \quad} \text{D-glucose} + \text{D-fructose}$$

For verification of the result as well as for increasing the sensitivity, fructose from [6] can be measured subsequently.

Determination of **maltose** is carried out after enzymatic hydrolysis [7] via glucose [1 and 2]:

$$\text{[7]} \quad \text{Maltose} + H_2O \xrightarrow{\quad \alpha\text{-glucosidase} \quad} 2\ \text{D-glucose}$$

Sucrose, which also has an α-glucosidic bond, is split as well as maltotriose, which is regarded as sugar and not as "starch".

Lactose is the most important carbohydrate in milk. Determination is carried out after enzymatic hydrolysis [8] either via galactose [5] or glucose [1 and 2]:

$$\text{[8]} \quad \text{Lactose} + H_2O \xrightarrow{\quad \beta\text{-galactosidase} \quad} \text{D-galactose} + \text{D-glucose}$$

Determination via glucose is carried out with hexokinase/glucose-6-phosphate dehydrogenase [1 and 2]. The glucose oxidase/peroxidase method cannot be recommended, as already mentioned earlier.

The enzymatic determination of lactose, e.g. according to DIN 10 344 or according to L. 01.00/17 of § 35 German Food Law method collection, is recommended for example as a reference method for infrared reflection analysis of milk products.

Determination of **raffinose**, e.g., in refinery sugar products, is carried out according to [9 and 5]:

$$\text{[9]} \quad \text{Raffinose} + H_2O \xrightarrow{\quad \alpha\text{-galactosidase} \quad} \text{D-galactose} + \text{sucrose}$$

An old topic in food analysis is the determination of starch. Prior to the measurement via glucose [1 and 2], the starch must be hydrolyzed. For that, acid hydrolysis is not suitable because of hydrolysis of other "polyglucoses" and undesirable side reactions. The enzymatic hydrolysis is carried out with amyloglucosidase; however, maltose as well as other oligoglucosides are also hydrolyzed [10]:

$$\text{[10] Starch + (n-1)H}_2\text{O} \xrightarrow{\text{amyloglucosidase}} \text{n D-glucose}$$

The starch must be dissolved during sample preparation. This is carried out by autoclaving or by using a mixture of HCl and dimethyl sulfoxide (DMSO).

Differentiation of starch and oligoglucosides (maltodextrins, glucose syrup, starch sugars) is accomplished by making use of their different solubilities in ethanol/water mixtures in which starch is insoluble.

15.6.2 Inorganic and Organic Acids

L-malic acid, a compound of the citrate cycle, is determined according to [11 and 12]:

$$\text{[11] L-Malate + NAD +} \xrightarrow{\text{L-malate dehydrogenase}} \text{oxaloacetate + NADH + H}^+$$

$$\text{[12] Oxaloacetate + L-glutamate} \xrightarrow{\text{GOT}} \text{L-aspartate + 2-oxoglutarate}$$

The determination of D-malate is described in the relevant literature, the necessary enzyme, however, is not commercially available.

Formic acid, produced by metabolism of fungi and bacteria, but also a preservative, is determined according to [13]:

$$\text{[13] Formate + NAD + H}_2\text{O} \xrightarrow{\text{formate-DH}} \text{bicarbonate + NADH + H}^+$$

Ascorbic acid, a vitamin for humans and a technological additive is determined in a chemical-enzymatic system [14]. To improve specificity, ascorbate is removed enzymatically in a sample blank determination [15]:

$$\text{[14] L-ascorbate (x-H}_2\text{) + MTT} \xrightarrow{\text{PMS}} \text{dehydroascorbate (x) + formazan}$$

$$\text{[15] L-ascorbate + 1/2 O}_2 \xrightarrow{\text{ascorbate oxidase}} \text{dehydroascorbate + H}_2\text{O}$$

Dehydroascorbate, which also has vitamin-C properties, can be determined after being chemically oxidized to ascorbate [16] via [14 and 15]:

[16] Dehydroascorbate + reduced DTT ⟶ L-ascorbate + oxidized DTT

Aspartic acid is determined according to [17 and 18]:

$$\text{[17] L-aspartate + 2-oxoglutarate} \xrightarrow{\text{GOT}} \text{oxaloacetate + L-glutamate}$$

$$\text{[18] Oxaloacetate + NAHD + H}^+ \xrightarrow{\text{L-malate DH}} \text{L-malate + NAD}^+$$

If only aspartic acid is to be determined from a sample containing amino acids, or if other amino acids are present in the sample in low quantities only or are not important for evaluation, the individual enzymatic determination of aspartic acid should be preferred to a complete amino acid analysis. The same is applicable for determination of glutamic acid. In the enzymatic test only the L-forms of aspartic acid and glutamic acid are detected.

Succinic acid, also a compound of the citrate cycle, is determined according to [19, 20 and 21]:

$$\text{[19] Succinate + ITP + CoA} \xrightarrow{\text{SCS}} \text{IDP + succinyl-CoA + phosphate}$$

$$\text{[20] IDP + PEP} \xrightarrow{\text{pyruvate kinase}} \text{ITP + pyruvate}$$

$$\text{[21] Pyruvate + NADH + H}^+ \xrightarrow{\text{L-lactate-DH}} \text{L-lactate + NAD}^+$$

Pyruvic acid, a central substance of the metabolism, is determined according to [21]. The enzymatic measurement has proven good for determina-tion of pyruvate in milk in the ppm-range a million times.

Determination of **citric acid**, a key substance in metabolism, is carried out according to [22]. The oxaloacetate formed is determined according to [18], pyruvate formed by chemical or enzymatic decarboxylization of oxoloacetate is determined according to [21]. The NADH consumed in [18] and [21] is equivalent to citrate.

$$\text{[22] Citrate} \xrightarrow{\text{citrate lyase}} \text{oxaloacetate + acetate}$$

The classical reference (pentabromoacetone) method for determination of citric acid takes almost two days, the chemical "rapid method according to

Rebelein" requires approximately 5 hours. In contrast, the enzymatic measurement is finished in 15 minutes. Titrating the acid with alkaline is quicker, but those results should not be compared to the results of the enzymatic method, even though this is sometimes done. In the titration method protons are determined, in the enzymatic method the citrate ion.

Acetic acid, a component of "volatile acids", is specifically measured according to [23, 24 and 25]:

$$[23] \; \text{Acetate} + \text{ATP} + \text{CoA} \xrightarrow{\text{ACS}} \text{acetyl-CoA} + \text{AMP} + \text{pyrophosphate}$$

$$[24] \; \text{Acetyl-CoA} + \text{oxaloacetate} + H_2O \xrightarrow{\text{CS}} \text{citrate} + \text{CoA}$$

Oxaloacetate is formed during a preceding indicator reaction [25]:

$$[25] \; \text{L-malate} + NAD^+ \underset{\longleftarrow}{\overset{\text{L-malate DH}}{\longrightarrow}} \text{oxaloacetate} + NADH + H^+$$

D-gluconic acid, produced by oxidation of D-glucose, is determined according to [26 and 27]:

$$[26] \; \text{D-gluconate} + \text{ATP} \xrightarrow{\text{gluconate kinase}} \text{D-gluconate-6-phosphate} + \text{ADP}$$

$$[27] \; \text{D-gluconate-6-P} + NADP^+ \xrightarrow{\text{6-PGDH}} \text{ribulose-5-P} + NADPH + H^+ + CO_2$$

Determination of **glucono-δ-lactone** is carried out according to the same principle [26 and 27], but after alkaline hydrolysis [28]:

$$[28] \; \text{D-glucono-}\delta\text{-lactone} + H_2O \xrightarrow{\text{pH 10-11}} \text{D-gluconate}$$

Glutamic acid, an important amino acid, is determined according to [29 and 30]:

$$[29] \; \text{L-glutamate} + NAD^+ + H_2O \xrightarrow{\text{glutamate-DH}} \text{2-oxoglutarate} + NADH + NH_4^+$$

$$[30] \; NADH + H^+ + \text{INT} \xrightarrow{\text{diaphorase}} NAD^+ + \text{formazan}$$

The color reaction is very sensitive, interfering reducing substances must be removed during sample preparation.

The determination of **hydroxybutyric acid** in eggs has gained much interest lately. The determination is carried out according to [31 and 30]:

$$[31]\ \text{D-3-hydroxybutyrate} + NAD^+ \xrightarrow{\text{3-HBDH}} \text{acetoacetate} + NADH + H^+$$

The enzymatic determination of **isocitric acid** [32] has been partly responsible for the success which the enzymatic methods have gained in fruit juice analysis.

$$[32]\ \text{D-isocitrate} + NADP^+ \xrightarrow{\text{ICDH}} \text{2-oxoglutarate} + NADPH + H^+ + CO_2$$

Lactones and esters of isocitric acid are determined according to the same principle [32] after alkaline hydrolysis:

$$[33]\ \text{D-isocitric acid lactone [-ester]} + H_2O \xrightarrow{\text{pH 9–10}} \text{D-isocitrate (+ alcohol)}$$

The stereospecific determination of lactic acid is one of the most important enzymatic methods. **L-lactic acid** is the final product of glycolysis in the animal kingdom, **D-lactic acid** is produced only by some microorganisms. The determinations are carried out according to [34 and 36] and [35 and 36], respectively:

$$[34]\ \text{L-lactate} + NAD^+ \xrightarrow{\text{L-lactate-DH}} \text{pyruvate} + NADH + H^+$$

$$[35]\ \text{D-lactate} + NAD^+ \xrightarrow{\text{D-lactate-DH}} \text{pyruvate} + NADH + H^+$$

$$[36]\ \text{Pyruvate} + \text{L-glutamate} \xrightarrow{\text{GPT}} \text{L-alanine} + \text{2-oxoglutarate}$$

The enzymatic determination of **nitrate** with a stable enzyme, according to [37], has been published first in 1986:

$$[37]\ \text{Nitrate} + NADPH + H^+ \xrightarrow{\text{nitrate reductase}} \text{nitrite} + NADP^+ + H_2O$$

Oxalic acid is component of kidney stones. Furthermore, it inhibits resorption of calcium. The enzymatic determination is carried out according to [38 and 13]:

$$[38]\ \text{Oxalate} \xrightarrow{\text{oxalate decarboxylase}} \text{formate} + CO_2$$

Due to possible interference, the oxalate oxidase method cannot be recommended for food analysis.

The enzymatic determination of **sulfite** according to [39 and 40] was described first in 1983:

$$[39]\ \text{Sulfite} + O_2 + H_2O \xrightarrow{\text{sulfite oxidase}} \text{sulfate} + H_2O_2$$

$$[40]\ H_2O_2 + \text{NADH} + H^+ \xrightarrow{\text{NADH peroxidase}} 2\,H_2O + \text{NAD}^+$$

The enzymatic sulfite determination in serie is much faster than all distillation methods.

15.6.3 Alcohols

Ethanol is formed by yeasts especially under anaerobic conditions. The enzymatic determination is the enzymatic method first employed in routine analysis. Today, the determination is carried out according to [41 and 48]:

$$[41]\ \text{Ethanol} + \text{NAD}^+ \xrightarrow{\text{alcohol dehydrogenase}} \text{acetaldehyde} + \text{NADH} + H^+$$

The enzymatic ethanol determination is very fast. For alcoholic beverages (wine, beer, spirits) a rapid method is recommended which takes less than five minutes from starting the reaction with the diluted sample to obtaining the final result.

Cholesterol, an important steroid, is measured for determination of the egg content e.g. in noodles and advocaat. The determination is carried out according to [42,43 and 44]:

$$[42]\ \text{Cholesterol} + O_2 \xrightarrow{\text{cholesterol oxidase}} \Delta^{4}\text{-cholestenone} + H_2O_2$$

$$[43] \quad H_2O_2 + methanol \xrightarrow{\text{catalase}} formaldehyde + 2\ H_2O$$

$$[44] \quad Formaldehyde + NH_4^+ + 2\ acetylacetone \rightarrow lutidine\text{-}dye + 3\ H_2O$$

The method is specific in the analysis of sample material of animal origin, if plant components are present, phytosterols may also react with a 3β-hydroxy group (except lanosterol).

The enzymatic determination of cholesterol is much faster and more reliable than the digitonin method, and therefore much more suitable for screening purposes, such as the calculation of the egg content of noodles in order to sort out "suspicious" samples for gaschromatographic examination.

Glycerol is widely found in nature and usually bound in the form of lipides. The determination is carried out according to [45, 46 and 21]:

$$[45] \quad Glycerol + ATP \xrightarrow{\text{glycerokinase}} glycerol\text{-}3\text{-}phosphate + ADP$$

$$[46] \quad ADP + PEP \xrightarrow{\text{pyruvate kinase}} ATP + pyruvate$$

There are several possibilities for determination of **sorbitol**. Determination according to [47] alone is not very satisfactory. Determination via fructose, that is [47] together with [3,4 and 2] proved worthwhile resulting in a specific determination of sorbitol.

$$[47] \quad D\text{-}sorbitol + NAD^+ \xrightarrow{\text{sorbitol-DH}} D\text{-}fructose + NADH + H^+$$

Another possibility is the combination of [47] and [30] which is recommended for samples with large amounts of glucose and fructose relative to sorbitol.

15.6.4   Other Analytes

Acetaldehyde is determined in wine or yogurt for example according to [48]:

$$[48] \quad Acetaldehyde + NAD^+ + H_2O \xrightarrow{\text{aldehyde-DH}} acetate + NADH + H^+$$

Ammonia is determined specifically according to [49]:

$$[49] \quad NH_4^+ + 2\text{-oxoglutarate} + NADH \xrightarrow{\text{glutamate-DH}} \text{L-glutamate} + NAD^+ + H_2O$$

Determination of **creatine** and **creatinine**, e.g., for calculating the meat content of soups, is carried out according to [50, 51, 46 and 21]:

$$[50] \quad \text{Creatinine} + H_2O \xrightarrow{\text{creatininase}} \text{creatine}$$

$$[51] \quad \text{Creatine} + ATP \xrightarrow{\text{creatine kinase}} \text{creatine-phosphate} + ADP$$

Urea, an important end product of protein metabolism, is specifically determined according to [52 and 49]:

$$[52] \quad \text{Urea} + H_2O \xrightarrow{\text{urease}} 2\,NH_3 + CO_2$$

There are several possibilities for determination of **lecithin**, an important phospholipide and component of emulsifiers. In practice, the choline kinase method has proven best. The determination is carried out according to [53, 54, 55, 46 and 21]:

$$[53] \quad \text{Lecithin} + H_2O \xrightarrow{\text{phospholipase C}} 1,2 \text{ diglyceride} + \text{phosphorylcholine}$$

$$[54] \quad \text{Phosphorylcholine} + H_2O \xrightarrow{\text{alkaline phosphatase}} \text{choline} + \text{phosphate}$$

$$[55] \quad \text{Choline} + ATP \xrightarrow{\text{choline kinase}} \text{phosphorylcholine} + ADP$$

Triglycerides are important constituents of lipides and can be determined enzymatically, after enzymatic saponification [56] via glycerol, according to [45, 46 and 21]. The enzymatic method is advantageous when only small sample amounts are available or when analyzing milk products with low fat content.

$$[56] \quad \text{Triglycerides} + 3\,H_2O \xrightarrow{\text{lipase/esterase}} \text{glycerol} + 3 \text{ fatty acids}$$

In most cases the formation or the consumption of NADH or NADPH, respectively, is measured. The systems usually "operate" without interference, the corresponding extinction coefficients are almost independent of the reaction conditions. Furthermore, they are very well-known. If spectral photometers are used, the measurements are done at the maximum absorption of light (340 nm). For spectrum line filter photometers, very often used in Germany, measurement at 334 nm is recommended.

In some cases, the NAD(P)H-system cannot be used or is only of limited use for the determination:

Due to the reaction equilibrium (i.e. in the determination of glutamic cid, hydroxybutyric acid and sorbitol), the NADH formed reduces INT to a formazan dye which is then measured.

A formazan is also measured when determining ascorbic acid. In this case a chemical reaction is made "highly specific" by means of an enzyme (ascorbate oxidase).

The determination of cholesterol is one of the rare cases in food analysis for which an oxidase is used (usually oxidases are not
recommended because of interferences due to reducing substances). For measurement of the hydrogen peroxide formed in the oxidase reaction, there are several indicator reactions, for determination of cholesterol in food the Kagejama reaction has proven best: here acetylacetone, ammonia and formaldehyde, formed by methanol and H_2O_2 in presence of catalase, produce a lutidine dye which is measured photometrically.

Substrate determinations in food analysis are normally carried out as end point methods. The substrate to be measured is transformed quantitatively and one of the reaction partners or products is measured photometrically. This means that the total determination period is not dependent on the "pure" working time (needed for pipetting and absorption measurements, for example) but only on the waiting period until the reaction, with decreasing speed, has reached the end point. The reaction time can be cut about in half when the incubation temperature is increased from 20-25 °C to 37 °C.

In clinical chemistry, the speed of analysis as well as high sample throughput is much more important than "high" requirements regarding accuracy and precision of the analytical results. Here enzymatic substrate determinations are also carried out by kinetic measurements, which are much faster than end point determinations.

Kinetic measurements are based on a relation between the substrate concentration in the analyte and the reaction speed of the enzymatic reaction. The measurements are carried out "very well controlled" since the reaction speed is highly dependent on the reaction conditions. The used enzymes must have a "very high Michaelis constant" regarding the substrate to be measured, contrary to the end point method. This high constant can be reached, e.g., by addition of competitive inhibitors, if necessary.

The concentration differences to be measured can be very small and the measurement times are only a few minutes, therefore, high demands are made regarding the instrumentation. Usually analysis automates, such as centrifugal analyzers, are used.

Kinetic substrate determinations are not yet employed for food analysis. In this field, no experience is available as to what extent the samples are responsible for matrix effects. These could possibly be compensated by means of internal standards.

Use of analyzers for kinetic measurements in food analysis should not be a problem of methodology rather than a financial one (e.g. purchase of expensive equipment), because the cost for analysis should stand in reasonable relation to the total cost. The same is also valid for instrumental non-enzymatic analysis which might be necessary for solving some problems, but in other cases has a very disadvantageous of cost/performance ratio.

REFERENCES

BERGMEYER, H.-U. (1984–1986): Methods ot Enzymatic Analysis, Volumes I–XII, Verlag Chemie, Weinheim, Deerfield BeachlFlorida

Methoden der biochemischen Analytik und Lebensmittelanalytik (1986), Boehringer Mannheim GmbH, D-6800 Mannheim 31

Die enzymatische Lebensmittelanalytik (Serie), Boehringer Mannheim GmbH, D-6800 Mannheim 31

HENNIGER, G. (1979): Enzymatische Lebensmittelanalytik, Zeitschrift für Lebensmittel-Technologie und Verfahrenstechnik (ZFL) **30**, 137–144 und 182–185

Immunochemical Methods

H.O. Günther

16 Immunochemical Methods

H.O. Günther, Augsburg

16.1 Introduction

Immunochemical or immunological methods are procedures with reactions between antigens (proteins with additional groups such as carbohydrates, lipides, or hapten conjugates) and antibodies. These antibodies are produced in the animal body and commercially available as antisera or even as pure gamma globuline fractions. Within a certain concentration range precipitation between the reaction partners occur. This can be made visible in a high polymer agarose gel by staining. In this way the reaction can be recorded permanently. The positive reactions are pictured as fine colored lines.

For formation of these lines or bands it is necessary that both reaction partners in the gel diffuse towards each other. This explains the term "double gel diffusion" (according to Ouchterlony [1]). For the immuno-electrophoresis, the proteins are separated by means of electrical direct current prior to starting the immuno-diffusion. Depending on their charge they migrate over different distances in the gel.

These two methods, necessary for identification of proteins with identity reactions, take several days before the results can be obtained. Therefore, for larger sample amounts faster methods are required.

Two methods are suitable: First the counter current electrophoresis according to Gocke and Howe [2] and second the enzyme immuno assay in form of the ELISA (sandwich-form) [3–5].

Although for the counter current electrophoresis common sample preparation is necessary, the immuno assay can be carried out with liquids directly from the diluted solution or with solid food from the extract without further clean-up. The counter current electrophoresis usually requires 6–8 hours, the immuno assays between 2 hours and maximal 2 days, depending on the preparation.

16.2   Methods

16.2.1   Counter Current Electrophoresis According to Gocke and Howe [2]

16.2.1.1 Principle

Filling holes are punched opposite of each other into a polysaccharide-containing gel (agarose) with high electro osmotic mobility. They can receive 20–50 µl of the sample or standard solution, respectively. The protein solutions are filled so that they migrate to the positive pole, while the antiserum solutions migrate to the negative pole. Both will meet in the middle between the punched rows and will form precipitation lines. Precondition for this method is a difference in electrophoretic mobility of sample and antibody protein. Alkaline proteins must be pretreated with KSCN to change polarity e.g. by carbamylization [10].

16.2.1.2 Performance

A solution of agarose with tris-tricine buffer (0.05 M; pH=8.2) is heated until the rising bubbles burst. After cooling down to 70–80 °C, the clear solution is poured quickly onto a Gel-Bond-Foil® FMC, Rockland, Maine (USA) which lies on a glass plate on a levelling table. For two hours the gel is hardened in a humid chamber, then a gel puncher or a corkscrew is used for punching holes into the gel. The distances between the rows of holes and the arrangement can be seen in Figure 1. Since the molecular weights of the antibodies are between M_r = 150,000–170,000 D, the weights of the sample proteins mostly are below M_r = 70,000 D, a pre-electrophoresis of the antisera for 60–90 min at 150 V is recommended.

The sample extracts migrate 90 min to 2 hours. In many cases the white precipitation line of the positive controls can be recognized with the naked eye. The plates are kept in the humid chamber overnight, the next morning the sample protein precipitates are visible, too. But often the lines appear after staining only.

First, however, the unprecipitated proteins must be removed from the gel. This is done by pressing the excess water off with a 1 kg weight and by rinsing with NaCl solution (1%). The solution is changed 3–4 times every 30 minutes, then pressing again and drying on air or with a blow-drier. After staining, the foils can be evaluated and then punched for documentation.

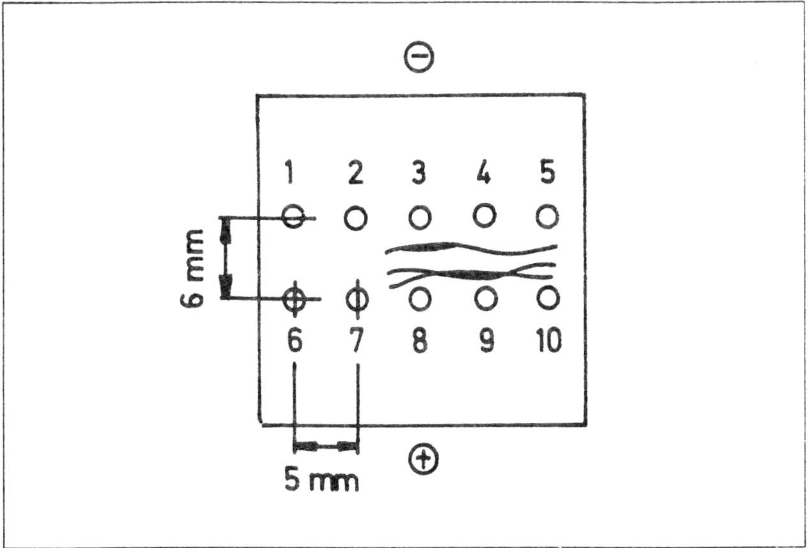

Fig. 1 Diagram of a Counter Current Electrophoresis with the Detection of Rice Protein in Foreign Beer.

```
     1 = Beer extract, negative to rabbit serum, normal (no specific antibodies)
     2 = Beer extract, negative to antiserum to rice
     3 = Rice protein, positive control
     4 = Beer extract, positive to rice protein
     5 = Rice protein, positive control
     6 = Rabbit serum, normal
  7–10 = Antiserum to rice protein of rabbit, No. 6118, Behring-Werke, Marburg
```

In many cases the bands obtained are successfully identified by punching the filling holes for the control (total protein) and the sample in distances of 3 mm. This causes the lines of identical samples to flow together. In order to avoid mis-evaluation a third hole, also filled with sample extract, should be placed next to the others. In case of a positive reaction the bands of the control should extend across to both sample bands.

Figure 1 shows a positive detection of rice in beer. If new antisera batches are employed, a negative control (rabbit antiserum without specific anti-bodies, e.g. Behring-Werke, Marburg, stabilized rabbit serum, ØTØR 04) should be applied.

16.2.1.3 Total Working Period

The following schedule can be employed for extracts from food or precipitated proteins from beer or other liquids:

Tab. 1 Schedule for a Determination Series

Time	Activity
12.00 am	Pouring the gel
2.00–3.00 pm	Punching; filling of antisera for pre-electrophoresis
3.00–4.00 pm	Filling of sample extracts; electrophoresis
5.00 pm	Placing into humid chamber overnight
Next morning:	
8.00 am	Pressing the gel
9.00 am	Rinsing in NaCl solution 4 x 30 min
11.00 am	Pressing the gel
12.00 am	Drying the gel plates
1.00 pm	Staining the gel plates
1.30 pm	De-staining
3.00 pm	Drying
4.30 pm	Evaluation

Tab. 2 Detection of Proteins in Various Food

Detection of proteins	Food
corn, rice, millet, pepsin	beer [6]
caseinate, whey protein, hazel nut, peanut, soybean	pasta products and confectionery [7]
w. bean protein, chicken protein, almond, cocoas, corn, rice, wheat	for chocolate products after removing tannins [8]
soy, caseinate, whey proteins	meat products [14]
poultry protein (troponin)	meat products [9]
cow milk protein after carbamylization	goat cheese [10]
corn, wheat, rice, poultry meat chicken protein, whey protein	soups and sauces [11]

Depending on the plate gel size, 10–50 samples can be electrophorated simultaneously. The exact instructions are published in the official collection of analysis instructions according to §35 LMBG and the relevant publications.

16.2.1.4 Applications for this Method

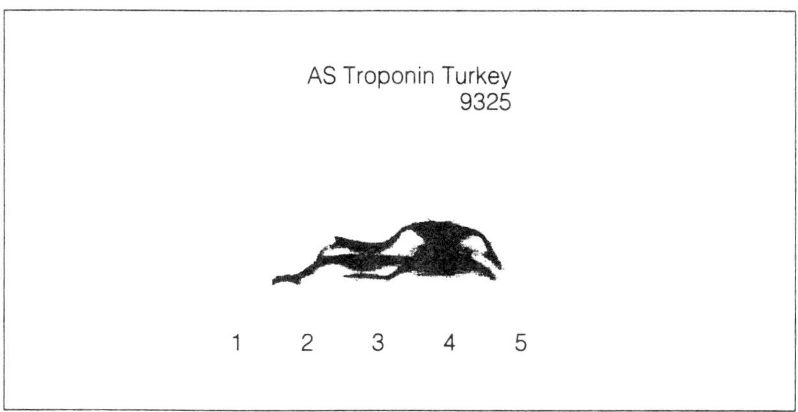

Fig. 2 Counter Current Electrophoresis with the Detection of Chicken Meat
    1 = Beef extract
    2 = Pork extract
    3 = Chicken, muscle extract, cooked in special pot (Rîmertopf)
    4 = Chicken, troponin, positive control
    5 = Chicken soup negative, meat fibers positive (not shown)
Upper row = Antiserum to turkey, No. 9325, Behring-Werke, Marburg

If no identification reaction, as shown in Figure 1, can be recognized, it is necessary for positive reaction to perform the official method of double gel diffusion for verification of identity between positive control and sample protein.

16.2.2   Enzyme Immuno Assay (= ELISA)

Another fast analysis method with a sample throughput of 50–100 samples in two days or two hours, respectively, are the enzyme immuno assays [3–5]. They must be run specifically for the substance (protein or hapten) or the substance group to be detected.

16.2.2.1 Principle

For the ELISA [2] (**E**nzyme **L**inked **I**mmuno **S**orbent **A**ssay) the solid phase surfaces of plastic materials are coated with a specific antibody for the antigen to be detected (here the protein papain). The antigen reacts with the antibody, present in excess, which is fixed to the wall of the carrier. In the second immuno reaction the second enzyme labelled antibody with specificity to the same antigen forms a sandwich complex in which the antigen is enclosed from both sides. The not bound, enzyme labelled antibody is removed by washing. In the last step, the enzyme labelled antibody causes a color reaction with a substrate and a chromogen. Its concentration is determined in the visible range, e.g. at 405 nm with internal standards, and compared to a calibration line.

16.2.2.2 Production

The reagents for an enzyme immuno assay must be prepared. These include an antiserum directed against the protein, as well as a antiserum conjugated with an enzyme label. The antiserum is prepared against a purified antigen, if possible. For that the enriched protein mixture, e.g., from nuts or soy beans, must be separated on a column into the individual components after ammonium sulfate precipitation. Since the main fraction usually remains after the food production process, it is isolated preparatively and used for immunization. The antiserum obtained is tested for specificity against all proteins which could be present in the food concerned. Unspecificities must be adsorbed and removed. Usually, the antiserum is fractionated; only purified antibodies or gammaglobuline fractions are employed.

For protein detection the antibody is conjugated with an enzyme. Within the last years, two methods have turned out to be quite favorable:

1. The oxidation of the carbohydrate portion of the enzyme, e.g., of the peroxidase with periodate [15]. The aldehydes formed are coupled to the amino group of the protein. A mixture of conjugate and non-reacting molecules is obtained, which also contains polymers, formed by self conjugation. It is therefore necessary to separate the conjugate fraction, recognizable by the absorption ratio at different wave lengths. The pure conjugate fraction obtained is diluted prior to performing the assay, e.g., 1 + 300 or even 1 + 500. With one milliliter of conjugate many samples can be investigated.

2. Coupling method with MBE (M-maleimidobenzoic acid N-hydroxy-succinimide ester) [15]. MBE is a hetero-bifunctional reagent which re-acts with the amino groups of the antibody (IgG). It can be separated by centrifugation and reacts further with the SH-groups of the enzyme, e.g., galactosidase. A bridge-like function is set up between the antibody and the enzyme. With that, yields and activities are high.

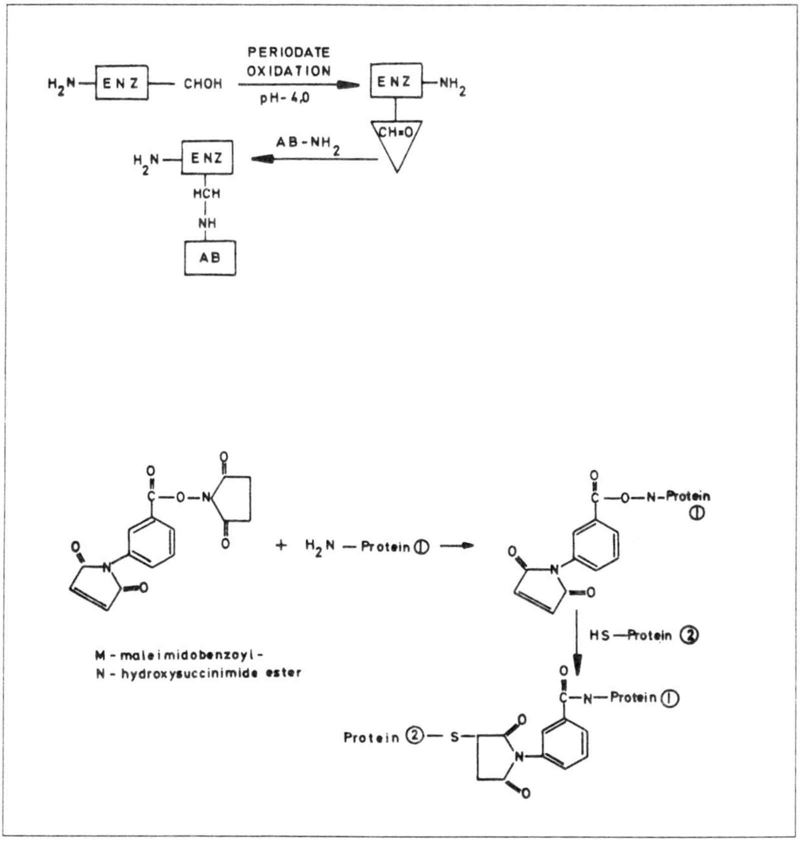

Fig. 3 Enzyme Antibody Conjugation
Top: Periodate method
Bottom: Hydroxy succinic acid method for enzymes carrying SH-groups

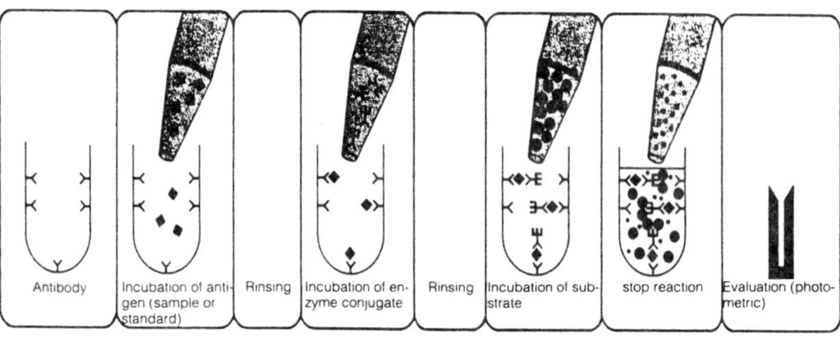

Fig. 4 Test Principle for ELISA (Sandwich form), Behring-Werke, Marburg (Source: Enzygnost-SP 1, Behring Diagnostika, Marburg, West-Germany)

16.2.2.3 Performance

Performing ELISA is relatively easy, as shown in Figure 4. After production of the antisera, the plastic carriers are coated with the purified antibodies (1μg/ml). After incubation overnight the filling holes are washed and dried in the exsiccator. After drying for one day, the plates, trays, or spheres are sealed in plastic foils under vacuum and stored at -40 °C. Durability is at least 6 months.

The following example is applicable for papain in beer [12].

Step 1: Fill 200 μl (depending on carrier) of the diluted sample solution into the filling holes of the plastic carriers, coated with anti-papain. In-cubate after closing with plastic foil for 60 min (for beer) at 37 °C in the water bath.

Step 2: Wash the emptied plates with washing puffer PBS with Tween 20.

Step 3: Fill 200 μl of the diluted conjugate solution and cover the carrier with plastic foil. Incubate overnight in the refrigerator at +4 °C (16 h).

Step 4: Wash the emptied plates 3 times with washing solution, once with bidistilled water.

Step 5: Add 200 μl chromogen (p-nitrophenly phosphate in substrate buf-fer). Incubate in the water bath at 37 °C for 30 min.

Step 6: Stop enzyme substrate reaction by adding 200 μl stopping solution (cold sodium bicarbonate). Determine the extinction by means of a reading unit (ELISA reader) and record the values. Or fill the 400 μl into 1 ccm glass cuvette and add 1.5 ml PBS buffer, measure at

405 nm against PBS blank value or 200 µl buffer and measurement in semi-micro cuvette.

Step 7:   Determine the papain quantity on the calibration curve.

Total working period, including the overnight waiting period = 20 hours.

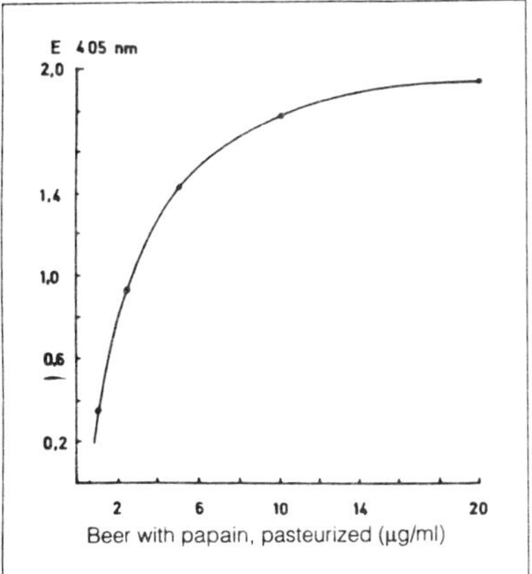

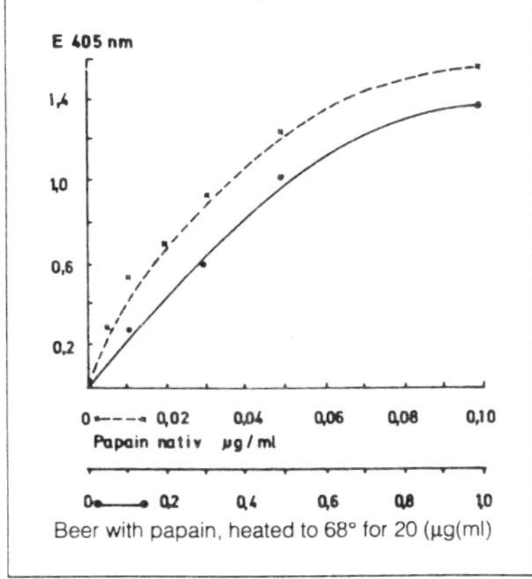

Fig. 5 Calibration Curves for Papain in Beer for Different Enzyme Immuno Assays
(Source: Bauwissenschaft. 39, H.-Carl Verlag, Nürnberg, 1986, West-Germany)

16.2.2.4 Evaluation

For this immuno assay, all sample extinctions ranging from 0 to 0.2 units are evaluated as negative, from 0.2 to 0.3 units as suspicious and above 0.3 units as positive. The background blank for each assay is to be determined by addition of the required protein. Without automatization 100 sample can be determined in one day, with instrumental reading unit 300–400 samples daily. The enzyme immuno assays is one of the fastest analysis methods known. However, as long as the absolute specificity cannot be guaranteed for each assay, the ELISA is recommended as screening test. The positive samples are further investigated using official analysis methods.

16.3   Outlook

With the application of monoclonal antibodies and intensification of the color reaction the assay can be performed within two hours. An enzyme immuno assay using monoclonal antibodies has been developed for detection of cereal proteins (gluten) from wheat, rye, barley and oats – even denatured – in food [13]. With it there are good prospects for detecting gluten absence in certain foods and raw fruit additions to beer in large number of samples.

REFERENCES

1. OUCHTERLONY, O.: Progress in Allergy **6**, 30, Karger Verlag, Basel 1962

2. GOCKE, D. und HOWE, C.: J. Immunology **104**, 1031 (1970)

3. VAN WEEMAN, B. und SCHUURS, A.: FEBS Letters **15**, 232 (1971)

4. GÜNTHER, H. O., VAAG, PIA und GIBBONS, G.: Lebensmittel Ger. Chemie **38**, 69 (1984)

5. DONHAUSER, S., VOGEL, H. und WINNEWISSER, W.: Monatsschrift für Brauwissenschaft **38**, 244 (1985)

6. GÜNTHER, H. O. und BAUDNER, S.: Monatsschrift für Brauwissenschaft **32**, 200 (1979)

7. GÜNTHER, H. O., MENGER, ANITA und BAUDNER, S.: Lebensmittel Ger. Chemie **38**, 114 (1984)

8. KLEIN, E.: persönliche Mitteilung 1986

9. SCHWEIGER, A., BAUDNER, S. und GÜNTHER, H. O.: Electrophoresis **4**, 158 (1983)

10. BERNHAUER, H., BAUDNER, S. und GÜNTHER, H. O.: Z. Lebensm. Unters. Forsch. **177**, 8 (1983)

11. GÜNTHER, H. O: nicht veröffentlicht

12. PARTANEN, P., WINNEWISSER, W., LÜBEN, G. und GÜNTHER, H. O.: Monatsschrift für Brauwissenschaft **39**, 293 (1986) u. amtl. Sammlung nach §35 LMBG–L36.007- Mai 1986

13. SKERITT, J. H.: Food Technology in Australia, **37**, 570 (1985)

14. KALTWASSER, E., BAUDNER, S., GÜNTHER, H. O.: Fleischwirtschaft **64**, 722 (1984)

15. Zitiert aus: NGO, T, LENHOFF, H. M.: Enzyme mediated Immunoassay, Plenum Press, New York - London (1985)

Theory and Application of Isotachophoresis in Food Analysis

P. Offizorz

17 Theory and Application of Isotachophoresis in Food Analysis

P. Offizorz, Rosdorf

17.1 Introduction

The isotachophoresis (Greek: iso, equal; tacho, speed) is an electrophoretic separation technique with high resolution. It allows separation and determination of different ion species in a complex mixture even in the nanogram range. This technique has the advantage of easy sample preparation, usually a simple membrane filtration. The analysis time normally is about 10–20 min.

The relevant literature lists many different descriptions regarding the technical layout [1,2,3,4], therefore only the carrier-free capillary technique, suitable for analytical determinations is described here. However, the isotachophoresis can, like other electrophoretic techniques, also be employed for preparative determinations, especially in the field of protein analytics.

For application of isotachophoresis in food analytics, especially for separating low-molecular charged food particles, the preparative technique is of little significance.

17.2 Instrumentation

Heart of an analytical isotachophoresis unit (Fig. 1) is a capillary column (0.5–1.0 mm diameter, 15–60 cm length) made of polytetrafluorethene (PTFE) or quartz glass. The capillary, sometimes divided into pre- and main column, is connected with the reservoirs for leading and terminating electrolytes containing the electrodes of the constant current power supply.

The capillary is thermostatable by air or kerosene bath. The equipment normally contains two detectors, an specific UV-detector as well as and universal detector, usually a thermometric-, conductivity-, or potential gradient detector. The sample is injected by micro-liter syringe through a suitable septum at the end of the column end at the boundary between leading and terminating electrolytes.

17.3 Basic Theory

An electrophoretic migration is called "isophoretic" if ion species of the same sign arrange themselves in an electric field in consecutive zones behind a preceding zone of higher mobility (leading ion) and in front of a zone of lower mobility (terminating ion). They have in common, an oppositely charged counter-ion for keeping the electrical neutrality, as well as a defined pH-value corresponding to each separation problem.

The current flow depends on the analyte. Contrary to the zone electrophoresis there is no background electrolyte acting as connection between anode and cathode as second-order conductor.

The principle of separation is based on different effective mobilities (m_{eff}) of ions present in the system.

17.3.1 Ionic Mobility

The absolute mobility of a supposedly sphere shaped particle depends on the intrinsic charge and the friction resistance (due to particle size and viscosity of the surrounding medium):

$$m_0 = \frac{Z^\pm \times e^0}{6\pi\eta \times r} \left[\frac{cm^2}{V \times sec}\right]$$

$Z^\pm$ = Charge number

e^0 = Electrical elementary charge

r = Particle radius

η = Viscosity of medium

For weak acids and bases the effective mobility depends additionally on the degree of dissociation of the analyte.

$$m_{eff} = \alpha \times m_0$$

The degree of dissociation α is regulated through the Henderson-Hasselbach equation of the pK-value of the surrounding buffer system.

$$pH = pK_a \pm \log(1/\alpha - 1)$$

Therefore, changing the pH-value of the buffer system can result in an increase (decrease) of the effective mobility.

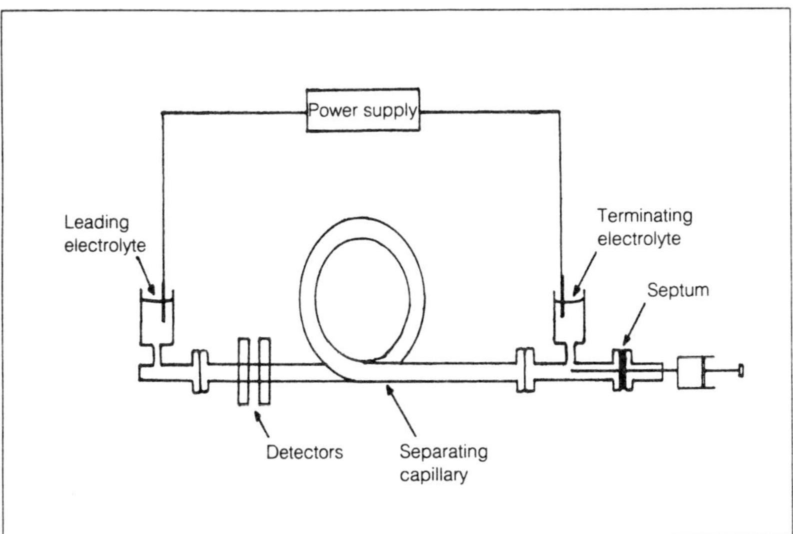

Fig. 1 Schematic Diagram of an Isotachophoresis Instrument

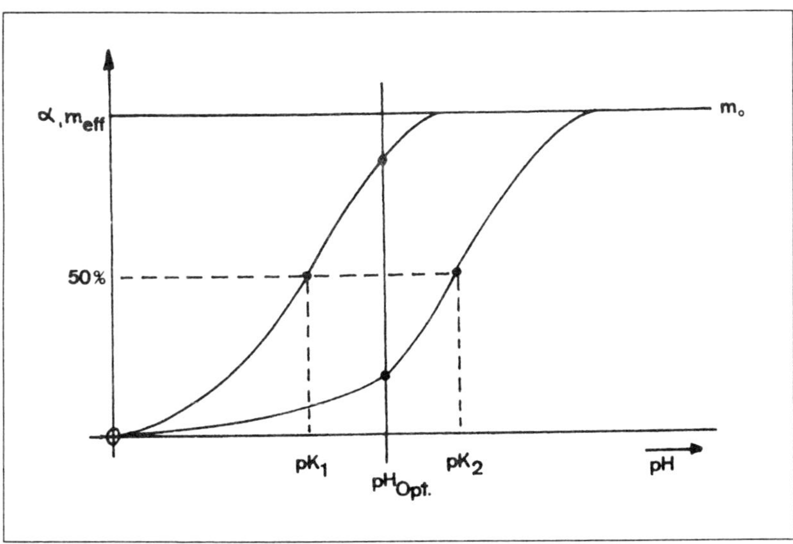

Fig. 2 The Effective Mobility of Two Weak Acids as Function of the pH-Value

Figure 2 shows the effective mobility of two weak acids as function of the pH-value of the surrounding buffer system. To obtain effective and uncomplicated separation, mobility differences as large as possible are desired. Therefore, the electrolyte system is adjusted to this pH-value by using the counter-ion in the so-called "pK-value method"

The electrophoretic migration velocity (v) of an ion depends on the effective mobility (m_{eff}) and the field strength (E), according to the formula

$$v = m_{eff} \times E$$

If the sample to be analyzed contains the ions A^- and B^- the electrolytes and the pH-value are selected so that the of the sample ions lie between the mobilities of leading and terminating electrolytes. After reaching an electrophoretic equilibrium, the sample ions migrate in separated zones with equal velocity through the capillary, following the leading electrolyte. The decreasing order of mobility from L^- to F^- is compensated by increasing field strength (Fig. 3).

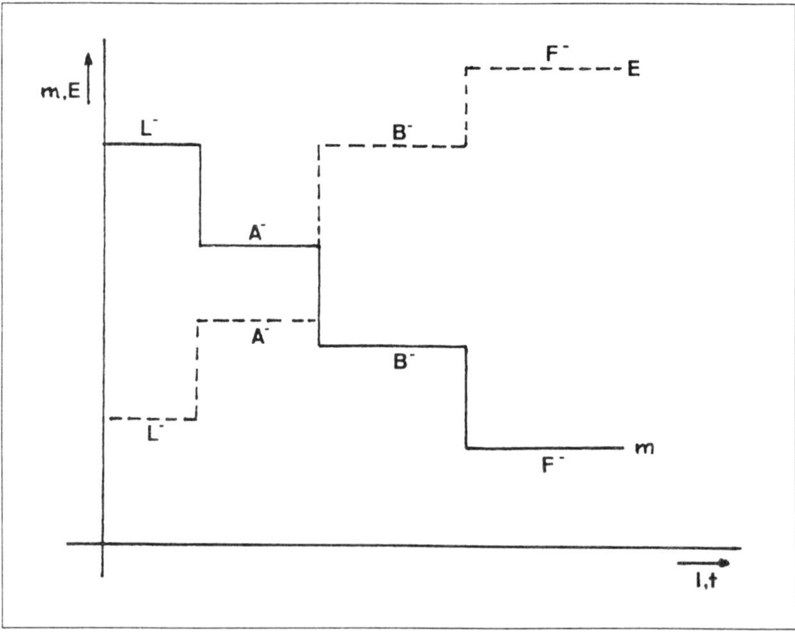

Fig. 3 Course of Mobilities and Field Strengths

17.3.2 Concentrating Effect

The existence of the so-called Kohlrausch's regulating function which can be simplified for a given buffer system with a defined constant pH-value to

$$C_L = C_A \times k$$

explains the possibilities of quantitative determination of an analyte by means of the iostachophoretic principle

Thus, the concentration of a sample ion within its zone is solely depending on the concentration of the leading electrolyte. Because the concentration within one zone is homogeneous and the diameter of the capillary does not change, the **zone length** is directly proportional to the amount of ions it contains and **not the intensity (height) of the signal.**

17.4 Detection

As can be seen from figure 1 the isotachophoresis instruments, usually available, have two detectors at the end of the separation column. A UV-detector as well as another unspecific detector, such as conductivity-, thermometric-, or potential gradient detectors, are commonly used.

17.4.1 UV-Detector

The UV-detector is the most sensitive detector available today. Under favorable circumstances ion concentrations in the nanogram range can be determined.

The concentration of sample ions is defined by the concentration of the leading electrolyte and is homogeneous within each zone. Therefore, no "UV-peaks" will be obtained but rather trapezoidal signals with absorption plateaus, similar to a Gaussian distribution curve.

The plateau height is, depending on the molar extinction coefficient, substance specific. For quantitative determinations the plateau length is significant.

17.4.2 Thermometric Detector

The joule heat product will change stepwise at each zone boundary due to changing field strengths.

With

$$I = E \times i$$

the heat occurring at constant current is directly proportional to the field strength and reciprocally proportional to the mobility of the analyte. Because the thermoelement is located outside of the capillary, the heat must first pass through the capillary wall. This delay causes a relatively indistinct signal which can only be used for qualitative determinations. Therefore, modern instruments are no longer equipped with thermometric detectors.

17.4.3   Conductometric and Potential Gradient Detectors

Both detectors are comparable regarding their resolution and they are more suitable for quantitative determinations in the µg-range than the thermometric detector. As in the thermometric detector, the changes of field strength in the individual zones are utilized. According to Ohm's law, the internal resistance in a zone at constant current is proportional to the field strength and reciprocally proportional to the mobility of a substance. The **step height** is the characteristic value for an analyte. The **zone length** is proportional to the amount of substance present (Fig. 4).

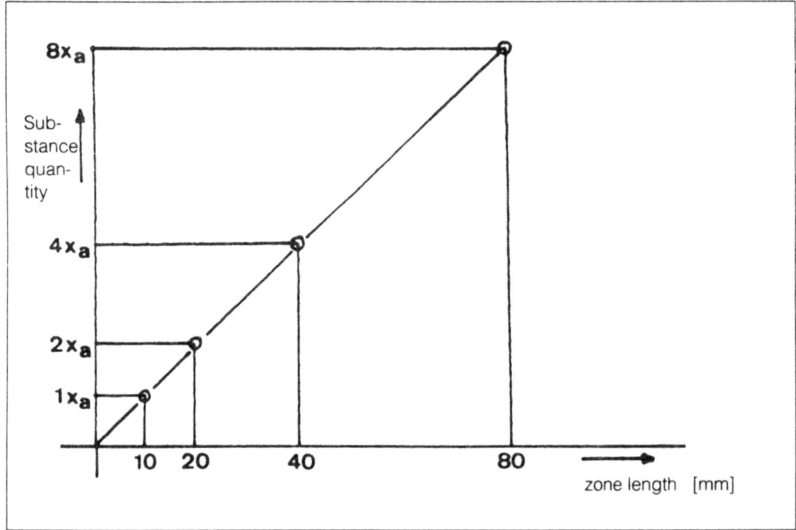

Fig. 4 Quantification by Measuring the Zone Length of the Conductivity and Potential Gradient Detector, resp.

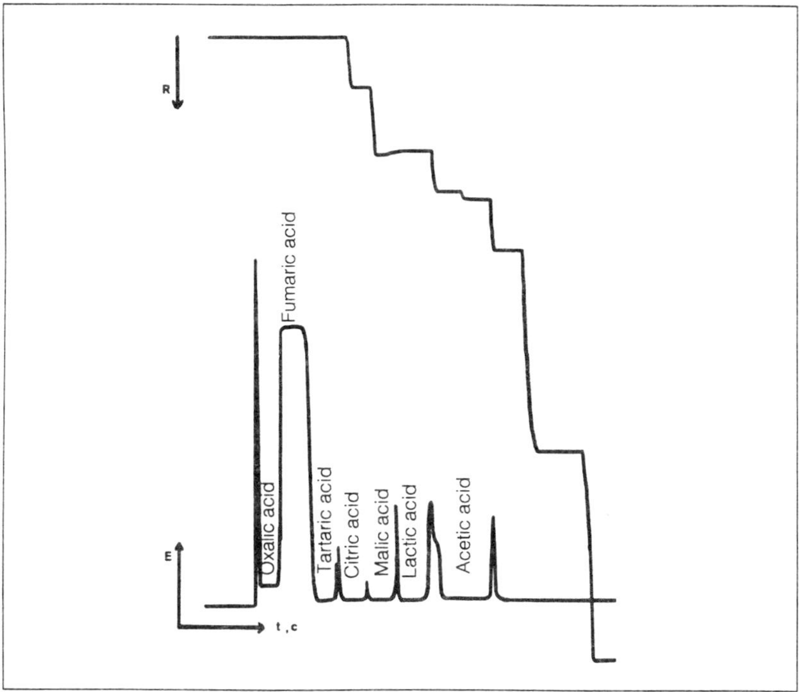

Fig. 5 Isotachophoretic Separation of Aliphatic Mono- and Dicarboxylic Acids
E = extinction, t = time, c = concentration,
R =resistance, zone length = approx. 600 ng

17.5 Applications of Isotachophoresis in Food Analysis

Isotachophoresis can successfully be employed for the analysis of food constituents as well as for the determination of additives or residues in food.

Ryslavy, Bocek, Deml, and Janak [5] used this method to determine anions in mineral waters. Organic acids in cherry juice, citrus juices and berry juices were examined by Everaerts et al. [6,7,8]. Scholze and Mayer [9] determined organic acids in coffee by means of isotachophoresis, using a computer based evaluation system for the quantitative determination. Kaiser and Hupf [10, 11] described in two studies the isotachophoretic examination of wine and fruit juices.

Figure 5 shows a typical isotachopherogram of organic acids which can be determined simultaneously in a single buffer system.

Histamine, a biogenic amine which can be used as a "freshness indicator" for fish, was also detected by isotachopheresis from several fish species [12].

The amount of the water-soluble vitamins Thiamine (B_1), Riboflavin (B_2), Pyridoxine (B_3), nicotinic acid, and nicotinic acid amide as well as ascorbic acid in vitamin preparations and vitamin supplement foods can be determined by isotachophoresis without further preparation [13, 14]. Figure 6 shows the simultaneous separation of the cationic vitamins Thiamine, Pyridoxamine, and Pyridoxol. Even quantities from 1n–100 ppm can be determined by means of the UV-detector.

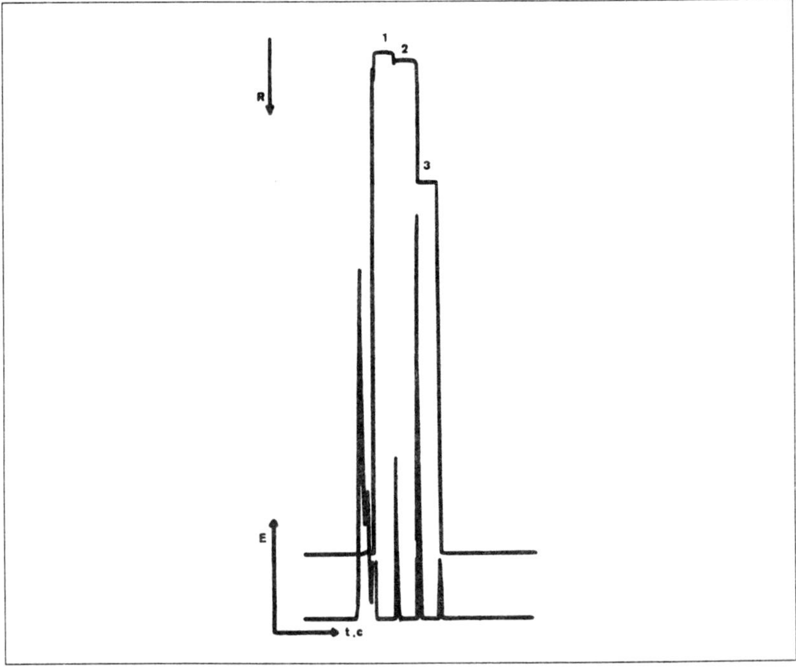

Fig. 6 Isotachphoretic Separation of Thiamine-dichloride (B_1) **1**, Pyridoxamine-dihydrochloride (B_2) **2**, Pyridoxol-hydrochloride (B_3) **3**
E = extinction, t = time, c = concentration, R = resistance

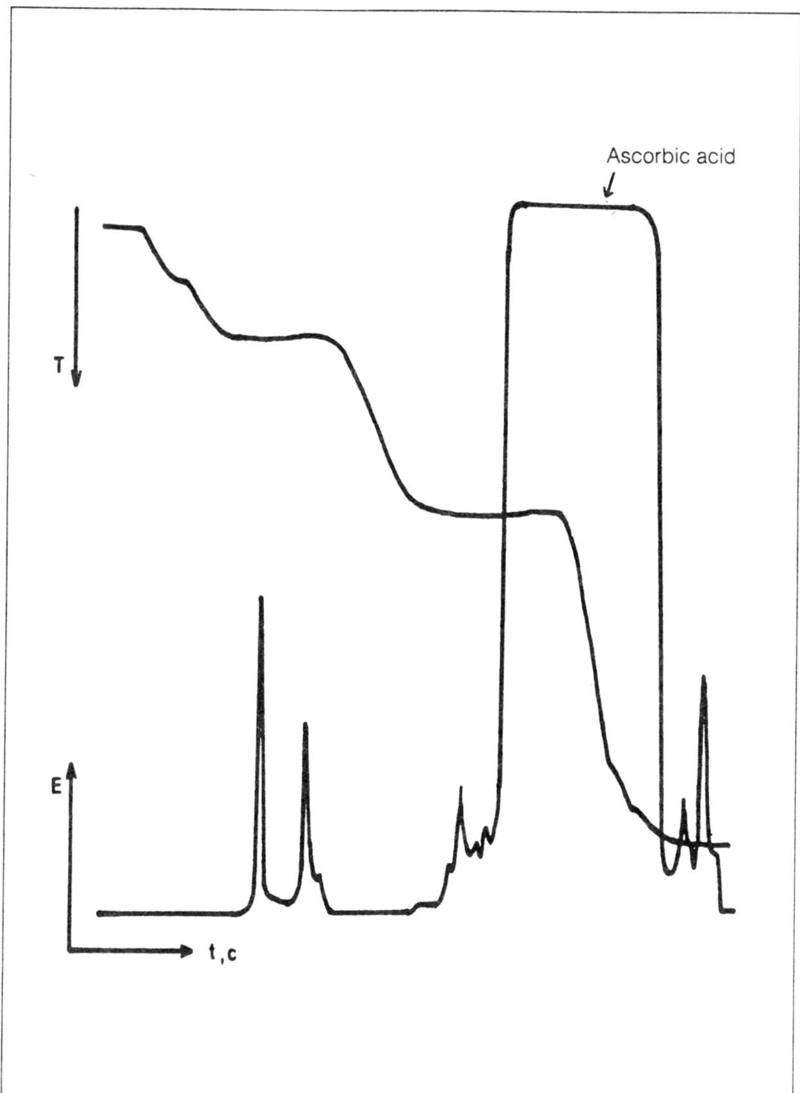

Fig. 7 Isotachophoretic Determination of Ascorbic Acid in a Reconstituted Orange Concentrate. The Signal Corresponds to a Quantity of 0.48 µg Ascorbic Acid
E = extinction, t = time, c = concentration, T = temperature

Small amounts of ascorbic acid used as antioxidant in bottled beer can also be determined by isotachophoresis in the ppm-range [15].

Ascorbic and dehydro-ascorbic acid in reconstituted citrus concentrates (Fig. 7) can be determined simultaneously by measuring the vitamin C contents before and after reduction by mercaptoethanol [16].

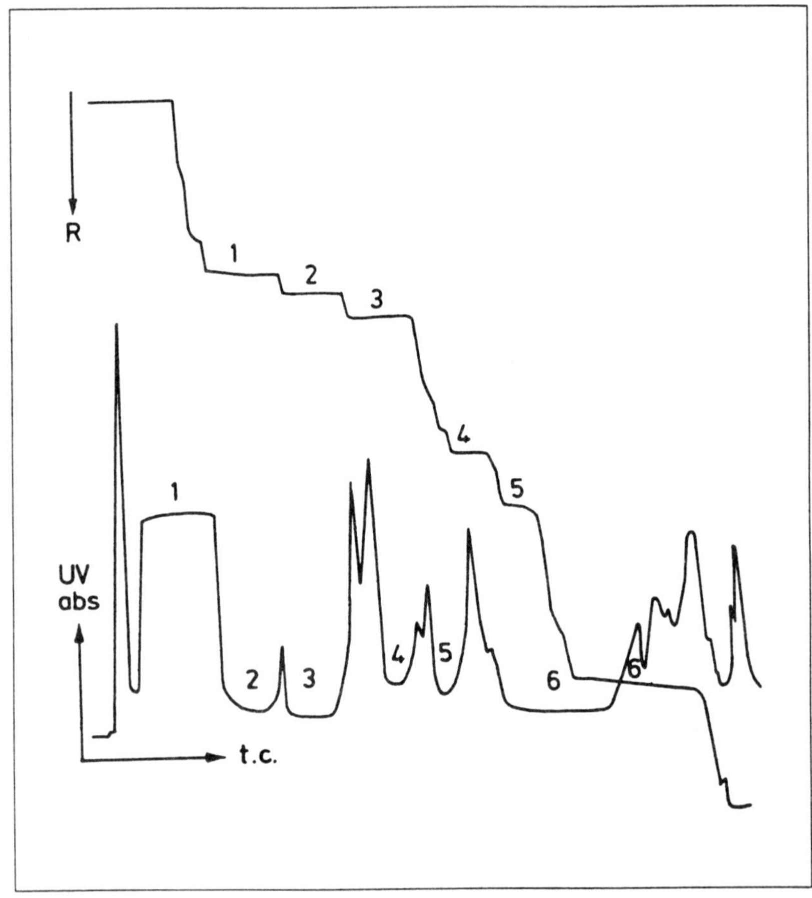

Fig. 8   Isotachophoretic Separation of Sweeteners in Sweet Mustard
Saccharine (1.28 µg) **1**, Phosphate **2**, Cyclamate (0.91 µg) **3**, Citric Acid **4**,
Lactic Acid **5**, Acetic Acid **6**
R = resistance, t = time, c = concentration

Offizorz and Rubach [17] described the possibilities for quantitative analysis of the sweeteners cyclamate and saccharine in various food items. Figure 8 shows an isotachopherogram of determination of above mentioned sweeteners in sweet mustard.

Rubach et al. [18] described a method for determination of the preservatives benzoic acid, sorbic acid, propionic acid, and PHB-ester in lemonade base, wine, bread, fish marinades, and other food. For determination of preservatives in fatty foods a sample clean-up by means of so-called Extrelut-columns (Merck) is recommended. Figure 9 shows a typical isotachopherogram for determination of preservatives in bread.

The preservatives o-phenylphenol and thiabendazol are used as surface treatment agents on citrus fruits. Those substances can be determined iso-

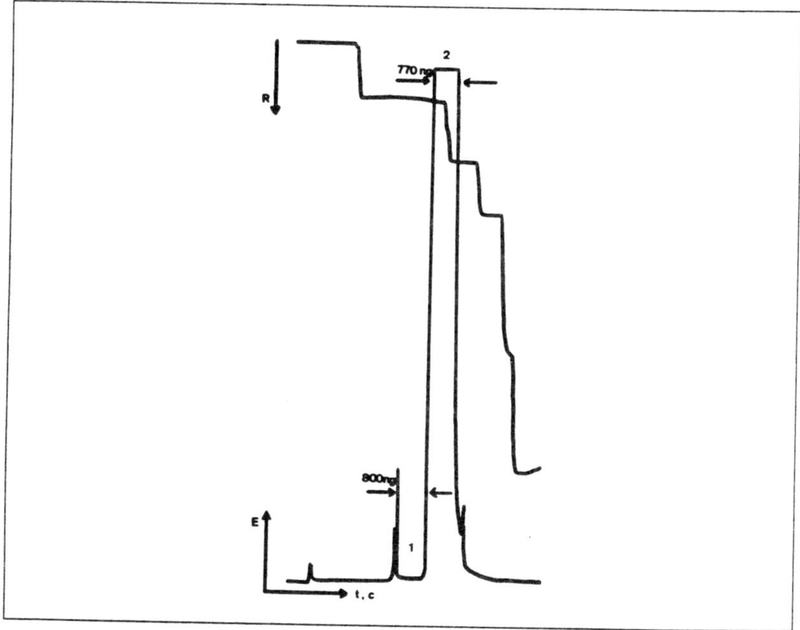

Fig. 9 Isotachophoretic Separation of Propionic Acid **1** and Sorbic Acid **2** in Bread
E = extinction, t = time, c = concentration, R = resistance

tachophoretically after cleaning and enrichment in an Extrelut-column even in quantities below 1 ppm [19].

Furthermore, in application notes of the Shimadzu company, Tokio [20, 21] the separation of amino acids in lemon and tomato juices and the determination of several anionic components in sake, tomato juice, wine, acid milk, tea, and drinking water are described.

17.6 Summary

Isotachophoretic analysis techniques have the advantage of avoiding excessive sample preparation. Time-consuming enrichment of the analyte, which might influence the accuracy of the result, is only necessary if the concentrations fall below a value of approx. 0.5–1 µg for the optimum injection volume.

Finally, it can be stated that capillary isotachophoresis should be seriously taken into consideration especially in the food analysis, in addition to other separation techniques such as HPLC or GC.

REFERENCES

1. EVERAERTS, F. M., BECKER, J. L., VERHEGGEN, TH. P. E. M.: Isotachophoresis Theory, Instrumentation and Applications, J. Chrom. Library Volume 6, Elsevier Scientific Publishing Company, Amsterdam t976

2. HJALMMSSON, S. G., BALDESTEN, A.: A critical Review of Capillary Isotachophoresis, Anal. Chem. 267 (1981) 261–352

3. BALDESTEN, A.: Science Tools 27 (1980) 1

4. HAGLUND, H.: Science Tools 17 (1970)1

5. RYSLAVY, Z., BOCEK, P., DEML, M., JANAK, J.: J. Chrom 147 (1978) 369

6. EVERAERTS, F. M., VERHEGGEN, TH. P. E. M.: J. Chrom. 91 (1974) 837

7. EVERAERTS, F. M., MULDER, A. J., VERHEGGEN, TH. P. E. M.: Am. Lab. p. 37 (1973); Int. Lab. p. 43 (1974)

8. EVERAERTS, F. M., PROSE, P., VERHEGGEN, TH. P. E. M.: Proteides of Biological Fluids, Vol. 22, Peeter, H. (Ed.) Pergamon Press., Elensford, N. Y. (1975), 721

9. SCHOLZE, A., MAYER, H. G.: Lebensmittelchem. gerichtl. Chemie 36 (1982), 111

10. KAISER, K.-P., HUPF, H.: Dtsch. Lebensm. Rundsch. 75 (1979), 300

11. KAISER, K.-P., HUPF, H.: Dtsch. Lebensm. Rundsch. 75 (1979), 346

12. Rubach, K., Offizore, P., Breyer, C. Z. Lebensm. Unters. Forsch. 172 (1981), 351

13. Röben, R., Rubach, K.. Analytical and Preparative Isotachophoresis, Holloway, C. J. (Ed.), Walter de Gruyter & Co., Berlin, New York (1984)

14. Offizorz, P., Rubach, K.: Fresenius Z. Anal. Chem. 317 (1984) 662–663

15. Röben, R., Evers, H., Krüger, E., Rubach, K.: Brauwissenschaft 37 (1984) 425

16. Rubach, K., Breyer, C.: Dtsch. Lebensm. Rundsch. 76, (1980) 228

17. Offizorz, P., Rubach, K.: Dtsch. Lebensm. Rundsch. 79, (1983) 88

18. Rubach, K., Breyer, C., Kirchhoff, E.: Z. Lebensm. Unters. Forsch. 170, (1980) 99

19. Weiss, C., Rubach, K.: Analytical and Preparative Isotachophoresis, Holloway, C. J. (Ed.), Walter de Gruyter & Co., Berlin, New York (1984)

20. Shimadzu I P Data Sheet, C A (1979)198

21. Shimadzu I P Application Data No. 1 C A 198—903 C

"Rapid Methods" in Sensory Analysis of Food

A. Fricker

18 "Rapid Methods" in Sensory Analysis of Food

A. Fricker, Karlsruhe

18.1 Introduction

The term "rapid method" in the chapter title was set in quotation marks because there are no real sensory analysis procedures that may be referred to as rapid methods. Nevertheless, existing procedures can be redesigned resulting in a significant acceleration of the procedure. With this, of course, losses regarding method accuracy and/or meaningfulness of results must be accepted. Here is an example: In general, a significance level of at least 5% is the goal for statistical evaluation of sensory results; but even 1% is possible if the investigations are extended correspondingly. If a significance level of only 10% is to be achieved, less testing is necessary and the result is obtained much faster. One thing is certain: the so-called "boss sensory test" – that is, when the boss or whoever thinks he's the boss takes part in a sensory evaluation during a telephone call and then makes a quality assessment – that is **not** a satisfactory method.

18.2 Methods of Sensory Analysis

How many different sensory tests available? With the answer to that question, it can be decided, which of them may be "changed" into a rapid method, even if certain disadvantages have to be accepted.

The following test methods are distinguished:

(1) Difference tests
(2) Description tests
(3) Evaluation tests
(4) Threshold tests

In context here, the quality evaluation of food is important, therefore, the threshold tests are eliminated because in these tests not the food quality rather than the quality of the tester is the main consideration. The description tests are also of little practical use because they are much too extensive and time-consuming, even though an evaluation is reached at the end. For example, take the best known description test, the "profile test" or "dilution test". The definition alone according to DIN 10 950, shows that this test cannot be changed into a rapid method: For the profile tests "the characteristical properties of one or more test samples are described comprehensively according to odor, taste, or texture components in the sequence of their appearance (character notes according to time) and according to their intensity (character notes according to intensity)". Even the "single description test" is not suitable because here the food is only **described**, not evaluated.

Only the difference tests and evaluation tests remain.

18.2.1   Difference Tests

At first glance one could assume that difference tests, too, cannot be used for quality **evaluation** because, according to DIN 10 950 they are defined as "difference tests for determination of the difference between two test samples". But considering, in how many cases only the question, if a quality difference of any kind can be detected between two samples is to be decided, then it is clear that evaluations are possible.

18.2.1.1 Paired Difference Tests (DIN 10 954)

This procedure may be called a rapid method, because it can be performed very easily and experienced testers can answer the following test questions very quickly:

a) Can a difference between two samples be recognized?
b) In which sample is the parameter in question more pronounced?
c) Which sample is preferred?

Experience has shown that a quick decision leads to better results. The questions themselves contain an evaluation, because if one sample is preferred, one assumes that this sample is "better". But it should be noted that the more intense perception of a property (question b) should not always be related to the term "quality improvement"; food can also be too sweet or too sour, only to mention two simple examples. The possible result that there is "no difference" can also be an evaluation if seen in connection with question c). This is the case when one test sample serves as defined standard or when samples are compared which have been stored for different periods.

18.2.1.2 Duo-Trio Test

The definition for this test method, which is not standardized by DIN, is according to DIN 10 950: "Simultaneous presentation of an individual sample and one or more sample pairs in which one sample is identical with the individual sample. The testers have to decide which sample of the sample pair corresponds to the individual sample." If a food with a certain desired quality standard is used as an individual sample and the question is changed to determining the deviating sample as well as kind and strength of the deviation, then this is a relatively quick and sensitive test for recognizing small quality differences. Fewer samples must be tested, since the individual sample remains the same, other than for the mostly employed difference test, the "triangle test".

18.3   Comment on Terminology

Several times now, the term "standards" has been used. In performing sensory test methods as rapid methods, the application of clear quality standards is of significant importance, and the question of the definition of the term "sensory standard" should be discussed.

Basically, this definition is very free as long as all persons concerned with sensory tests – really all of them – agree on the definition. But this is an illusion. It would be most useful, especially for rapid methods, to take a sample containing the desired properties for that special product as a sensory standard. The optimum total quality as well as, depending on aim and purpose of the test, the optimal presentation of the individual characteristics should be striven for. In other words: In one case, e.g., the texture or the taste or the odor can be considered as individual attribute, in another case the acceptability of the food.

But it is clear that for other problems other characteristics can be applied as sensory standards.

18.4   Evaluation Tests

As the name says, these are sensory tests in which the food is evaluated for quality. According to DIN 10 950 the evaluation test is aimed at evaluating the test sample as a whole or of individual characteristics of it. Classification in the sense of allocation into quality categories, classes, or ranges is also an evaluation.

18.4.1   Evaluation Test with Scale

This sensory test is the classical method for determining the food quality. It includes an evaluation of the sensory impressions by kind and intensity. This method is laid down in DIN 10 952, part 1 "evaluation test with scale: test methods" and part 2 "evaluation method with scale: preparing test scales and evaluation schemes". According to the fundamentals of this standard the Deutsche Landwirtschaftgesellschaft (DLG) has performed extensive quality tests and set detailed specific evaluation criteria for the individual food groups. The standard is included in the "Amtliche Sammlung von Untersuchungsverfahren nach   35" LMBG (part of the German food law). A detailed description of this method cannot be given here (cf. the standard publications). However, it should be mentioned that different scales or scale ranges can be used. Table 1 from the standard shows how such scales are structured and how certain properties are related to certain "numerical values".

Tab.1   Example of a Non-Specific Evaluation Scheme

k = 9	k = 6
9 perfect, optimal	6 perfect, fulfills all requirements
8 typical, faultless	5 slight deviations
7 typical, with minor deviations	4 noticeable deviations
6 noticeable deviations	3 clear fault
5 noticeable impairments, minor fault	2 strong fault
4 clear fault	1 totally changed (cannot be evaluated)
3 strong fault	
2 very strong fault	**k = 3**
1 totally changed	3 perfect to minor deviations
	2 noticeable deviations to clear faults
	1 strong faults to totally changed

As can be seen for a relatively exact description a scale with nine values is used. The DLG works with a six point system which is sufficient for their purpose. This non-specific evaluation scheme can be adapted specifically to each food item, i.e. for a defined food item with "x" points a very special quality definition can be made or vice versa. Thus a product specific evaluation scheme can be provided. If a tester finds certain properties in the food tested he must give this property the points according to the scheme. Individual attributes such color, form, odor, taste, texture are evaluated separately according to these schemes. For statistically significant results many tests with many testers must be performed. But there are two possibilities to change the evaluation test with scale to a relatively rapid method. First, the performance as a hedonic test, that is, only the popularity (How does the product taste to me?) is taken as the criteria and/or only three numerical values are used, as already provided in the table. However, in the latter case the development of a specific evaluation scheme will be difficult, but the use of the more general remarks seems to be sufficient. Also, depending on the purpose it is possible to limit the testing to individual attributes such as texture.

18.4.2 Ranking Test (DIN 10 963)

Another evaluation test which can also be performed very quickly is the ranking test. For this test three or more test samples should be "sorted into a order of rank according to a preset criteria (e.g. general impression, specific characteristics, specific property)" (according to DIN 10 950). This method is used for "quick classification of products, e.g., according to kind or intensity of individual attributes". Experienced testers can perform these tests in a very short time, if the number of samples is not too large (3–6). The evaluation obtained is very meaningful. Here too, the employment of sensory standards is useful, as previously described. Even though the meaningfulness can be significantly increased by a statistical evaluation method, for a rapid method this extensive calculation is not needed. The individual result are usually sufficient to obtain a rough idea as to the quality classification.

18.5 Summary

With simplifications, increased employment of sensory standards and other changes in sensory test methods, a relatively quick determination of food quality can be obtained. The following methods were described: paired difference test, duo-trio test, evaluation test with scale, and ranking test. It must be emphasized that losses in accuracy and meaningfulness of the tests must be expected.

Physical Methods for Rheology, Consistency and Particle Size Measurements

E. Windhab

19 Physical Methods for Rheology, Consistency and Particle Size Measurements

E. Windhab, Quakenbrück

19.1 Rheology – Consistency

Two different groups of measuring methods for rheological parameters can fundamentally be differentiated:

- absolute rheological methods
- empirical methods

With the help of absolute rheological methods, parameters that are independent of the measuring device can be determined. Empirical methods yield consistency parameters consisting of several superimposed absolute physical parameters. If measurements are used only for the relative comparison, empirical values are most often suitable. For the design of technical apparatus (pipes, mixing and stirring systems, etc.) and their scale up as well as for the description of rheological phenomena (viscoelastic effects) absolute rheological parameters are necessary. More over it has to be recognized that superposition of absolute rheological parameters in empirical measurement techniques may lead to compensation effects followed by a loss of relevance for the valuation of the rheological behaviour.

19.1.1 Rheological Fundamentals

The important rheological parameters for most of the practical purposes are:

The **flow function** $\tau\,(\dot{\gamma})$ describes the functional dependency of the fluid shear rate and the resulting shear stress τ in the fluid.

The **viscosity function** $\eta\,(\dot{\gamma})$ can be calculated from the flow function by Newton's law $(\tau\,(\dot{\gamma}) = \eta\,(\dot{\gamma}) . \dot{\gamma})$.

The **yield value** τ_0 is the minimum shear stress that must be overcome to cause a deformation of the fluid (important for suspensions, dispersions, etc.).

The **normal stress functions** γ_1, γ_2 $(\dot{\gamma})$ describe the elastic fluid behaviour dependent on the shear rate.

The **dynamic moduli** G' (ω); G'' (ω) and the **dynamic viscosity** coefficient η' (ω) or the complex viscosity η^* (ω) that describe the viscous (G'' (ω); $\eta'(\omega)$ η^* (ω)) and the elastic (G' (ω)) flow behaviour measured under definite unsteady flow conditions (oscillatory test).

The **slip function** τ (υ_g) is not a pure fluid parameter but also dependent on the wall conditions (adhesion forces, roughness, etc.) (v_g = slip velocity). For an accurate measurement of rheological parameters the slip function at the walls of the measuring gap must be known if slippage cannot be avoided. For concentrated suspensions and dispersions wall slippage has to be taken into account.

The **slip limit** τ_1 is the minimum shear stress to overcome the adhesion forces to the wall. τ_1 is dependent on the parameters characterizing the interaction between fluid and wall.

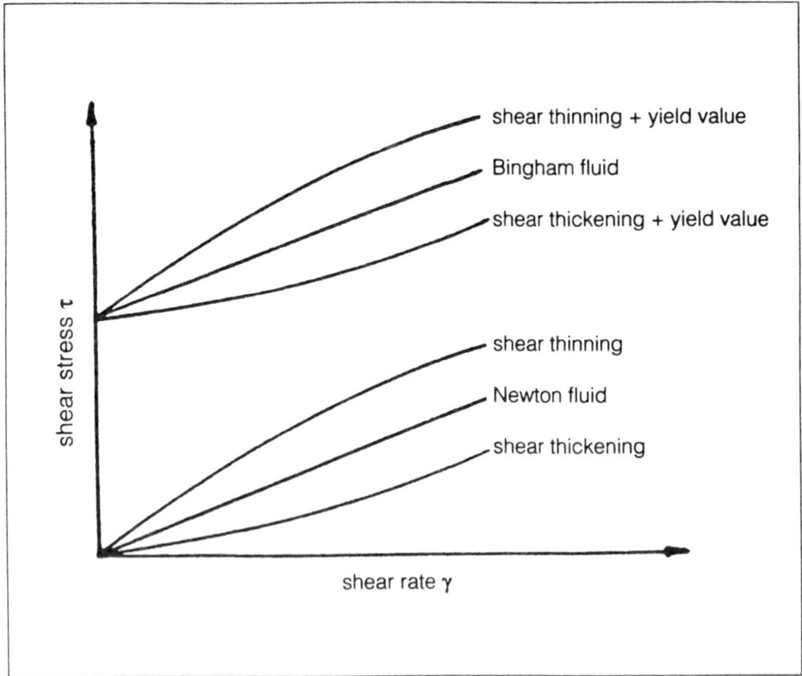

Fig. 1 Typical FLow Functions

The parameter that is most often the easiest to be measured is the **flow function** $\tau\,(\dot{\gamma})$ respectively the viscosity function $\eta\,(\dot{\gamma})$.

Figure 1 shows typical flow functions. The slope of the flow function corresponds to the viscosity. "Shear-thinning" behaviour means decrease of viscosity with increasing shear rate. The opposite is called "shear-thickening". Newtonian flow behaviour means constant viscosity (independent of the shear rate). A yield value may be superimposed (plastic behaviour, Bingham fluid, etc.).

Time dependent flow behaviour

The rheological parameters described above may be a function of the measuring time. Figure 2 shows qualitatively the time dependency of the viscosity respectively shear stress. Increase of viscosity with time at **constant** shear rate is called **rheopexy**, the opposite, which occurs more often, is called **thixotropy**.

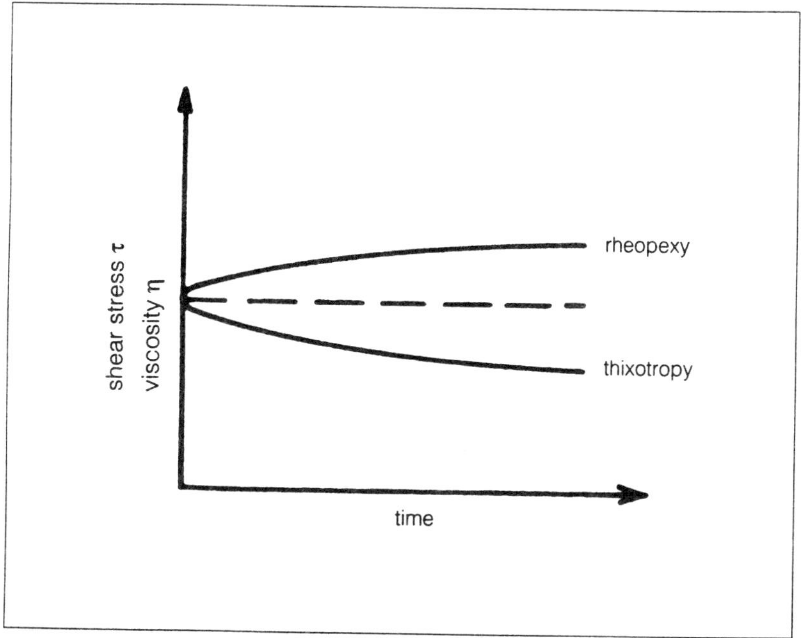

Fig. 2 Time Dependent Behaviour of Flow Functions

Thixotropy, rheopexy as well as shear-thinning and shear-thickening behaviour are caused by structural changes in the fluid.

19.1.2 Measurement of Rheological Parameters

19.1.2.1 Absolute Rheological Parameters

19.1.2.1.1 Presumptions

Following presumptions have to be fulfilled for measurement of absolute rheological parameters:

- rheometric flow (plane, laminar, parallel flow layers)
- adhesion to the walls
- homogeneity

19.1.2.1.2 Experiments

The typical experiments to measure absolute rheological parameters are described in Figure 3.

experiment	rheological parameter (measured)
– flow test	– flow function viscosity function – normal stress functions – slip functions
– creep test given stress and measured strain	– flow function viscosity function – yied value – slip function – slip limit – compliance
– oscillation test given oscillating stress/ strain and measured oscilation strain/stress	– dynamic viscosity coefficient – complex viscosity – storage modulus – loss modulus

Fig. 3 Experiments to Determine Absolute Rheological Parameters

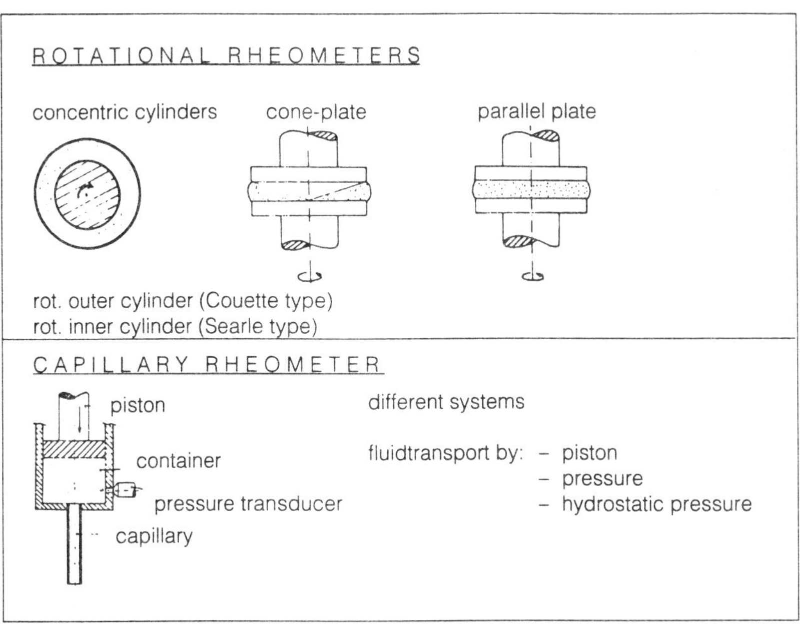

Fig. 4 Gap Geometries for "Absolute" Measuring Rheometers

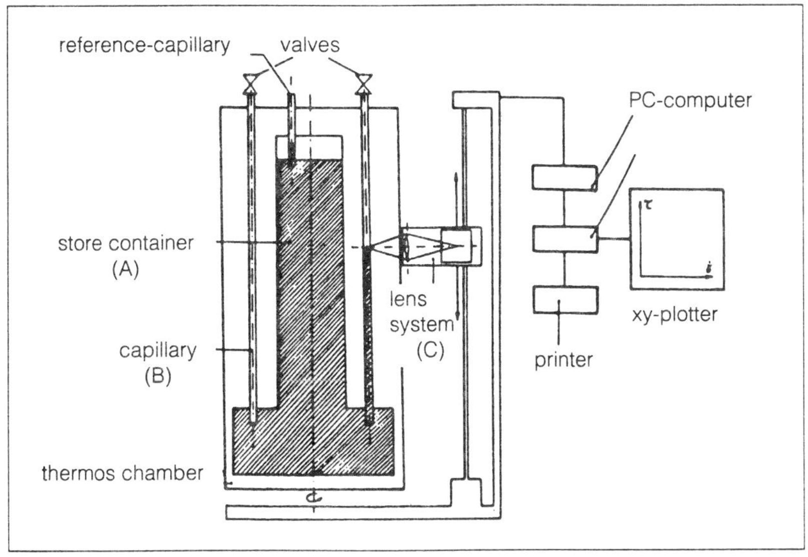

Fig. 5 Hyrheometer

19.1.2.1.3 Measuring Geometries

The most often used geometries for measuring the gap are shown in Figures 4 and 5. Generally, rotational viscometers are used for low viscous to high viscous fluids ($\eta \approx 1\text{--}10^6$ mPas), capillary rheometers (high pressure capillary rheometers) allow to measure even higher viscosities ($\eta \approx 1\text{--}10^9$ mPas).

19.1.2.1.4 Measuring Equipment

A general view of available viscometers (rheometers) is given in Tables 1 and 2. Apparatus that allow to carry out only the flow test have prime costs of 5,000 to 10,000 \$. Measuring equipment which is able to realize further experiments (creep test, oscillatory test) have prime costs in the range of 10,000 to 50,000 \$ (few of them are even more expensive).

Tab. 1 Measuring Devices I

type of rheometer	geometry	manufacturer	flow test	creep test	oszill- test	normal force
rotational	concentric	Brabender	X		X	
rheometers	cylinders	Bruss	X	X		
		Carrimed	X	X	X	
		Bohlin	X	X	X	
		Brookfield	X			
		Contraves	X		X	
		Haake	X		X	
		Viscom.-UK LTD	X			
		Physica	X			
		Rheometrics	X	X	X	
	cone-plate	Brabender	X		X	X
		Bohlin	X		X	X
		Carrimed	X	X	X	
		Contraves	X		X	
		Haake	X		X	
		Rheometrics	X	X	X	X
	parallel	Bohlin	X		X	
	plate	Carrymed	X	X	X	
		Haake	X		X	
		Rheometrics	X	X	X	

Tab. 2 Measuring Devices II

type of rheometer	geometry	manufacturer	flow test	creep test	oszill- test	normal force
capillary rheometer	capillary	Ceast	X			
		Göttfert	X			
		LTG-Karls- ruhe		X		
		Paar		X	X	
		Schott	X*			
		Lauda	X*			
falling ball viscometer	ring gap	Paar	X*			
		Haake	X*			

* Only Newtonian fluids

19.1.2.1.5 New Measuring Principle

A new measuring principle developed by the Deutsches Institut für Lebens-mitteltechnik (German Institute of Food Engineering, DIL, Quakenbrück, West-Germany) is shown in Figure 5. With this stress-controlled capillary rheometer viscosity and flow function ($\eta \approx 50\text{--}10^6$ mPas) as well as the yield value τ_0 can be measured with high accuracy ($\pm$ 0.01 mPas). Furthermore, the slip function $\tau\,(\gamma)$ and the slip limit τ_1 are measurable. Measuring the height of the ascending fluid level in the measuring capillary enables a higher sensitivity than torque measurement does in rotational rheometers.

19.1.2.2 Measurement of Empirical Rheological Parameters

19.1.2.2.1 Measuring Principles

For the measuring equipment suitable for measurements of empirical con-sistency parameters, the following classification into five groups is possible:

1. rotational methods
2. penetration methods
3. flow out methods
4. sag methods
5. tearing methods

These methods are described in the Figures 6-10 with their measuring geo-metries, measuring parameters and range of application.

R O T A T I O N A L   M E T H O D S

geometry: rotational stirrers without defined shear gap

measuring parameter: torque or power input of the motor

application: low viscous to medium viscous fluids

examples (stirrers)

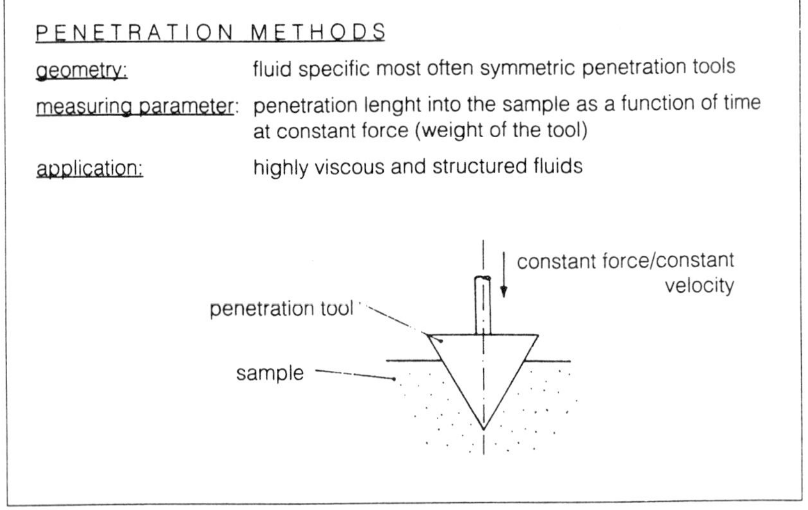

Fig. 6   Rotational Methods

P E N E T R A T I O N   M E T H O D S

geometry: fluid specific most often symmetric penetration tools

measuring parameter: penetration lenght into the sample as a function of time
at constant force (weight of the tool)

application: highly viscous and structured fluids

Fig. 7   Penetration Methods

<u>FLOW OUT METHODS</u>

<u>geometry:</u> fluid container with defined outlet (pipe)

<u>measuring parameter:</u> flow out time of a defined volume or mass

<u>application:</u> low viscous to medium viscous fluids

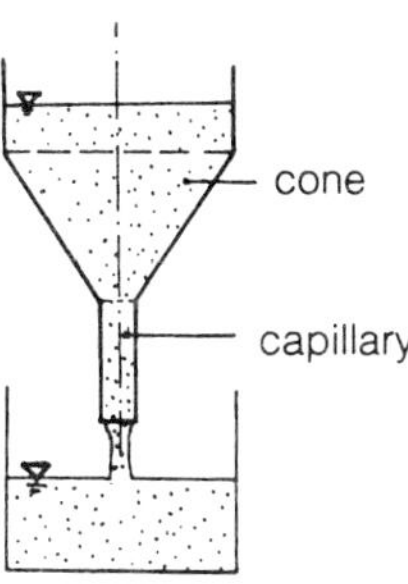

Fig. 8 Flow out Methods

<u>SAG METHODS</u>

<u>geometry:</u> cylindrical or conical shape of the sample

<u>measuring parameter:</u> sag height after specified time

<u>application:</u> structured fluids (gels)

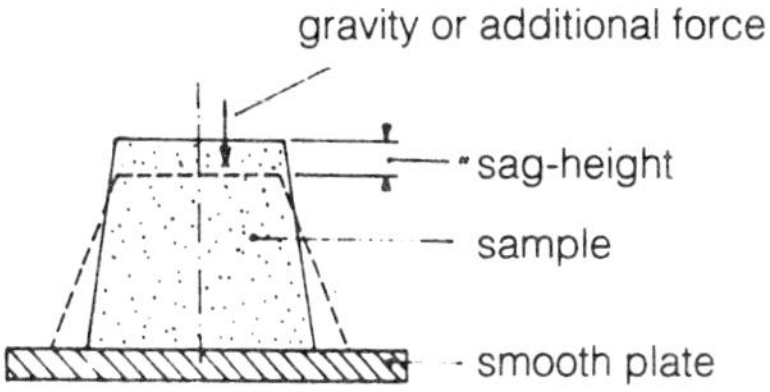

Fig. 9 Sag Methods

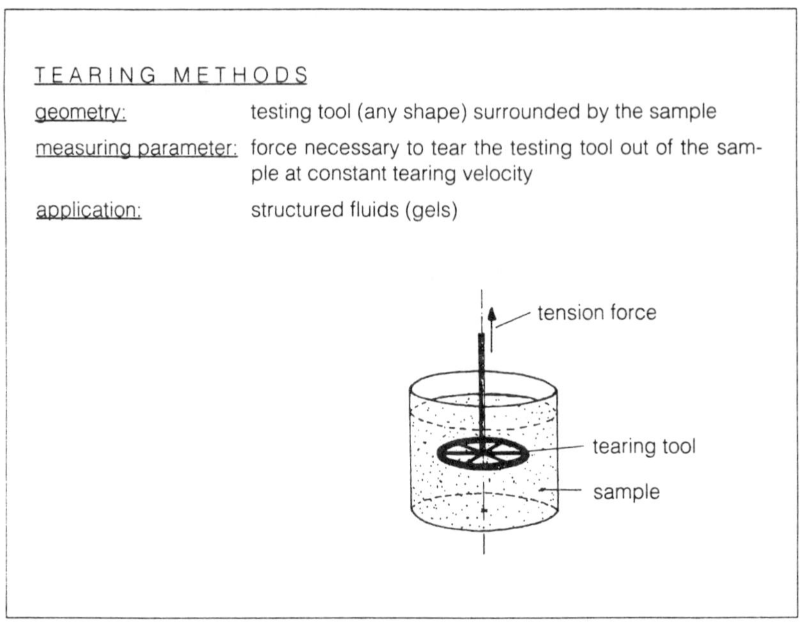

Fig. 10 Tearing Methods

19.1.2.2.2　Measuring Devices (Manufacturers)

Table 3 shows some manufactures of empirical measuring devices for consistency parameters.

19.2　Particle Size Measurement

Many foods as well as the "raw materials" are suspensions, powders, dispersions or emulsions. Their physical behaviour is mainly influenced by the properties of the disperse phase. One of the important parameters to characterize the disperse phase is the particle size respectively the particle size distribution.

Tab. 3 Empirical Rheological Measuring Devices

type of device	manufacturer/device
rotational methods	Brabender Brookfield Bruss Coesfeld F.I.R.A./ jelly tester Fann Vickometers UK
penetration methods	Instron J. J. Lloyd Instruments LTD Shimatzu Stevens Zwick SUR (Berlin)
flow out methods	Fann Erichsen SUR (Berlin) Haake/Konsistometer
sag methods	Kinematica Waegeningen – Sag Tarr – Baker IFT – Sag
tearing methods	Herbstreith Luers – Lochmüller all tension/pressure testing apparatures

19.2.1   Fundamentals of Particle Size Measurement

For the particle size characterization different particle properties may be considered:

Geometrical properties:

- length (diameter, feretdiameter, etc.)
- areas (projection area, surface, etc.)
- volume

<u>cumulative frequency distribution</u>

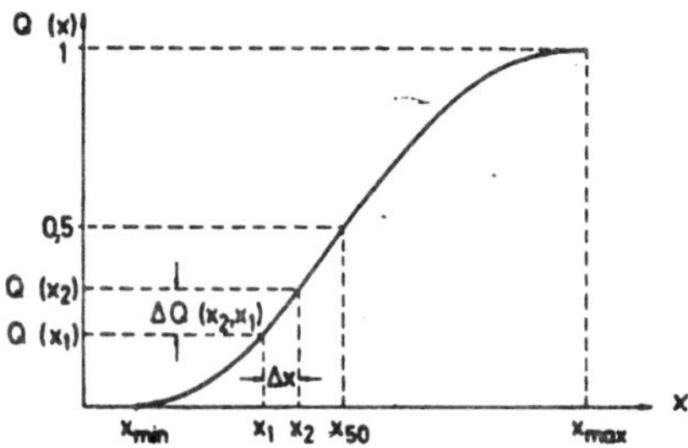

<u>absolute frequency distribution</u>

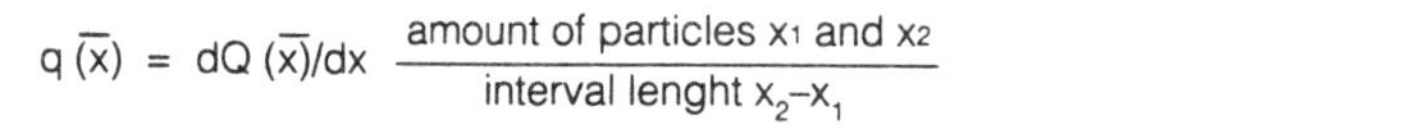

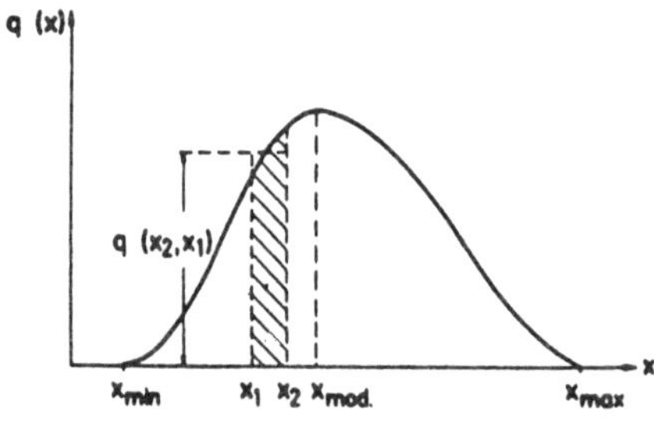

Fig. 11 Particle Size Distribution

Equivalent diameters:

- diameter of sphere with equal surface
- diameter of sphere with equal projection area
- diameter of sphere with equal volume

Other particle properties:

- sedimentation velocity
- extinction
- density
- light scattering
- light diffraction

For these characteristical parameters the frequency distributions can be determined. Two different methods of description for the particle size distribution are commonly used, the accumulated distribution and the incremental frequency distribution:

Accumulated distribution $Q(x_1)$:

$$Q(x_i) = \frac{\text{amount of all particles between } x_1 \text{ and } x_2}{\text{total amount of particles}}$$

Incremental distribution $q(\overline{x}_i)$:

$$q(\overline{x}_i) = d\,Q(x)/dx = \frac{\text{amount of particles between } x_1 \text{ and } x_2}{\text{Interval length } x_2 - x_1}$$

19.2.2 Measuring Methods

The measuring methods differ dependent on particle size beginning with conventional sieving analysis to modern optical methods (light scattering, light diffraction).

Figure 12 shows a general view of the usual methods and their range of application (particle size range). Most of these methods use equivalent diameters as characteristic parameter dependent on the measured parameter.

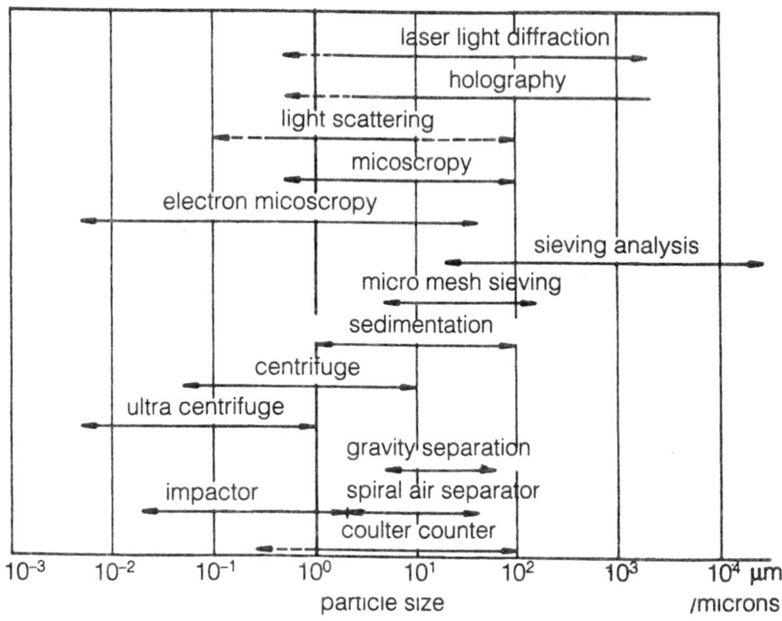

Fig. 12 Range of Application of Difference Measuring Methods Dependent
on Particle Size

19.2.2.1 Measuring Devices

Table 4 shows examples of measuring apparatus and the manufacturing
companies. The prime cost range extends from about 1,200 $ (sieving ana-
lysis) to more than 50,000 $ (optical methods, light scattering, light diffrac-
tion, electron microscopy).

Tab. 4 Measuring Devices for Particle Size Measurement

measuring principle	particle size/ μm	measuring parameter	manufactures
sieving analysis	5–125 000 5–100 (ultra-sonics)	diameter	Endecotts LTG–Karlsruhe Retsch, Humbold-W. Rheowum, Siebtech. GmbH
sedimentation	0.5–100 (gravity field) 0.1–150 (centrifugal field)	sedimentation velocity	Micromeritics Schimadsu
air seperation	0.3–100	mass	Retsch (Laborimpaktor) Alpine
electrokinectic methods	0.3–1000	electrical resistence	Particle Data Inc. Coulter
optical methods	0.5–1800	light diffraction light scattering	Cilas Malvern Leads and Northrup Sympatec Polytec Kratel Royco ETI (Dap 2000)
	dependent on enlargement	picture analysis	Leitz Al-Tectron Kontron (Zeiss) Hamamatsu Nikon Tracor Oympus
	0.001–5	photon cor-relation	Malvern Coulter

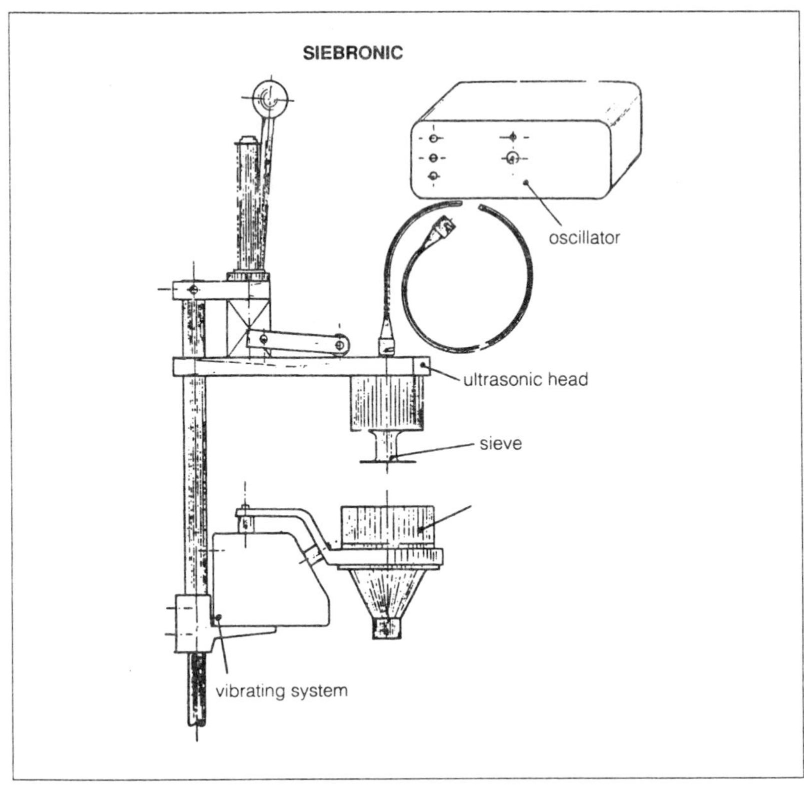

Fig. 13 Measuring Device "Siebtronic"

19.2.2.2 Short Time Sieving Analysis

The sieving analysis is the cheapest way ($\geq$ 1,2000 $) of particle size measurement. Sieving of suspended particles (micro mesh sieving) is possible down to 5 microns. A combination of vibrating sieves with an ultrasonic oscillating device, which is positioned parallel to the sieve, avoids the build-up of a filter cake, enables short sieving time (some minutes). Figure 13 shows such a measuring device called "Siebronic", developed by the company LTG (Gesellschaft für meßtechnische Entwicklung und Beratung mbg, Karlsruhe, West-Germany).

Microbiological Rapid Methods

R. Zschaler

20 Microbiological Rapid Methods

R. Zschaler, Hamburg

20.1 Direct and Indirect Count Determination Methods

The microbiological quality and with it the durability of perishable or even preserved food of plant or animal origin is determined mainly by the amount of living, multiplicable microorganisms present in food.

Determination of viable germs is of central importance. Direct and indirect count determination methods are employed for this purpose. They can be divided into very rapid methods (the results will be obtained in less than one hour), rapid methods (the results will be obtained within 4–8 hours) and longer lasting methods (the results will be obtained after 12, 48, or even 72 hours).

Figure 1 shows a survey of the individual methods, divided into these three groups. The methods marked with a cross will be discussed shortly in the following.

Figure 2 shows the mentioned methods, but they are in different order, that is, they are sorted into **"non-cultural"** and cultural methods. The individual possibilities are numbered and can be assigned to the categories 1, 2, and 3.

20.1.1 Direct Epifluorescence Filter Technique

First, the **D**irect **E**pifluorescence **F**ilter **T**echnique (DEFT) should be mentioned. It was developed by Pettipher et al. in 1980 and is an improved microscopic total count determination. It was developed mainly for liquid products, especially for raw milk. But today, even solid food can be tested by means of this technique. For that the product must be liquefied by homogenization, prefiltration, treatment with enzymes and tensides, if necessary.

After filtration through 0.5 µm Nucleopore membrane filters the individual microorganism cells retained on the filter are stained with a fluorochromatic dye, e.g. acridine orange, and counted under the fluorescence microscope. The counting is very difficult. A significant improvement is the analysis by means of a monitor. After gaining some experience with this technique the results can be obtained after 20–30 min.

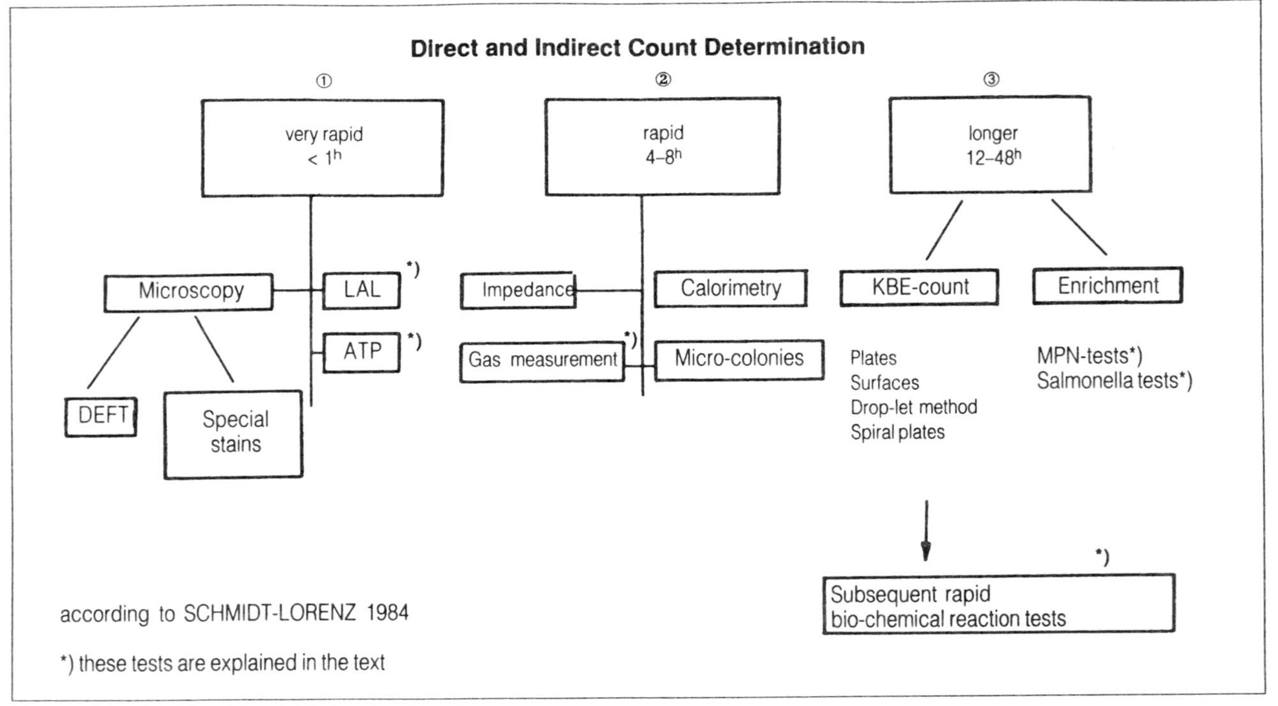

Fig. 1 Survey on Individual Methods

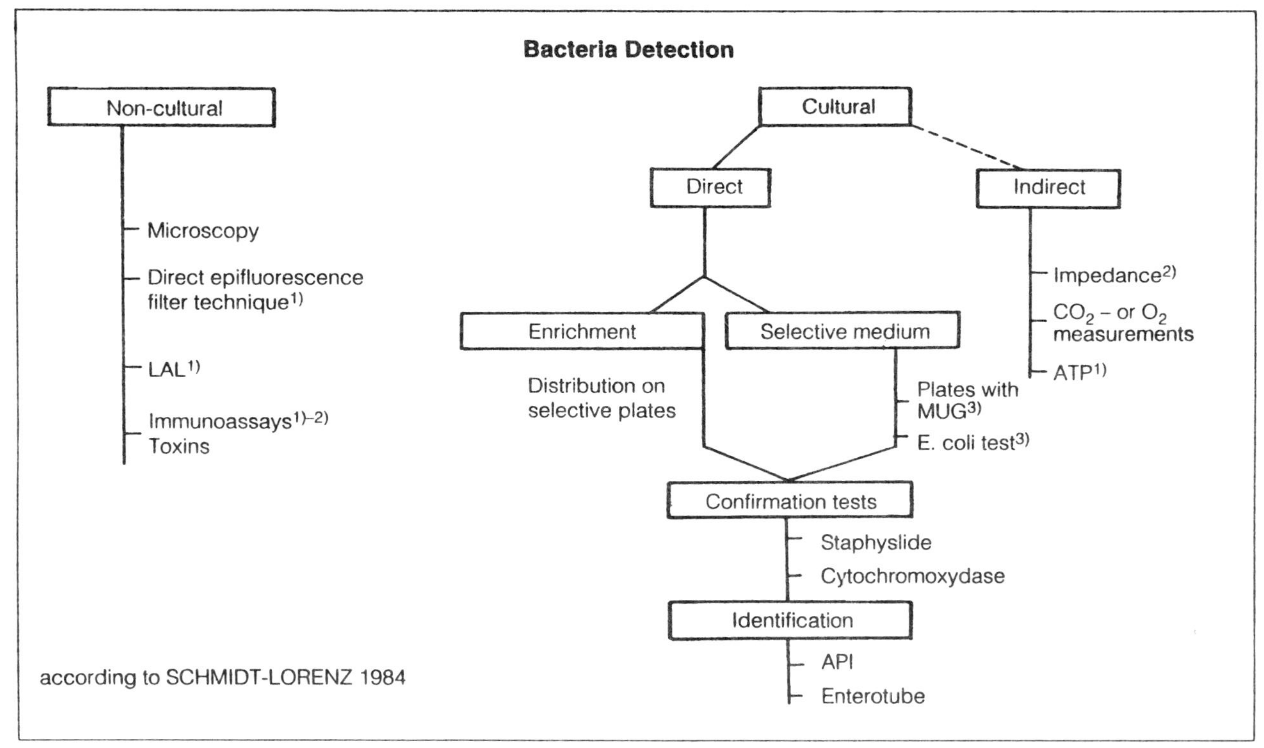

Fig. 2 Detection of Bacteria by Non-Cultural and Cultural Methods

20.1.2   Limulus Amoebocyte Lysate Test

Among the very rapid methods is the so-called **LAL-Test,** an abbreviation
for **L**imulus **A**moebocyte **L**ysate **T**est. It is a very sensitive detection test for
lipopolysaccharides in the cell walls of gram negative bacteria. Therefore, it
can only be applied for detection of gram negative germs. This test is
based on the principle that the blood of the horseshoe crab coagulates
when infections with gram negative germs take place. A gel is formed between the cell constituents, the lipopolysaccharides of such bacteria and the
amoebocytes, the only blood corpuscle of the limulus blood. The amoebocytes contain pro-enzymes and agglutination enzymes. If lipopolysaccharides are present, the pro-enzymes are activated. This gel formation is used
for detection of endotoxins. A linear correlation exists between the endo-
toxin concentration and the number of gram negative bacteria present.

Today, ready-made microtitration plates can be purchased for this test, either by Mallinckrodt Diagnostic or by bioMÇrieux. The so-called limulus test
is applied in practice for the examination of egg noodle products in commercial plants. Since the test result is available within 1–2 hours, it is possible to detect infection with gram negative germs in a container with whole
egg or egg yolk within this time. If the internal limit is exceeded, acceptance
of the container can be refused.

20.1.3   Enzyme Linked Immuno Sorbent Assay

The list of non-cultural methods should be concluded with the example of
the **ELISA** test (**E**nzyme **L**inked **I**mmuno **S**orbent **A**ssay). Here, not the
germs are detected indirectly, but the toxins formed by certain germs. The
so-called **Fey test**, described here, is a semi-quantitative indirect ELISA test
for which plastic beads, coated with specific antibodies, are used. With the
employment of this test the direct detection of enterotoxins of Staphylococcus aureus in food is possible. Disease symptoms can be triggered by enterotoxin quantities between 1.0 and 10.0 µg in 100 g food. For the Fey test,
the sensitivity ranges from 0.1–1.0 ng/g food (depending on extraction and
type of food). The ELISA technique, according to Fey, allows much faster results (within 24 hours) than the commonly used precipitation method.

The course of examination is shown in Figure 3. The suspicious sample is
extracted. The toxin-containing extract obtained is first incubated with anti-
body-coated plastic beads and control beads and then the conjugate is
added. After addition of dye the samples are evaluated after a waiting period.

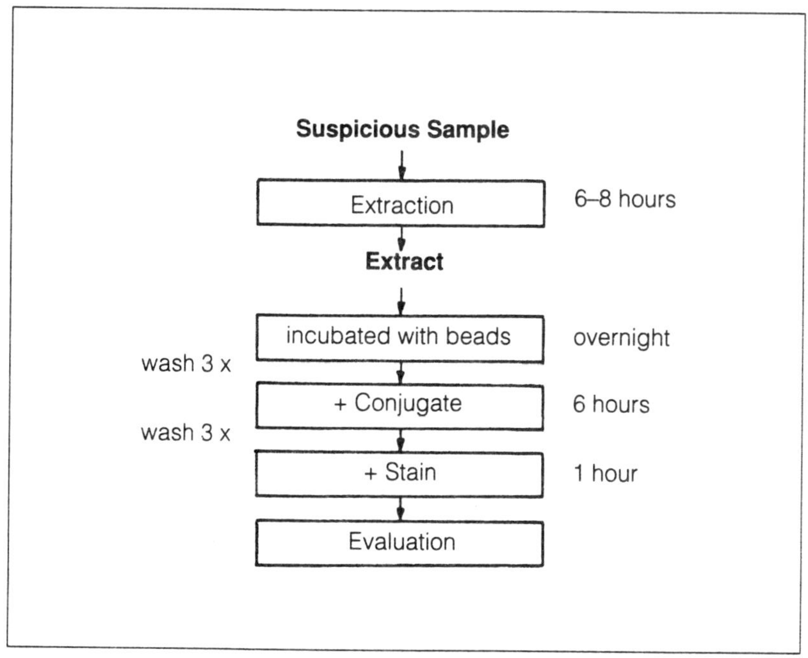

Fig. 3 FEY-ELISA Examination Scheme

Extraction: Water is added to the food which is homogenized and then centrifuged at 10,000 rpm for 20 min at 5 °C. The pH-value of the remaining liquid phase is adjusted to 4.50 as measured with a pH-electrode. Another centrifugation under the same conditions follows. The pH-value of the liquid phase is then increased to 7.50.

For removing an eventual fat layer, filtration through moist hydrophilic cotton is done. For fatty food it might be necessary to extract the fat by adding 10% chloroform.

Tween 20, with an end concentration of 0.25%, and rabbit normal serum, 2.5%, is added to the solution to bind the protein A, formed by almost all Staphylococcus species, which otherwise could react with the toxin anti-bodies. With that, the extraction is concluded and the **practical execution** of the test starts. The test kit, available on the market, contains the following reagents and materials:

4 tubes with enzyme conjugate (phosphatase compounds) A-D,
4 tubes with plastic beads, anti-staphylococcus enterotoxin A-D
(color coded) +
1 tube with 20 control beads
substrate buffer
washing solution, 10fold concentrated
normal rabbit serum
bottle with tablets for color reaction.

First, the food extract is poured into a glass bottle and 4 test and 4 control
beads are added.

In a shaker the bottle is shaken in horizontal position overnight at room temperature. The next morning, the beads are washed three times. The last
washing is carried out in the plastic tube, also included in the test kit. The
corresponding enzyme conjugate (antibody coupled with phosphatase) is
then added.

If a toxin is present, it will be bound within the 6-hour incubation period at
room temperature. For determination of the toxin the beads are put into a
colorless nitrophenyl phosphate solution after washing them three times.
The bound phosphatase forms a yellow nitrophenolate. This reaction occurs
within one hour. Evaluation follows. An approximate reading is possible with
the naked eye. For more exact evaluation a photometer is used and the extinction is measured at 405 nm.

Each sample with an extinction value equal or above the mean value plus
3x the standard deviation for the negative control is considered positive. For
interpretation in our laboratory the following evaluation criteria are used:
all samples with a factor < 1 are negative
all samples with a factor >2 are positive
the range between both values is considered questionable or suspicious.

20.2   Cultural Methods

20.2.1   Impedance Measurement

For the **cultural** methods, first an indirect method must be mentioned: the
impedance measurement. In the meantime, easy to handle, mainly automated equipment consisting of an incubator (for 1-512 samples in 2-8 incubation chambers), computer including software, color monitor and printer
has become available for impedance measurement.

The plastic modules with 16 chambers are γ-radiated disposable articles and relatively expensive (DM 12.00). The method of impedance measurements is probably the most tested procedure and seems to be suitable for the practise. It is not widely used in Germany, but in the Netherlands and Great Britain its use is expanding constantly. Mainly two equipment types are used, the Malthus Meter from the Kratel Company and the Bactometer from the Bactomatic Company (price approx. DM 195,000.00).

Principle: During the growth stage the metabolism products are enriched and cause a change of electrical resistance in a liquid medium. The conductivity of the medium increases or conversely the resistance decreases. The difference can be noted only at cell counts above 10^6-10^7 T/ml. This means that a detectable change of resistance is determined more easily, the higher the original count of the investigated product is. If products with low germ counts are examined, longer incubation periods in the chamber of the supplied incubator must be considered. Also, the culture medium for the incubation of the sample material must be selected corresponding to the expected bacteria type. Supporters of this method see its major advantage in the processing of large sample amounts and in the gain in time (for examination of UHT-milk, the result is obtained after 5 instead of 10 days).

- total count determination
- selective detection of microorganisms
- durability forecast
- antibiotics resistence determination
- desinfectant tests

Fig. 4 Fields of Application for Impedance Measurements

20.2.2 Measurement of Gas Changes

Another indirect method is the count determination by **measurement of gas changes,** especially **measurements of CO_2 or O_2.** The method is based on the measurement of gas changes in liquid or diluted samples. The measurement value consists of
1. the microbial oxygen consumption and
2. the microbial carbon dioxide production.

Decreasing oxygen concentration is determined polarographically with an oxygen selective electrode according to the Clark principle. The following relation is applicable: The quicker the dissolved oxygen in a gas-tight sealed food sample is reduced, the higher the number of actively respiring microorganisms.

This indirect count determination by oxygen consumption with only short measuring periods of approx. 1 hour has become accepted for end product control in the production process. The oxygen determination has been introduced as an especially suitable method for "on delivery" control of egg products, pasteurized milk, pasteurized liquid whole egg and other food. Precondition for this method is that about 10^4/ml bacteria are present in the product, otherwise the method cannot be applied quickly enough. This method was developed in the Fraunhofer-Institute in Munich.

20.2.3 ATP Measurement

Another indirect method is the measurement of the so-called adenosine triphosphate content, the so-called **ATP measurement.** Here, a luciferin-luciferase preparation is employed. During the reaction, ATP, which is present in live cells in almost constant quantity, is transformed into adenosine monophosphate and light. The intensity of the light is produced directly proportional to the cell concentration of ATP. This procedure as an indirect method can be applied only for germ counts above 10,000/ml.

In our laboratory, this method was tested for examinations of personal products and detergents. Since the original counts for these products usually are very low, 0.1 g of these products have been pipetted into a culture bouillon and incubated overnight in a shaking water bath at 30 °C. After 12-14 hour the ATP measurement was taken by means of the luminometer LKB. Compared to a standard for control purposes, the ATP was measured and registered. The germ count was determined using software, developed in our institute. Each increase in ATP content, compared to the standard, was evaluated positive. This allows very precise results regarding present contaminations in sample material of 0.1 g weight. After confirming the results of 2,500 samples the false positive samples were below 0.4%. This method seems to be faster, more efficient and cheaper than the counting method.

20.2.4 Direct Detection Method for E. coli

As one of the direct cultural methods, the detection of E. coli should be described in detail here.

In 1975, Baird-Parker and Anderson in England developed a test, called the **direct plate method for enumerating E. coli** in food by means of **membrane technique.** This method has also been accepted in Switzerland for testing drinking water.

Today, it is a standard method in many laboratories. The result can be obtained within 20–24 hours, and it is very selective. The rapid detection of E. coli includes lactose negative species as well as the possibility of detecting damaged bacteria.

Figure 5 shows the steps necessary for detection of E. coli by the direct plate method.

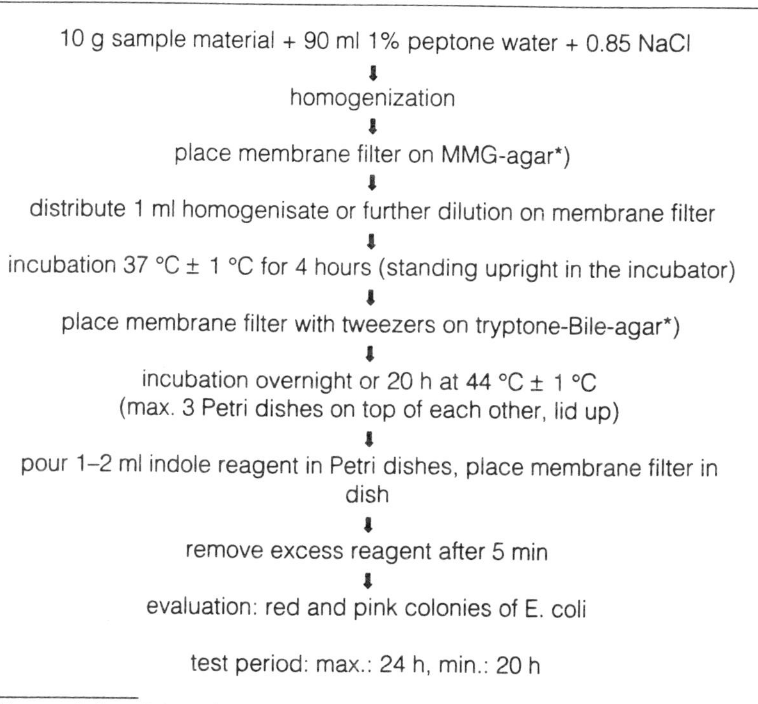

Fig. 5 Direct Plate Method for Detection of E. coli

The colonies of E. coli grow on cellulose acetate membranes which are placed on a Tryptone Bile Agar. We use a membrane filter made by the Oxoid, Sartorius, or Schleicher and SchÅll Companies.

The indole reagent is prepared according to Vracko and Sherries (1963) and consists of 5.0 g 4-dimethylaminobenzaldehyde in 100 ml 1 n HCl.

Figure 6 (see page 386) shows a filter with E. coli colonies after staining.

Recently, another method for detection of E. coli had been published. This method, described by Glaeser, Fachhochschule Sigmaringer in 1986 as well as by Hahn, Bundesanstalt fÅr Michforschung, Kiel, is an optical fluorescence technique.

This method is based on the fact the E. coli, contrary to almost all other enterobacteria, possesses the enzyme ß-glucuronidase. This enzyme sets 4-methylumbelliferone free from 4-**methylumbelliferyl-ß-D-glucuronide** (MUG). This substance fluoresces when exposed to radiation of long-wave UV-light. MUG is added to the bouillon such as Brillant-Green for determination of the coliform titer, but is also added to selective culture media for detection of enterobacteria. Fluorescence in the test tube when exposed to long-wave UV-radiation clearly indicates the presence of E. coli, correspondingly on plates or microtiter wells. The method is very safe and can be carried out with very little effort (96% recovery).

Figure 7 (see page 387) shows a positive reaction in a liquid culture medium after addition of MUG.

20.3 Other Identification Methods and Confirmation Tests

Besides the previously mentioned identification methods, there are a number of identification possibilities and confirmation tests. They can be described in short as follows.

20.3.1 Staphyslide-Test

First, there is the so-called **Staphyslide-Test.** With this test it is possible to state within seconds whether a suspicious colony on Baird-Parker agar or blood agar is Staphylococcus aureus or not. Staph. aureus possesses a protein receptor capable of combining with fibrinogen. 96% of human pathogenic Staph. aureus strains have this receptor. Stabilized sheep red blood cells, sensitized with fibrinogen, cause agglutination of the test strain, if it is Staph. aureus.

This test is easy and quick. First, remove the suspicious colony with the platinum loop from the medium. Perform the following reaction: The colony is spread carefully with a Pasteur pipette or a platinum loop into one drop of sensitized red blood cell suspension as well as into one drop of control red cells. Large clumping appearing within **15** seconds in the sensitized red cell suspension indicates the presence of Staph. aureus.

Other agglutination tests, as laid down in the BGA-Method Collection to LMBG § 35 (German Food Law) require about 6–8 hours prior to reading (this method is currently applied for examination of ice-cream)

20.3.2 Aminopeptidase Test

Another helpful simplification for laboratories which usually could not perform complicated microbiological examinations is the replacement of the gram staining by the **aminopeptidase test.** This test was developed in 1978 by Cerny. It is based on the differences in composition of cell walls of bacteria. Examinations on a series of bacteria proved that predominately gram negative bacteria contain the enzyme **L-alanine-aminopeptidase** in large quantities while practically all gram positive or gram variable bacteria show no or only low enzyme activity.

Method Principle:

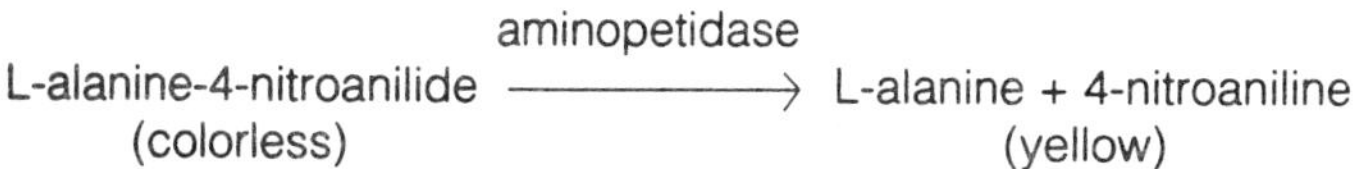

The test is easy to employ due to the test strips developed by the Merck Company. The colony to be tested is removed from a plate with a platinum loop. Preferably a culture media without color indicator is used for growing so that the colony remains colorless. The colony is placed into a test tube which contains between 0.2 and 1.0 ml dist. water. The suspension should be homogeneous and the strip is placed in the test tube so that the reaction zone is covered by water. The incubation of the test tubes is done either in the water bath or in the incubator at 37 °C for approx. 10 min, but no longer than 30 min. The strip is then read. The reaction is definite: The gram negative strains color the dist. water yellow, with the gram positive strains the water stays colorless. Figure 8 (see page 390) shows the positive and negative reaction during the test.

Fig. 6 E. coli Colonies after Staining

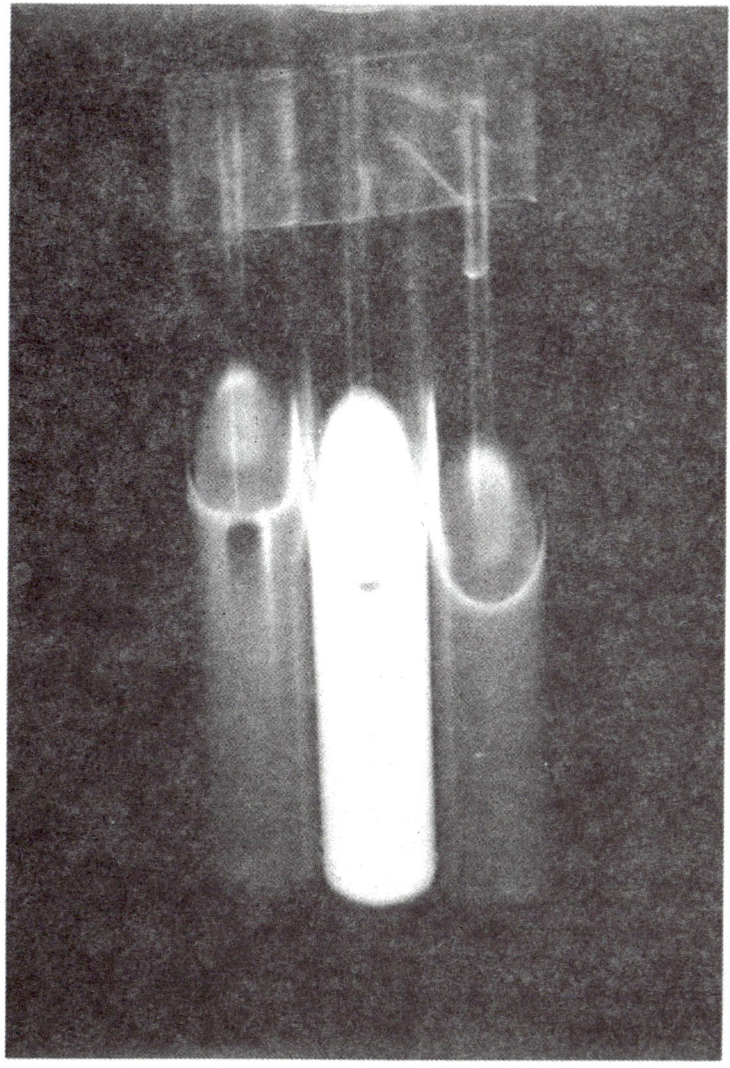

Fig. 7 Flourescence Test with Addition of MUG in Brilliant Green Tubes. Center tube E. coli positive

Dr. Hahn, Kiel kindly allowed reproduction of this picture

20.3.3   Detection of Cytochromoxidase

Another rapid method, which simplifies the differentiation of gram negative bacteria, is the detection of **Cytochromoxidase**. This test is very significant for the identification systems for enterobacteria, as described in the following. It allows, among others, the differentiation between the large groups of enterobacteria and pseudomonas. This test is based on a color reaction. Formation of indophenole blue, which is indicated by blue coloring of the test material, from **NN-dimethyl-p-phenylendiamine, α-naphtol and oxygen** is catalyzed when cytochromoxidase is present. Two system are available on the market:

The system of the Oxoid Company, which employs sticks, and the system of the Pathotec Company, using strips.

The sticks of the Oxoid Company are stored in the refrigerator. When starting a test, the cooled sticks are pressed onto one individual colony, so that a small amount of the colony remains on the surface of the test area of the stick. The stick is then placed for approx. 30 sec. between lid and bottom of a Petri dish. The reaction is then concluded. A dark-blue coloring of the stick surface indicates a positive reaction.

Performing the cytochromoxidase test with strips is even more simple. A small portion of a colony is taken with a platinum loop or a glass spatula and distributed on the reaction zone of the strip. The reaction can be read within 30 sec in the zone.

Figure 9 (see page 390) demonstrates the performance of the cytochromoxidase test with strips.

20.3.4   Identification of Enterobacteria, Non-Fermenter, Yeasts, Staphylococci and Clostridia

To conclude this survey on today's rapid methods in microbiology some well-known rapid tests for identification of **Enterobacteria, Non-Fermenter, Yeasts, Staphylococci and Clostridia** should be mentioned.

For some years now, we in our laboratory have gained much experience with the so-called **API-**systems of the bioMérieux Company. As an example, the API 20-E system for identification of enterobacteria as described below:

This system contains 20 micro tubes in which the different dehydrated substrates are stored. The substrates are dissolved when the bacteria test suspension is added. The durability of the test strips is 18 months when stored at 2–8 °C. For preparing the test, the incubator tubs, made of plastic, are numbered on the lengthened strap and 2 ml dist. water (not necessarily sterile) is place on the tub bottom assuring a humid atmosphere during incubation of the test strips. The API-strip is removed from the sealed package. One strip each is placed diagonally into the incubation tub. If suspension medium (sterile aqua dest.), supplied by API in ampules, is used, the top of this ampule must be opened by thumb pressure. If these ampules are not used, it is possible to work with test tubes which contain 5 ml sterile aqua dest. with a pH-value between 5.0 and 7.0. A bacteria suspension is produced by distributing a selected colony of the material to be tested in the 5 ml water and mixing thoroughly. The bacteria suspension is carefully applied onto the API 20-E strip by means of a Pasteur pipette. Air bubbles in the microtiter tablets are to be avoided. Finally, the lid is placed onto the tub and the strips are incubated for 18–24 hours at 35–37°C. Then, the reac-tions are read. For some test reactions additional reagents are necessary: Kovacs-reagent, oxidase reagent, nitrite reagent, Voges-Proskauer reagent and iron-III-chloride. Reading the test strips is relatively simple. It is done either by means of the supplied identification tables or, if available, by using the API-profile index which makes things significantly easier.

The API coding table, as shown in Figure 10 (see page 391), considers all 21 reactions simultaneously and converts them into a seven-digit number by sorting them into groups of three and, for each of the positive results, provides a specific number the so-called profile. This profile is then looked for in the index and, if present, the identification is made. If the determined number cannot be found in the index, it is possible to call the computer service of the bioMérieux Company for information on how to classify the strain.

Other miniature test systems as **Enterotube** of the LaRoche Company, Mini-tek of BBL, Pathotec of the Boehringer Company or systems from Abbott Laboratories as well as from the Becton & Dickenson Company are available. However, application possibilities are predominately in the medical fields because the microorganisms coded in this program come mainly from this field. Furthermore, a pre-condition is the use of pure cultures, and they are expensive.

In conclusion can be stated that for the last 15 years, distinct progress regarding the development of rapid detection methods in the field of micro-biology can be seen. This progress was made possible with totally new meth-ods such as the ATP- and impedance measurements, but also with improvements in and perfection of existing methods.

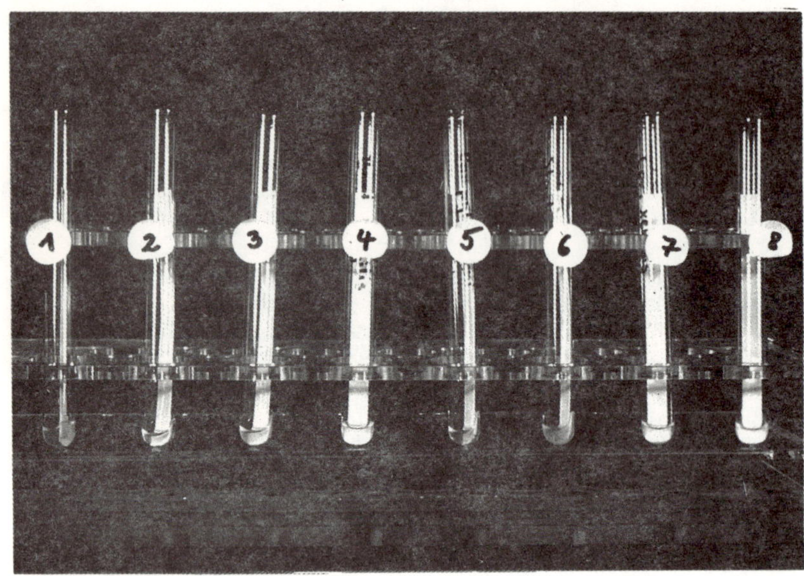

Fig. 8 Aminopeptidase Test*
  1–3 Enterobacteria = yellow   = gram negative
  4–8 other germs   = colorless = gram positive

---

* The Merck Company kindly allowed reproduction of this picture.

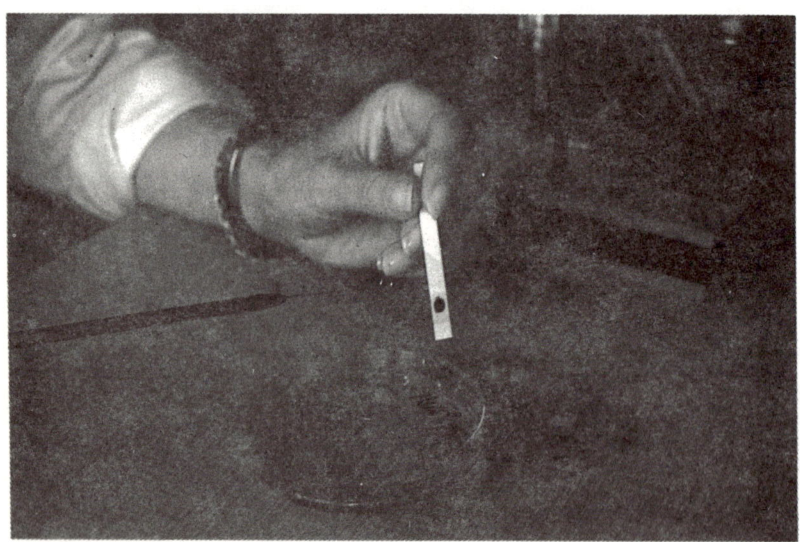

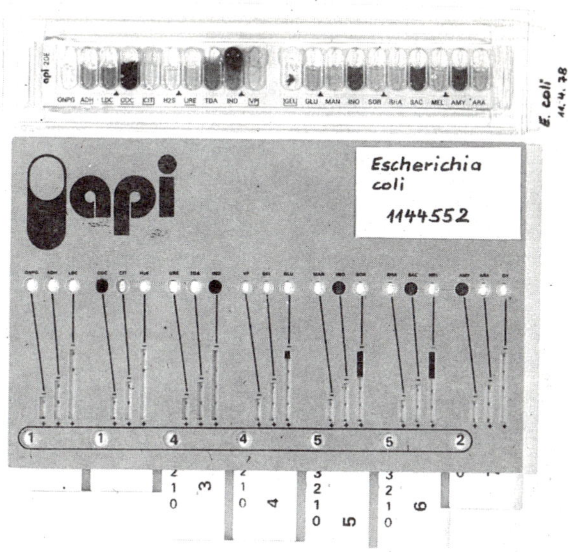

Fig. 10   API-Coding Table and Strips with Enterobacteria

REFERENCES

1. ANDERSON, J. M., BAIRD-PARKER, A. G.: A rapid and direct plate method for enumerating Eschericia coli Biotype I in food, J. appl. Bact. **39**,111—117, 1975

2. BAUMGART, J.: Mikrobiologische Untersuchung von Lebensmitteln, Behr's Verlag, 1986

3. COSTIN, J. D., KAPPNER, M., SCHMIDT, W.: Differenzierung von gram-positiven und gram-negativen Bacterien mit dem L-Alanin-Amminopeptidase-Test, Forum Mikrobiologie **6**, 351–352,1983

4. DICK, E., WIEDEMANN, R., HAMMES, W. P.: Schnellnachweis mikrobieller Infektionen in Bier, Chem. Mikrobiol. Technol. Lebensm. **10**, 37–41, 1986

5. FELS, P.: Zukunftsaspekte der Inprozeßkontrollen am Beispiel der ATP-Bestimmung mittels Biolumineszenz-Verfahren, Concept-Symposium "Steuerung der Arzneimittelqualität durch Inprozeßkontrollen", 23.–24. März 1982, Bad Nauheim

6. FEY H.: Nachweis von Staphylococcen-Enterotoxinen in Lebensmineln, SGHL, Heft **13**, 51–67, 1983

7. FRIESENBERG-EDEN, R., TRICANCO, M. K.: Impedimatic Determination of Total, Mesophilic and Psychrotraphic Counts in Raw Milk, J. Food Sci. 48 (6), 1750–1754, 1983

8. GLAESER, H.: Schnellnachweis von E.coli in Weichkäse mit Hilfe eines einfachen Fluoreszenztests.

9. SCHMIDT-LORENZ, W.: Demonstrations-Tagung SGHL, Sept. 1984

10. ZSCHALER, R.: Methode zum Nachweis von E.coli in Lebensmitteln, Lebensmitteltechnik, 12, 697–699, 1985

11. ZSCHALER, R.: Nachweis von Staphylococcen-Enterotoxin in verschiedenen Lebensmitteln, Hygiene und Medizin **11.,** 300–303, 1986